Canis dingo

Understanding Australia's Apex Predator

Bradley P. Smith

CSIRO
PUBLISHING

A catalogue record for this book is available from the National Library of Australia

ISBN: 9781486318483 (pbk)
ISBN: 9781486318490 (epdf)
ISBN: 9781486318506 (epub)

How to cite:
Smith BP (2026) *Canis dingo: Understanding Australia's Apex Predator*. CSIRO Publishing, Melbourne.

Published by:

CSIRO Publishing
36 Gardiner Road, Clayton VIC 3168
Private Bag 10, Clayton South VIC 3169
Australia

Telephone: [+613] 9545 8555
Email: csiropublishing@csiro.au
Website: www.publishing.csiro.au
Sign up to our email alerts: publishing.csiro.au/earlyalert

Front cover: At first light, a wild dingo watches from the long grass of Australia's high country. (Photo by Charles Davis)
Back cover: A dingo makes its way down a sand dune on the coast of New South Wales. (Photo by Charles Davis)

Edited by Dr Kerry Brown
Cover design by Cath Pirret Design
Typeset by Envisage Information Technology
Printed by Ingram Lightning Source

CSIRO acknowledges the Traditional Owners of the lands that we live and work on across Australia and pays its respect to Elders past and present. CSIRO recognises that Aboriginal and Torres Strait Islander peoples have made and will continue to make extraordinary contributions to all aspects of Australian life including culture, economy and science. The use of Western science in this publication should not be interpreted as diminishing the knowledge of plants, animals and environment from Indigenous ecological knowledge systems.

Jun26_RP_ILS

Contents

PART 6: EMBRACING THE WILD 217

Preface

A dingo stepped towards the edge of the enclosure, its movements smooth but wary. Behind it, others hovered in the background, their sandy coats nearly disappearing into the dry earth. But this one held my focus. It watched me intently, never taking its eyes off me as it moved, every step measured, ready to react. The air smelled of dust and discarded bones, remains of their daily feedings, left to pile up and bleach in the sun. I stood still, struck by the sense that the dingo was sizing me up just as much as I was observing it.

It was 2006 and I found myself at a small wildlife park outside Adelaide, fresh from completing my undergraduate psychology degree and ready to explore a question that had captivated researchers around the world: could these animals understand human gestures? Domestic dogs, as had recently been shown, excelled at interpreting human signals like pointing or gazing towards an object, skills their wolf ancestors seemingly lacked, shaped by thousands of years of domestication. But dingoes? These canids, free-ranging across Australia yet historically entwined with human life, represented a puzzle. Were they dogs gone wild or something altogether different? Their unique evolutionary story, shaped by long-term isolation on the Australian continent and ongoing debates about their origins, made them compelling subjects for exploring how communication between humans and animals might have evolved.

I had always been drawn to peering into the minds of non-human animals, trying to see the world through their eyes. With its unique evolutionary history, Australian wildlife particularly fascinated me and dingoes seemed the perfect research subjects. At the time, their psychology was largely unexplored, and I was eager to find out what made them tick.

Though I'd spent years studying the behaviour of a variety of species (such as orangutans, African wild dogs, and sea lions) held in zoo exhibits, dingoes were largely unfamiliar. And until now, all my animal studies had been conducted from the safety of the other side of the fence. Here I was, with my student colleague Jo Davis, venturing into something entirely new. The hollow clang of metal gates echoed through the enclosure as the wildlife park staff let us in, the rattle of chain and fading of footsteps marking their departure to other duties. The experiment we were there to attempt had seemed straightforward on paper: present the dingoes with two buckets, point to the one containing food and record their choices. We'd done it successfully with domestic dogs. But as I stood there, clipboard in hand, watching the dingo's cautious approach, tail low, ears sweeping back, I began to realise this would be anything but straightforward. The dingoes were far from the bold, fearless predators I had envisioned. Instead, their distrust of our unfamiliar figures and objects was clear. Their every move seemed measured, each step imbued with an intentionality that hinted at a deeper awareness.

I reached out instinctively to those brave enough to approach, my hand moving to pat as I would any pet dog. Their response was a revelation: they didn't shy away, but neither did they lean in. They existed alongside my touch, tolerating it with neither pleasure nor displeasure. In that moment, the difference between them and their domesticated cousins was clear: dogs seek us out, eager for approval, while dingoes remain apart, wild, indifferent and independent.

That first study never yielded its intended results. The dingoes, with their deep wariness, simply wouldn't engage with our carefully designed experiments. Although I would later successfully complete these experiments years later as part of my PhD research (covered in Chapter 7), that initial experience opened my eyes to something much more significant. Everything about them hinted at a distinctive evolutionary story. That the dingo was something unique, and uniquely Australian.

At the time, I thought I was simply exploring an interesting scientific question about animal cognition. I had no idea I was stepping into one of Australia's most contentious wildlife debates. I didn't yet realise the depth of division over the dingo's place in Australia or that battles were being fought in courtrooms, on farms and in the mind of the public. That would only become clear as I delved deeper into my research, speaking with scientists, farmers, conservationists and Australia's First Peoples.

Looking back now, I realise that this first encounter marked the beginning of what would become a lifelong journey into understanding these enigmatic creatures. It wasn't until 2009, as I faced reviewer comments on my first scientific publication, that I truly grasped the complexity of this path. The criticism wasn't about my methodology or findings, but something more fundamental: the scientific name I had used for my study subjects. I had written *Canis lupus dingo*, following what I thought was accepted taxonomy. One of the two reviewers insisted on *Canis dingo* instead, arguing for their status as a unique species rather than a wolf subspecies suggested by the former. Years later, I would face the same debate in reverse, with other scientists arguing these animals should be classified as *Canis familiaris* or domestic dogs. To call them this, however, felt like an oversimplification and ignored their fundamental nature. Dingoes are far removed from the domesticity and dependency we associate with our canine companions; their essence shaped by natural forces.

That first encounter foreshadowed the complexities of studying a species that defies simple categorisation. Dingoes exist at the crossroads of science, culture and ecology, each layer steeped in debate. Even their scientific name, a tool meant to provide clarity, has become a symbol of this contention. Through it all, I often recall the dark amber eyes of that dingo – watchful, soulful and hard to forget. That look reminds me that these animals have been misunderstood for generations. It pushes me to approach my research with more than just data in mind, but also with humility and curiosity.

This book reflects 20 years immersed in dingo science: exploring how they think and behave, how they are made, how they shape ecosystems and how we humans understand and manage them. In it, I aim to reconcile what science tells us about dingoes with what the public believes about them. To find common ground between evidence and emotion, between research and rhetoric, but it's also about updating the conversation. A wave of new research is reshaping our understanding of dingoes, yet public opinion hasn't kept pace. Too often, outdated ideas continue to influence how we manage and talk about these animals. This book is my attempt to help close that gap.

The urgency of this mission becomes clear the moment you ask any Australian about dingoes. The response is invariably passionate, often contradictory, and rarely neutral. From city dwellers who have never seen one in the flesh to outback residents who encounter them daily, from wildlife managers to suburban dog owners, everyone has a story to tell or a position to defend. I witnessed this firsthand during a 2019 dingo forum in Sydney. I had just finished presenting my research on non-lethal management approaches when a farming industry representative stood up during question time. 'Yes, but you don't have any skin in the game', he declared, his voice carrying years of frustration. The room erupted. People on both sides jumping in, voices rising as they argued over who had the right to make decisions about dingo management.

A few years earlier, while assisting a mining company with managing the dingoes that live around the site, I found myself presenting to a team of specialist underground miners during a pre-shift meeting. The goal was straightforward: to discuss what we had learned about the dingo population, updates relating to dingo management and provide guidance on how to behave safely around dingoes.

As I and the environment officer worked through the updates, one of the crew members piped up, his tone sharp and dismissive. 'They keep jumping into my ute [utility vehicle] tray and nicking my lunch box', he said. 'They're a bloody pest. Why don't we just get rid of the lot of them?'

His words hit me like a sucker punch. It was the first time I'd encountered such open hostility towards dingoes and I wasn't prepared for it. I froze, caught between confusion and disbelief. How could anyone feel so antagonistic towards such an extraordinary animal? I realised then that I'd spent so much time immersed in dingo research, surrounded by others who admired and valued them, that I'd almost forgotten these views existed, let alone that they were, in fact, the dominant view in many circles.

Before I could gather my thoughts, another miner in the room cut through the tension and said firmly, 'Well, they were here first.' 'This is their country.'

The moment hung in the air for a beat, then the conversation shifted, as if the challenge had quietly settled the matter. I left the meeting relieved that not everyone shared the same view, but also acutely aware of how polarising dingoes can be.

The dingo has become more than just an animal; it's a lightning rod for debates about identity, conservation and our relationship with the natural world. Farmers and rural communities feel the immediate impact of dingo management decisions, but the consequences reach far beyond property lines. The dingoes in our national parks belong to all Australians. Every taxpayer has a stake in their future. When we allow lethal control in these shared spaces to benefit a single group, we alter delicate ecological balances that affect us all. A city resident might never see a dingo, yet they live with the ripple effects of management decisions made hundreds of kilometres away: altered ecosystems, diminished biodiversity and the gradual erosion of our natural heritage. Whether in the outback or the suburbs, we share both the responsibility for and the consequences of how we manage these predators.

Throughout my research career, I have wrestled with the unique challenges of studying such a contentious species. My background in psychology taught me to question assumptions, including my own, ensuring that evidence, not emotion, guides understanding. Yet, like many scientists who study charismatic animals, I've learned to navigate between scientific rigour and the deep appreciation that comes from countless hours of direct observation and interaction. These experiences have shaped my perspective, while the training to question assumptions has helped ensure they don't interfere with the pursuit of truth. Everything I present in this book is grounded in the best available evidence.

Across the planet, from the grey wolves of North America to the Ethiopian wolves of Africa, wild canids share both an ancient lineage and a troubled relationship with humans. They are persecuted for taking livestock, feared for their predatory nature and pushed to the margins of an increasingly human-dominated world. In a rapidly changing Australia, dingoes face their own uncertain future as the boundaries between wild and managed spaces continue to blur. Yet understanding their story, one of resilience and adaptation, may hold the key to their conservation. From that first encounter in the wildlife park to years of research and observation, each discovery has revealed these animals to be more complex and fascinating than we imagined.

Cultural sensitivity warning

Readers are warned that there may be words, descriptions and terms used in this book that are culturally sensitive, and which might not normally be used in certain public or community contexts. While this information may not reflect current understanding, it is provided by the author in a historical context.

This publication may also contain quotations, terms and annotations that reflect the historical attitude of the original author or that of the period in which the item was written, and may be considered inappropriate today.

Aboriginal and Torres Strait Islander peoples are advised that this publication may contain the names and images of people who have passed away.

1
Introduction

AS DAWN BREAKS OVER THE VAST AUSTRALIAN DESERT, A LONE DINGO STEPS from the shadows, its tawny coat almost indistinguishable from the red sand. The cool air clings for a moment longer, though the rising sun already carries the promise of heat. Each step is deliberate and silent. Its eyes scan the horizon, tuned to the slightest twitch of movement among the spinifex and saltbush. The dingo moves with purpose, attuned to the landscape in a way that speaks of generations spent surviving here.

The sun climbs higher and the dingo makes its way to a secluded soak, a shallow depression where water seeps to the surface, sustaining life in the arid desert. These sites, known to desert-dwelling Aboriginal people for millennia, are a vital oasis in a land where water is scarce. The dingo drinks quickly, its tongue flicking into the muddy pool, its senses never switching off. This moment, having a drink and a good look around, captures the essence of the Australian saying 'a dingo's breakfast', a phrase rooted in the animal's reputation for vigilance and survival in the bush.

Raising its head, droplets of water still clinging to its muzzle, the dingo catches sight of movement in the distance. A ripple of motion, subtle but unmistakable, washes across the scrub as a mob of kangaroos shifts. The animals graze and hop lazily, their pale coats blending with the sparse foliage, but the dingo's sharp eyes pick out the details instantly.

Instinctively, the dingo lowers its body, moving silently through the vegetation. Its muscles coiled, eyes focused intently on its favoured prey.

In a sudden burst of speed, the dingo launches forward. The kangaroos scatter, powerful legs propelling them in great leaps. The dingo homes in on a juvenile, singling it out as the most vulnerable member of the mob. The chase zigzags through scattered bushes, a frantic ballet of predator and prey. Fine sand kicks up with each bounding leap, flaring behind the panicked kangaroo as it darts away, its ears twitching sharply, locked onto the threat closing in.

The juvenile, smaller and more nimble than the dingo, takes a sharp turn through a dense patch of spinifex grass. The dingo follows closely but hesitates momentarily, its paws brushing against the sharp, spiky foliage. In that split second, the kangaroo gains ground and bounds towards the safety of open terrain.

The dingo pauses, ceasing the chase, its eyes following the fleeing kangaroo as it disappears into the shimmering horizon. Its breath comes in sharp bursts, muscles quivering from the effort. There is no frustration, only the calm acceptance of a predator accustomed to the odds being against it.

As the dingo lopes across the burning sand to rejoin its pack waiting in the shade of the dunes, it seems to capture the essence of the desert itself (Fig. 1.1). It's a simple, timeless scene:

Fig. 1.1. The star of the story: a desert dingo caught mid-stride on the red sand of the Australian desert. Source: Lea Scadden/iStock

an animal doing what it has always done to survive. This is the dingo at its truest – a native predator, part of the land. Yet that truth is often overlooked. Instead, the dingo is hounded, treated as a problem simply for living its life alongside us. To some, it's not a symbol of wild Australia, but a pest that needs to be controlled. A predator caught in constant controversy and one of Australia's most divisive animals.

THE DINGO DIVIDE

For thousands of years, Aboriginal people understood and respected the dingo as both creator and kin. Dingoes were not simply wild animals to be feared or controlled, but integral members of the ecological and cultural landscape, woven into the very fabric of life on Country. With the arrival of European settlers, this balanced relationship was shattered. To the newcomers, dingoes represented the untamed, a problem to be eliminated. The stark contrast in perspectives between Aboriginal and European views of nature, and of dingoes in particular, has long been unmistakable, a theme returned to in later chapters. The consequences of colonisation fell heavily on both people and animals. The parallels between dingo persecution and Aboriginal dispossession are striking. Just as Aboriginal people were systematically removed from their lands, dingoes were systematically eliminated from pastoral areas. Both were seen as obstacles to European-style agriculture and progress. The current classification of dingoes as wild or hybrid dogs reflects the same colonial denial of Aboriginal sovereignty and connection to Country. The notion of purity makes these parallels hard to ignore: colonists sought to erase

Fig. 1.2. Painting titled 'King Dingo (on horseback 1)' by Vincent Namatjira (2024). This artwork reimagines the dingo in royal regalia, challenging colonial narratives that marginalised dingoes and Aboriginal culture. By elevating the dingo to a position of dignity, Namatjira reflects on themes of displacement, resilience and the reversal of colonial power dynamics. Source: courtesy of the artist, Iwantja Arts and Ames Yavuz. Private Collection Australia.

Aboriginal culture, showing no tolerance for so-called half-castes. Similarly, dingoes that did not fit the idealised image of pure wildness and were deemed inferior, further justifying their extermination.

This shared history of displacement and erasure is vividly reimagined in Aboriginal artist Vincent Namatjira's 2024 exhibition, King Dingo. In his series, Namatjira depicts dingoes, his totem animal, in British royal regalia, a bold inversion of colonial narratives (Fig. 1.2). 'These new paintings are about a reversal of colonial power and ownership', Namatjira explains. By portraying dingoes as regal figures, he subverts their historical framing as pests, instead elevating them to a position of dignity and respect.

The first European encounter with a dingo in 1606 set an ominous tone. Spanish explorer Don Diego del Prado recorded seeing a dog in the Torres Strait, describing it only as 'of good size' with flesh that tasted better than venison. That the first dingo to meet a white person was consumed for sustenance would prove darkly prophetic.

The true breaking point came in 1788 with the arrival of the First Fleet and their cargo of sheep. Sheep, destined to become the backbone of colonial Australia's economy, embodied everything the settlers valued: progress, civilisation and prosperity. The dingo, as an apex predator, represented an existential threat to this vision. The catchcry at that time, 'Australia rides on the sheep's back', reflected the centrality of the wool industry to the nation's development and anything that jeopardised it, especially dingoes, was deemed an enemy.

Perhaps surprisingly, no one captured this colonial mindset more precisely than the American writer Mark Twain during his tour of the British Empire in the 1890s. Visiting Adelaide's Zoological Gardens in South Australia, he observed the dingo with his characteristic insight: 'He was a beautiful creature – shapely, graceful, a little wolfish in some of his aspects, but with a most friendly eye and sociable disposition'. Yet in the same passage, Twain acknowledged the animal's fate: 'But in an evil hour he got to raiding the sheep-runs to appease his hunger, and that sealed his doom. He is hunted, now, just as if he were a wolf' – a nod that this was not an entirely new problem and one reflected in predator stories across the world. Twain concluded with brutal clarity: 'The world was made for man – the white man.'

Walter Beilby gave voice to this ruthless calculus in 1897 when he stated, 'It will be a blessing for the squatters when the brutes are extinct.' Such sentiments were not merely personal opinions, but the crystallisation of colonial policy. And so, by the late 1800s, the dingo's fate had been sealed, driven not by science or necessity, but by the relentless demands of pastoral expansion. The sophisticated predator that had evolved in perfect harmony with the Australian landscape was now cast as a menace, while sheep grazing spread across vast regions.

Ironically, in a land that today is plagued by invasive species (cane toads, camels, European carp, foxes and feral cats) – all wreaking ecological havoc, and all introduced by colonialists – the most vilified creature is a native predator. The dingo was viewed solely through the narrow lens of its threat to livestock. Unlike other native animals which could be hunted or displayed as curiosities, the dingo offered no perceived value to settlers. It was instead marked for elimination as a necessary step towards progress.

The contradictions in our relationship with dingoes were starkly illustrated in 2016 on Pelorus Island, off the coast of north Queensland. Wildlife managers captured two wild dingoes from the mainland and released them with a calculated purpose: to eradicate a destructive goat population devastating the island's endangered ecosystem. Yet even as these dingoes restored balance to the fragile environment, they were treated as disposable tools. Each was implanted with a lethal 'time bomb', poison capsules programmed to release once their work was done. Hailed as ecological saviours in one breath and condemned as pests in the next, their fate embodied a divided approach to native predators. Their worth was measured purely in terms of their utility to humans. Once their mission was complete, they were eliminated, disposable instruments in our narrow vision of environmental management.

The vilification of the dingo reached its cultural apex with the 1980 disappearance of infant Azaria Chamberlain at Uluru, an event that amplified European fears about the 'savage' wilderness. Media sensationalism, followed by incidents such as the fatal attack on a young boy on the popular tourist island of K'gari in 2001, fundamentally shaped wildlife management policies (it was known as Fraser Island, but its name was officially restored to K'gari (pronounced *Gurr-rie*), meaning 'paradise', in 2023). These rare but highly publicised events justified expanding lethal control programs and reinforced the perception of dingoes as threats to human safety. On K'gari, where dingoes had once maintained relatively natural behaviours, the response was swift: increased killing, stricter regulations and fencing that further eroded their ecological and cultural roles.

The battle lines drawn on Australian soil with the arrival of the first sheep continue to shape dingo management policy today. The dingo, cast as 'public enemy number one', remains only a shadow of its true role as both a regulator of ecosystems and a cultural icon.

LEGACY OF COLONIAL THINKING

The systematic elimination of dingoes continues unabated. Each year, thousands of dingoes are killed across the continent, through a persistent, almost ritualistic approach to destruction, calcifying into an institutionalised war.

Curiously, this destructive approach emerged despite, or perhaps because of, a grudging colonial admiration for the dingo's capabilities. Settlers documented with mixed frustration and respect how dingoes outsmarted their traps, evaded their hunting parties and adapted to new threats. In their accounts, condemnation of the dingo's predation on livestock often sat alongside admiration for its intelligence and endurance. This duality reflected a broader colonial pattern: even as settlers penned lyrical descriptions of the continent's natural wonders and marvelled at its wildlife, they participated in its systematic alteration and destruction. The

Fig. 1.3. A section of the Dingo Fence cutting across Sturt National Park in far-western New South Wales. Built to exclude dingoes and protect livestock, it is one of the longest structures in the world. Source: photo by John Carnemolla/iStock.

dingo, like the landscape itself, was both respected for its mastery of the Australian bush and condemned for that very same prowess. Ultimately, this grudging respect only strengthened settlers' resolve to assert control over the landscape and its wildlife through ever more draconian measures.

No structure better exemplifies this determination for control than the Dog (henceforth Dingo) Fence (Figs 1.3 and 1.4). Stretching 5,614 km across the south-east of the continent, from coastal South Australia to Queensland, it is three times longer than the Great Wall of China, a physical barrier that does more than separate land. From space, the fence appears as a stark line dividing the continent, a physical manifestation of the boundary between two different visions of the Australian landscape.

Today, the methods of dingo eradication remain deeply rooted in a resource-centric approach to wildlife management. Up to four times a year, pieces of meat laced with deadly poison are scattered across public and private lands, by hand or from aircraft, with no way of controlling which animals take the bait. This poison, 1080 (sodium fluoroacetate), banned almost everywhere around the world except Australia and New Zealand due to its toxicity and inhumane effects, has no antidote. Trapping, shooting and government-sanctioned bounties supplement this chemical warfare, continuing a practice of extermination that prioritises economic interests over ecological balance.

The cultural language surrounding dingoes further reveals the depth of this destructive thinking. Calling someone a dingo in Australia is considered highly offensive. It is a term that suggests cowardice and treachery, traits supposedly exhibited by dingoes. This linguistic derision has even found its way into political discourse. In 1973, member of Federal parliament Doug Anthony declared of then Prime Minister Gough Whitlam, 'He has shown cowardly instinct by trying to restrain this form of action. All I say, in Australian language, is that he is nothing more than a dingo.' A decade later, in 1986, Labour minister John Dawkins was taunted in parliament as 'Dingo Dawkins', accused of cowardice during a censure motion. When he refused to respond, opposition MPs reportedly howled like dingoes across the chamber, disrupting proceedings in a bizarre display of political theatre.

The dingo's complex position in Australian culture is perhaps most strikingly illustrated by its absence from national symbolism. Despite being fast, intelligent and cooperative, traits we typically celebrate in sporting mascots and national icons, the dingo has never been embraced as a symbol of pride. As historian Merryl Parker notes, its long association with cowardice in Australian slang makes it unthinkable that anyone would shout, 'Up the Dingoes!' from the stands. The term itself has become a slur, not a source of inspiration.

Unlike the wedge-tailed eagle, which appears proudly on military insignia and commercial logos, the dingo remains excluded. And while other nations have elevated their apex predators, like the wolf in North America or the bear in Russia, Australia continues to sideline its own. The dingo's absence from these symbolic spaces reflects a deeper cultural discomfort, reducing it to a contested figure rather than a celebrated emblem.

At the time of writing, jurisdictions such as South Australia explicitly aim to 'eradicate the dingo' in their management plans. The Dingo Fence bisects the state: below the fence, dingoes are classified as pests and landholders, councils and even park rangers are legally required to carry out lethal control or risk prosecution. In stark contrast, north of the fence, dingoes are recognised as native species and protected under conservation laws.

This contradiction is not unique to South Australia. Across the country, the dingo's legal status shifts dramatically from one jurisdiction to the next. Depending on where they are found and how they are classified, dingoes can be legally protected in one area and targeted for destruction just metres away across a boundary line. These inconsistencies reflect deeper cultural and political divides over how the dingo is understood and managed. Table 1.1 summarises the current legal treatment of dingoes across Australian states and territories.

Table 1.1. High-level summary of how dingoes are classified and managed across Australian jurisdictions as of 2025

The term 'wild dogs' is widely used in law to include dingoes, dingo–dog hybrids and feral dogs. These categories and rules vary considerably between regions, and local conditions, such as land tenure, buffer zones and conservation areas, and can affect how control measures are applied.

State/ Territory	Legal classification	Protection status	Control measures
South Australia	Inside the Dingo Fence: Declared pest under the *Landscape South Australia Act 2019*. Outside the Dingo Fence: Unprotected native animal under the *National Parks and Wildlife Act 1972*	Inside the Dingo Fence: No protection; destruction/control is legally required, and state policy aims for eradication over time. Outside the Dingo Fence: No formal protection but ecological and cultural value recognised	Inside the Dingo Fence: Landholders must implement control measures, and penalties apply for non-compliance. Outside the Dingo Fence: Control is discretionary, except within a 35 km buffer zone north of Dog Fence where baiting is required
Queensland	Classified as 'wild dogs,' a restricted invasive animal under the *Biosecurity Act 2014*; also recognised as native wildlife under the *Nature Conservation Act 1992*	Protected within national parks and other protected areas, though park authorities may carry out control measures when wild dogs threaten biodiversity or visitor safety. No protection outside these areas	Landholders have a legal duty to control wild dogs under biosecurity law using methods such as baiting, trapping and shooting
New South Wales	Dingoes are considered 'wild dogs' and are listed as pest animals under the *Biosecurity Act 2015*; listed as native fauna under the *Biodiversity Conservation Act 2016* but explicitly excluded from protection	No protection anywhere in the state; wild dog control is permitted statewide, including in national parks where they may threaten wildlife or safety	Landholders have a legal duty under biosecurity law to control wild dogs using coordinated methods such as baiting, trapping and shooting; widespread aerial and ground baiting programs are carried out by parks authorities and landholders across public and private land

continued

Table 1.1. (Continued)

State/ Territory	Legal classification	Protection status	Control measures
Victoria	Listed as a threatened species under the *Flora and Fauna Guarantee Act 1988*; protected under the *Wildlife Act 1975*; unprotection orders allow lethal control in specific areas	Protected in the north-west (around the Big Desert park), but unprotected across much of the east, where culling is permitted on both public and private land within a 3 km buffer of public-land boundaries	Baiting, trapping and shooting are authorised in unprotected zones through large-scale wild dog control programs operated by the Victorian Government; a bounty is also offered for the scalps of wild dogs taken in eligible areas
Western Australia	Declared pest under the *Biosecurity and Agriculture Management Act 2007*	No formal protection; managed as pest animals across the state, including on public lands	Landholders are required under biosecurity law to control wild dogs on their properties and may be levied to fund regional programs; coordinated baiting and trapping are conducted through government and biosecurity group initiatives
Northern Territory	Dingoes are classified as protected wildlife under the *Territory Parks and Wildlife Conservation Act 2000*	Fully protected as native wildlife, though permits may be issued for control where they threaten livestock or safety	No broad-scale control programs; only localised management occurs under permit in response to stock conflict or problem animals
Australian Capital Territory	Dingoes (listed as 'waraguls') are declared pest animals under the *Pest Plants and Animals Act 2005*	Currently unprotected and recognised as pest animals under biosecurity law	Managed mainly around rural fringes to reduce livestock impacts. Policy reform is under review to consider future protection under the *Nature Conservation Act 2014*
Tasmania	Dingoes are not present and have never occurred in Tasmania	Not applicable	Feral domestic dogs are managed under the *Dog Control Act 2000* and controlled through local council and biosecurity programs to protect livestock and native fauna

Still, the killing continues, funded by millions of taxpayer and industry dollars each year. With every passing season, the effort to eradicate dingoes increases. Paradoxically, landholders seeking funding argue that the dingo problem is worsening. After decades of intensive lethal control, however, a worsening problem suggests that these strategies are not working. These management practices seem immune to evidence of their own futility.

Journalist Lionel Hudson observed while raising a dingo he ironically named Noxious, 'The dingo has had nothing but a raw deal from its countrymen...worst of all, we show no respect for him as a wild animal.' This lack of respect has become institutionalised through generations, continuing to shape Australia's approach to its native apex predator.

As Parker observes in her thesis on the dingo's place in Australian culture, the animal stands somewhere between wolf and dog, a kind of missing link. It's neither fully wild nor truly domestic, not quite a symbol of pride, but not forgotten either. Whether Australians see the dingo as a dangerous outsider, a clever bush dog or a sacred ancestor depends on which story they've grown up with. That uncertainty has made the dingo uniquely vulnerable to fear, myth and misunderstanding.

THE SCIENTIFIC ECHO

The colonial mindset that positioned the dingo as an enemy also became deeply embedded in the foundations of Australian scientific research itself. When Australia's premier research organisation, the Commonwealth Scientific and Industrial Research Organisation (CSIRO), initiated its first systematic dingo studies in the 1960s, their mission statement, once printed in a scientific journal, made their position clear: 'know thine enemy'. This wasn't objective scientific inquiry, but rather opposition research, funded almost exclusively by the Australian Meat and Livestock Corporation. The goal was not to understand the dingo's ecological role or biological significance, but to find more effective ways to eliminate it.

This adversarial approach to research continues to cast long shadows over contemporary scientific discourse, with funding sources often shaping both the language and focus of studies. Researchers supported by livestock industries tend to frame their work around 'wild dogs' requiring control measures, while conservation-focused studies discuss 'dingoes' as apex predators worthy of protection. This linguistic divide, far from being incidental, reflects deeper fractures in how these animals are conceptualised within the Australian ecosystem. A review of the scientific literature relating to dingoes, led by Tracey Kreplins (2018), all but confirmed this. Her analysis, covering 1952–2018, revealed that terminology choices, specifically the use of 'dingo' versus 'wild dog', correlate strongly with funding sources, operational contexts and publication periods. These data provide evidence of how institutional biases shape scientific discourse as well as the broader narratives about the dingo's role in the environment.

A persistent challenge in dingo research and management is that many fundamental questions about these animals are based not on objective evidence or scientific consensus but rather on strongly held opinions and controversial arguments. In many cases, those with the loudest voices often dominate discussions and once established, these descriptions or beliefs become difficult to shake. Persistent myths range from claims about their origins (brought by boats, domestic dogs gone wild), to their behaviour (killing for sport, not barking) and appearance (being pure only if red/ginger with white tips). Despite advancing scientific evidence challenging these assumptions, such beliefs persist, perpetuated by vested interests and creating divisions that continue to influence both public perception and scientific research.

The long-standing tensions within dingo research finally came to a head in 2023, when a groundbreaking study led by Valerio Donfrancesco used a systematic approach of multiple rounds of structured discussions and feedback, known as the Delphi method, to bring 28 dingo researchers, including me, together in dialogue. The process was revealing not just in its findings, but in its very structure: here, for the first time, were dingo experts from different camps systematically confronting their disagreements.

CURRENT BATTLEGROUNDS

The Donfrancesco study exposed five major battlegrounds (indicated in italics) in modern dingo science and management, each shaped by the colonial legacy that preceded it. As my own research has evolved within this complex landscape, I've watched these battlegrounds shift and intensify, while remaining fundamentally unchanged in their core tensions.

At the most fundamental level lies the question of *identity*. What exactly is a dingo? The debate around taxonomy extends beyond scientific classification to touch on deeper questions of belonging. Is it a distinct species unique to Australia, a subspecies of wolf, or something else entirely? The controversy intensifies around hybridisation. Some argue there are no 'pure' dingoes left, while others question whether genetic purity should even matter. These aren't just technical questions; they reflect ongoing colonial attitudes about what belongs in the Australian landscape.

The dingo's *ecological role* sparks equally heated debate. As an apex predator, it can significantly impact both native wildlife and introduced species. Some research suggests dingoes help control invasive species and maintain ecosystem balance. Other studies emphasise their threats to livestock and native species. The evidence itself becomes contested territory, with different groups interpreting the same data through markedly different lenses.

Management approaches remain perhaps the most visible battleground. While many regions implement lethal control programs, others explore coexistence strategies. Tourist destinations like K'gari struggle to balance dingo conservation with human safety. These conflicts often play out in brutal headlines, when dingoes are killed after incidents with tourists, when farmers report stock losses or when conservationists document the impacts of poison-baiting programs.

Beneath these surface disputes, however, lie deeper divisions about our fundamental relationship with nature. Conflict over *values* arises from differing underlying values and principles that guide experts' opinions and judgments. Some researchers and conservationists argue for preserving natural ecosystems in their pristine state, viewing dingoes as essential top predators maintaining ecological balance. Others advocate for active management, particularly where dingoes interact with human activities like farming, prioritising economic and safety concerns over ecological roles. This management approach often results in a so-called compromise: allowing dingoes to persist in areas with little human economic activity, while controlling them in regions used for agriculture. Yet all the compromise falls on the dingo. It's never a mutual trade-off: dingoes don't get a say. Their freedom, safety or survival is always the cost.

The conflict over values manifests in legislation that simultaneously protects dingoes in some areas while mandating their destruction in others. It appears in management plans that try to balance conservation with agricultural interests, often satisfying neither. And it emerges in public debates about dingo management on tourist sites like K'gari, where different value systems clash over how to handle dingo–human interactions.

These contemporary battles, while fought with modern tools and terminology, still carry the imprint of colonial attitudes that first cast the dingo as the enemy. The same fundamental question persists: is nature something to be controlled and managed for human benefit or are we part of a broader ecological community that includes predators like the dingo?

The final battleground relates to *evidence* – and emerges from disagreements over research rigour, objectivity and the weighting of evidence used to form arguments around the dingo. Research findings that might seem conclusive to one group are often viewed with scepticism by another, especially when funding sources align with particular value positions. What one researcher sees as proof of dingoes' ecological benefits, others view as flawed methodology or biased interpretation.

This persistent colonial influence shows most clearly in research funding patterns. Despite independent studies increasingly demonstrating the dingo's ecological importance as an apex predator, industry-funded research continues to prioritise lethal control methods and livestock protection. It's a pattern that shows how economic interests still shape scientific inquiry, even as our understanding of ecosystem dynamics grows more sophisticated. The result is a self-perpetuating cycle: research funded by agricultural interests tends to support lethal control, which in turn justifies continued funding for similar research, all while alternative approaches and ecological perspectives struggle for recognition and resources.

CURRENT STATE OF DINGOES IN AUSTRALIA

Today, dingoes face relentless persecution across much of the continent. Although some populations remain stable in remote or protected regions, mostly above the Dingo Fence, even those pockets near mine sites, large agricultural stations or tourist areas are not entirely safe. Elsewhere, dingoes have been driven to local extinction, particularly in areas of intensive agriculture, urban development, and even some national parks. In a sense, wherever people live or work, dingoes are rarely welcome (Fig. 1.4).

Although once listed as 'vulnerable' by the International Union for Conservation of Nature (IUCN), dingoes have since been removed from the list following shifts in how they're classified, particularly with the recent reframing of dingoes as domestic (feral) dogs suffering severe hybridisation. Nevertheless, threats to their survival remain severe. As a member of the IUCN National Dingo Working Group, established in 2020, I've seen firsthand how fragile their status is, how misunderstood they are as a species, both nationally and internationally, and how the laws and protections vary wildly between states and territories.

As mentioned earlier with the State of South Australia, across each Australian state and territory, there is a patchwork of contradictory policies that reflect our conflicted relationship with this apex predator (Table 1.1). It's not unusual for dingoes to be protected in one area and shot just metres away across an invisible boundary. The result is confusion, inconsistency and a deeply fractured national approach.

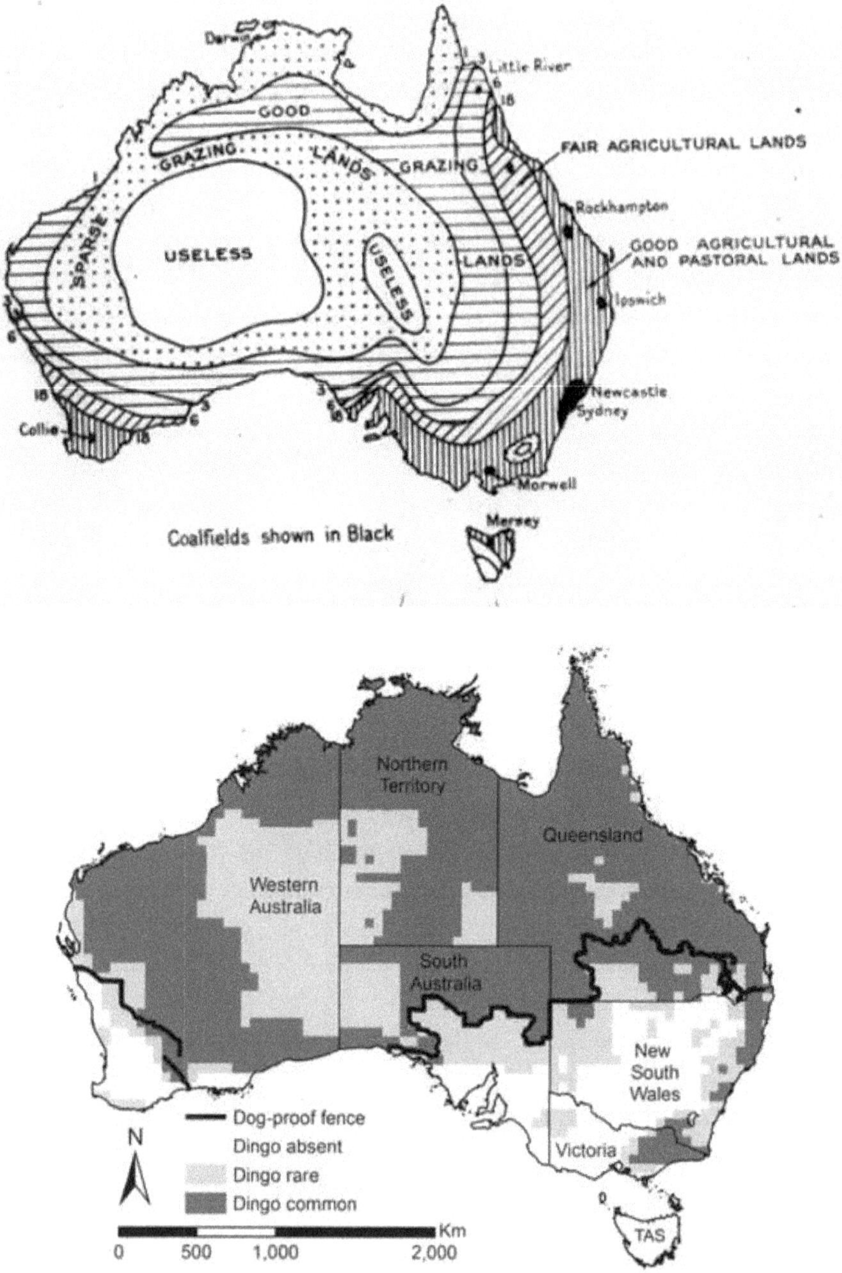

Fig. 1.4. (Top) 1956 habitability map of Australia by Griffith Taylor, illustrating land suitability for human settlement and agriculture. Much of the continent was classified as unsuitable or marginal for development. (Bottom) Modern distribution of dingoes in Australia, overlaid with major state-maintained dingo-proof fences (thick black lines). The areas where dingoes are most abundant today largely align with regions once deemed 'useless', highlighting how the dingo's continued existence is tolerated primarily in places where human interests are minimal. Sources: (Top) National Archives of Australia, A1200, L7597B; (Bottom) taken from Newsome *et al.* (2015).

UNLEASHING THE DINGO'S STORY

This book is not just about a predator. It's about an animal caught in some of Australia's deepest tensions: between wildness and control, science and story, past and future. Through the dingo, we catch a glimpse of a bigger struggle over how nature is understood, managed and remembered.

The chapters ahead follow the dingo's trail across deserts and ranges, farms and forests. We'll look at dingoes' mysterious origins, their place in Aboriginal cultures, their relationship with the landscape, their contested status in modern science and policy – and finally, how we might learn to live alongside them. Along the way, you'll hear from graziers and rangers, researchers specialising in fields such as ecology, archaeology, anatomy, genetics, psychology and taxonomy, and Traditional Owners, each with their own perspective on what the dingo is and what it means.

To look closely at the dingo is to confront the myths, the fears and the system that turned many Australians, and the world, against it. This book is my attempt to strip all that back. To show you the evidence, the history and the stories that reveal the dingo for what it really is and always has been: Not a pest or a problem, but one of the most remarkable animals on the planet.

Part 1

A new arrival

2
Origins

GIANT BEASTS ONCE ROAMED THE ISLAND CONTINENT OF AUSTRALIA: enormous diprotodons lumbered through prehistoric woodlands while marsupial lions hunted in the shadows. Though the megafauna vanished long ago, their smaller kin inherited the continent. Kangaroos bounded across open plains, their powerful tails propelling them forward. Beneath the soil, wombats carved vast tunnel systems. In the treetops, possums and gliders moved silently through the night. Even the predators, quolls and the tiger-like thylacine, belonged to this marsupial cast.

Nowhere else on Earth has ever looked like this. While the rest of the world was home to placental mammals, lions, wolves, apes, monkeys and deer, Australia had taken a different path. Long ago, it had formed part of the ancient supercontinent Sahul, which connected it to New Guinea. Many of the animals now thought of as uniquely Australian once roamed those tropical forests too. When Australia split from its northern neighbour, it was set adrift, geologically and biologically. Over millions of years, its animals evolved in near-total isolation. What emerged was an entire alternative version of what mammalian life could be.

Australia was, and still is, a land of extremes. Crushing droughts could give way to sudden floods that turned deserts into inland seas. Nutrient-poor soils and fire, not just as a force of destruction, but of renewal, shaped life at every turn. Yet in this harsh and unpredictable environment, Australia's peculiar creatures have adapted, evolved and endured. The balance held.

Then came the dingo.

Sure, there had been a few earlier arrivals. Native rodents and bats that managed to make the crossing from Asia millions of years before. The dingo was something else entirely. It was larger, social, physically stronger, and arguably more intelligent, capable of coordinated hunting and complex communication. It howled: eerie, synchronised calls that had never before echoed across this continent. This set it apart from the native predators it would eventually replace.

To the First Peoples, the original inhabitants of this land who had thrived here for over 50,000 years, the dingo was a revelation. Integrated into hunting practices and spiritual life, the dingo became a companion, a guardian, a being that blurred the line between wild and tame.

As the dingo became embedded into both human and ecological histories, it stirred questions that remain just as compelling today: Where did this creature come from? Who or what were its ancestors? And how did it reach the island continent of Australia? These mysteries have sparked decades of scientific debate, fuelled by archaeological discoveries, oral traditions and cutting-edge genetic research. Each new finding brings answers, but also uncovers new complexities, adding depth to a story that is still unfolding.

TRACING THE DINGO'S ORIGINS

In the quiet corridors of the University of Sydney, amidst the bustle of medical students and the precise preparations of surgeons-in-training, Professor Neil 'Mac' Macintosh was quietly consumed by a mystery. Officially, he was a professor of anatomy, dedicated to training the next generation of doctors. But his heart lay elsewhere, in unravelling the enigma of Australia's most contentious animal: the dingo.

For Macintosh (Fig. 2.1), the dingo was an intellectual obsession. He turned a corner of the university into a living laboratory, establishing a dingo breeding colony on campus grounds. He examined skeletal remains and reconstructed what was then the most complete early dingo skeleton, excavated at Fromm's Landing in 1960 and dated to ~3,200 years ago. He toured the outback, speaking with pastoralists, collecting specimens and conducting an ambitious landholder survey on dingo impacts, behaviour and management. His dedication knew no bounds. In 1974, in an audacious experiment, he even built a primitive boat and attempted to sail from Timor to northern Australia, dingo on board, to test whether such a voyage could explain the animal's arrival. He had to be rescued mid-journey, but the attempt captured the public imagination and exemplified his determination to solve the riddle of the dingo's origins.

Fig. 2.1. NWG Macintosh with one of his captive dingos on the grounds of the University of Sydney, circa 1954. Source: Macintosh archive, J. L. Shellshear Museum.

In 1975, he captured the uncertainty that had surrounded dingoes for some time, in a single, prescient statement:

> *The greatest problem in trying to write about the dingo is that one has no proof of the animal's identification, ancestry, affinity, place of origin or precise time of arrival in Australia; even the questions whether it was on arrival a partly-domesticated animal, subsequently becoming feral, or whether it is truly a wild dog, is disputed by many writers, and on almost completely polemical grounds.*

For all his ingenuity, Macintosh, like many scientists before him, was limited by the tools of the time. Researchers relied on two-dimensional skull measurements using callipers and basic morphological comparisons. No access to the internet, no genomic databases and none of the collaborative networks that researchers now take for granted. Those methods offered only the most rudimentary insights into the animal's past, but they almost always hinted at one thing: the dingo's Asian origins. Much else remained out of reach.

Theories about the dingo's origins have evolved over time, shaped by technological advances, archaeological discoveries and a growing understanding of how canids have dispersed around the world. In the 19th century, early naturalists such as Charles Darwin suggested that dingoes descended from grey wolves (*Canis lupus*), consistent with the prevailing view that all domestic dogs traced back to wolf ancestry. Around the same time, British palaeontologist Richard Lydekker proposed a connection to Indian wolves (*Canis lupus pallipes*), based on geographical proximity and physical similarities.

In the 20th century, Australian researchers began to explore the idea that dingoes had been introduced to the continent by people from South-East Asia. Klim Gollan completed his doctoral thesis on the topic, hypothesising that they shared ancestry with pariah dogs from the region. Laurie Corbett from CSIRO later refined these ideas, pointing to a stronger link with South-East Asian village dogs. Corbett, after travelling through South-East Asia and noting how similar the village dogs were to dingoes, also speculated about possible connections to Indian or even Arabian wolves. In earlier literature, some researchers went further still, suggesting that the dingo might be a type of jackal, noting its fox-like features and slender build.

Scientific understanding has advanced dramatically since Macintosh's time. The questions he raised about ancestry, arrival and identity, are no longer shrouded in quite so much uncertainty. Today, researchers have tools he could scarcely have imagined. Three-dimensional morphometrics and advanced DNA sequencing. Where once it was only bones and educated guesses, now there are entire genomes and CT scanners. Researchers can peer deep into the dingo's genetic code, tracing its evolutionary story in unprecedented detail, revealing ancient divergences, migration pathways and connections long hidden in the DNA.

WHEN THEY ARRIVED

The story of the dingo's arrival in Australia unfolds like an ancient mystery. The earliest concrete evidence comes from fossils unearthed in the limestone chambers of Madura Cave and other southern sites, dated to roughly 3,000–3,500 years ago. These remnants, bones, teeth, skulls and jaws, were intermingled with stone tools, discarded meals and the charred remains of ancient hearths, suggesting a bond between humans and dingoes. Cut marks on some bones may indicate they were used as tools or food, reflecting how deeply dingoes had become integrated into human survival and culture upon their arrival.

Interestingly, the earliest dingo fossils have been found along southern Australia's coasts. If dingoes entered from the north, as is most likely, their spread southward, a distance of some 3,000 km, would have taken time. Natural constraints, such as limited water sources, high mortality rates and their reproductive biology, may have slowed their movement. Dingoes breed only once a year and within packs, only the dominant pair typically reproduces. This combination of seasonal breeding and social suppression would have limited population growth and dispersal. Even with favourable conditions, it may have taken several hundred years for dingoes to spread across the continent.

Some researchers have attempted to estimate dingo dispersal rates by comparing them to the rapid spread of foxes and cats across Australia in the late 1800s. These introduced predators expanded across vast distances in just a few generations (~70–100 years), aided by environmental changes wrought by European colonisation, such as livestock, artificial water points and extensive land clearing. But using their spread as a model for dingoes is potentially misleading. The Australia that foxes and cats encountered was radically transformed. In contrast, the world the dingo entered had been shaped by very different forces.

If human assistance played a role in their movement, it was likely through Aboriginal exchange systems, subtle, decentralised and largely invisible in the archaeological record, but no less effective. These networks, sometimes described as 'cultural highways', stretched across the continent, connecting even remote regions through systems of exchange, ceremony and kinship. Through these enduring routes, goods, stories and animals could travel swiftly over vast distances. Dingoes, once introduced, may have moved far beyond their original entry point with the help of these pathways.

Taken together, the evidence suggests dingoes were present in Australia at least 3,000–4,000 years ago, and likely earlier. Of course, fossils can only provide a minimum age, not the precise moment of arrival, and gaps in the archaeological record mean the true timeline could stretch back even further.

Geography offers another important clue. Fossil evidence reveals no trace of dingoes in Tasmania, the large island to the south of mainland Australia. During the last Ice Age, Tasmania was connected to the mainland by a land bridge across Bass Strait. Around 12,000 years ago, rising sea levels submerged this connection, isolating Tasmania permanently. The complete absence of dingoes there strongly suggests they arrived in Australia after this land bridge disappeared, firmly anchoring their arrival to the mainland and placing it within the last 12,000 years.

The absence of older dingo remains, however, raises intriguing questions. If dingoes arrived earlier than the current fossil record suggests, why don't we see fossil evidence of them before 4,000 years ago? Several factors likely contribute. The preservation conditions in northern

Australia, where dingoes presumably first arrived, are poor. Tropical climates, acidic soils and scavengers, such as Tasmanian devils, accelerate the breakdown of bone. Rising sea levels since the end of the last Ice Age may have submerged some of the earliest coastal sites where dingoes first landed, placing important evidence out of reach. Dingoes' nomadic habits and low initial population densities would have further reduced the likelihood of remains being deposited in ways that would later fossilise. In general, animals must exist at a certain density over time before they're likely to leave a trace in the fossil record.

It is true that fossils of many other, even older, species have been found. However, this does not necessarily mean we *should* have found earlier dingo remains, at least not yet. If dingoes had arrived 20,000 or 30,000 years ago, we might expect to have uncovered evidence by now. Distinguishing between a 4,000-year and, say, a 6,000- or 8,000-year arrival is far more difficult, especially given the patchiness of Australia's archaeological record. Fossils provide certainty that dingoes were present by 3,500 years and that they were already well established across the continent by that time. What fossils cannot tell us with confidence is the precise timing of their arrival.

Despite these challenges, existing fossils reveal some fascinating details. Early dingoes were smaller than modern ones, with narrower skulls and less robust teeth. These traits suggest they initially preyed on smaller marsupials and reptiles, adapting their hunting strategies as they became more established in their new environment.

The absence of earlier cultural evidence aligns with the fossil record. To date, no Aboriginal rock art or artefacts clearly depicting dingoes have been reliably dated to earlier than 4,000–5,000 years ago. Although some images may be older, most have not been directly dated, and the current evidence suggests that dingoes only begin to appear in the cultural record after they appear in the physical one.

This same period, around 5,000 years ago, was also a time of significant transformation in Aboriginal societies, often referred to as a period of 'intensification'. It was marked by increasingly complex subsistence strategies, population growth and the development of new tools and technologies. The arrival of dingoes may have been both a reflection of and a contributor to this broader shift. Tim Flannery (1994) famously described it as a 'dingo-driven revolution', suggesting that dingoes may have influenced hunting strategies, mobility and ecological relationships. Although the scale of their impact remains debated, there is little doubt that dingoes became deeply integrated into both Aboriginal life and the Australian landscape during this transformative era.

By piecing together archaeological evidence, ecological data, cultural records and geological clues, researchers have converged on a likely window for the dingo's arrival in Australia: sometime after 12,000 years ago, and most likely before 4,000 years ago. This evidence can only take us so far. To dig deeper into the dingo's origins, we must turn to genetics.

GENETIC ORIGINS AND IDENTITY

The advent of genetic techniques opened a new frontier in dingo science and revolutionised our understanding of their origins and evolutionary identity. A key figure in this transformation was Alan Wilton, whose pioneering work in the 1990s laid the foundation for modern dingo genetics. In 1999, Wilton and colleagues published the first study using microsatellite markers, short, repeating sequences in the genome that vary between individuals, to differentiate dingoes from domestic dogs. This research showed that dingoes had significantly less genetic variation than dogs and that certain microsatellite alleles could help distinguish the two groups. This was groundbreaking. It provided the first robust, DNA-based framework for assessing dingo purity, supporting conservation efforts and challenging assumptions about hybridisation.

Building on this work, Wilton published a 2001 follow-up study focused on practical applications for detecting hybrids. He demonstrated that a set of diagnostic microsatellites could be used to assign individuals to populations, dingo, dog or hybrid, with relatively high confidence. This approach remained the basis for many assessments of dingo purity and hybridisation over two decades.

Wilton then collaborated in early 2000 with Swedish geneticist Peter Savolainen, a leading figure in canine mitochondrial DNA (mtDNA) research. Together, they analysed the mtDNA control region, which is a part of the genome passed down through maternal lineages, to trace the ancestry of Australian dingoes. Their 2004 study, based on samples from 211 dingoes, revealed a striking finding: over half of the individuals shared a single mtDNA haplotype, A29, which also occurred in dogs from South-East Asia and New Guinea. The remaining dingo mtDNA sequences clustered tightly around A29, differing by, at most, one or two mutations.

This created what geneticists call a star-like pattern, in which many closely related sequences radiate out from a single central ancestor, in this case, A29. Such a pattern typically forms when a small founding population rapidly expands, with little time for major genetic changes to accumulate. It suggested that all Australian dingoes descended from a very limited number of maternal ancestors, consistent with a genetic bottleneck. The estimated time of arrival, based on molecular clocks, was ~4,600–5,400 years ago, with a possible upper limit of around 10,800 years, depending on the mutation rate used. Importantly, the fact that A29 and its variants were only found in East Asia, island South-East Asia and New Guinea strongly supported a South-East Asian origin for the dingo, not India, as some earlier morphological studies had suggested.

Around this time, as a broader collaborative Savolainen and Wilton teamed up with Mattias Oskarsson and expanded the mitochondrial DNA research to include a larger regional dataset across mainland and island South-East Asia. Published in 2012 the study analysed 232 dingoes and hundreds of regional village dogs, confirming that all dingoes carried the A29 haplotype or closely related variants. The researchers estimated the dingo's arrival in Australia between 4,600 and 18,300 years ago, depending on mutation rates. Although broad, this range reinforced a key point: dingoes had diverged from other dog populations long before they entered Australia. The work also showed that A29, and other related haplotypes, were present in South-East Asia and New Guinea, but absent from Taiwan and the Philippines, further supporting a mainland South-East Asian origin for dingoes, distinct from Austronesian dog dispersal routes.

Together, these studies, first using nuclear microsatellites, then mitochondrial DNA, established the dingo as a genetically distinct population of ancient dogs.

Kylie Cairns, a conservation biologist and leading expert on dingo genetics at the University of New South Wales, has played a pivotal role in advancing this field. A former student of Alan Wilton, Cairns has carried forward his legacy and helped usher dingo genetics into the era of high-throughput sequencing and whole-genome analysis. Her work has made significant strides in refining our understanding of the dingo's ancestry, identity and conservation status.

Using complete mitochondrial genomes, Cairns and her collaborators have narrowed the timeline for the dingo's divergence from domestic dogs to between 8,000 and 11,000 years ago, a period that predates the rise of agriculture and the selective breeding that transformed dogs into the companions we know today. Supporting this, single nucleotide polymorphism (SNP) studies have shown that dingoes lack key domestication markers found in domestic dogs. Genes linked to starch digestion (AMY2B), tameness and coat variation are either absent or far less common in dingoes. This reinforces the idea that dingoes are not simply feral dogs, but part of an ancient lineage of dog-like canids that became genetically and behaviourally distinct before domestication fully took hold. We'll return to this question in more depth later in the book, where evidence from diet, behaviour and genes such as *AMY2B* further challenges the idea that dingoes are just domestic dogs gone wild.

Although mtDNA and microsatellite studies have provided critical early insights, they had limitations. Microsatellites analyse only small sections of nuclear DNA and mtDNA focuses solely on maternal lineage, offering an incomplete picture of the dingo's genetic history. Advances in genomic technology, particularly SNP analysis, have since provided a far more comprehensive understanding. SNPs, which examine single-point variations in DNA, allow researchers to analyse hundreds of thousands of genetic markers, dramatically increasing precision.

'SNP technology allows us to pinpoint exactly how dingoes diverge from domestic dogs, showcasing their unique evolutionary history', Cairns explains. These findings confirm that dingoes lack key domestication markers seen in domestic dogs, emphasising their status as a wild canid rather than a feral offshoot of domesticated dogs.

Notably, genomic studies have also revealed deep population structures within Australian dingoes. In 2016, Cairns and Wilton first identified two major mitochondrial lineages, south-eastern and north-western, suggesting either multiple introduction events or early divergence after arrival. Results from follow-up studies using Y-chromosome data and SNPs reinforced this pattern, showing clear geographical subdivision between dingoes in the south-east and those in the rest of Australia.

In 2018, Cairns and colleagues analysed over 60,000 SNPs to analyse dingoes from five bioregions across Australia. Their findings confirmed the existence of distinct regional populations, shaped by natural barriers such as arid zones and mountain ranges. More recent SNP-based work by Cairns in 2023 has extended this further, identifying clear biogeographical clustering and low levels of admixture in most populations, supporting the idea of long-standing genetic isolation.

Together, these studies suggest that although the dingo's arrival was likely the result of one or a small number of introduction events, what followed was a complex pattern of regional adaptation, ecological separation and genetic drift, one that continues to shape dingo populations today.

Additional context comes from whole-genome studies led by Shao-jie Zhang in 2020 and Anders Bergström in 2021, which estimate that dingoes diverged from South-East Asian dog lineages between 8,000 and 11,000 years ago. This divergence predates their earliest known presence in Australia, but the relationship between dingoes and New Guinea singing dogs (NGSDs), which share a close genetic ancestry, raises the possibility that dingoes may have entered Sahul earlier than the fossil record currently shows. Whether there was a delay between divergence and arrival, or whether the early record is simply missing due to preservation biases, remains an open question.

One of the longstanding criticisms in dingo genetics has been the lack of ancient DNA, samples that predate European colonisation. Until recently, almost all genetic studies relied on modern dingo populations as a baseline. This left a lingering question: could the genetic distinctiveness of dingoes simply reflect recent isolation or even contamination from domestic dogs brought by settlers? It was a caveat that shadowed every study. But not anymore.

Ancient DNA studies have now reinforced these findings. Yassine Souilmi and his team at the Australian Centre for Ancient DNA based at Adelaide University (2024) analysed dingo remains ranging from 400 to 2,746 years old. Crucially, these samples predate European colonisation, addressing the concern that modern dingo DNA may have been influenced by hybridisation with dogs introduced by settlers.

The results were clear: even thousands of years ago, dingoes were genetically distinct from domestic dogs. More importantly, by at least 3,000 years ago, the genetic structure seen in modern dingo populations was already emerging. This supports the idea that regional differentiation, first uncovered through SNP studies, has deep historical roots. It also highlights the dingo's long-standing adaptation and resilience within Australia's varied and often harsh environments.

Inspired by the work of archaeologists Loukas Koungoulos and Melanie Fillios in 2020, I synthesised archaeological and the most recent genetic findings to help refine the timeline of the dingo's arrival in Australia. The fossil evidence places dingoes in Australia by 3,000–3,500 years ago, but the molecular estimates are more variable, ranging from around 4,000 up to 11,000 years ago, depending on the methods and mutation rates used. However, the majority of genetic studies converge around 5,000–8,000 years ago, aligning closely with the earliest physical evidence. This would suggest a likely arrival around 6,000 years ago, followed by rapid population expansion and stabilisation across the continent by 3,000 years ago (Fig. 2.2).

The evidence now points to a genetic story that is not one of recent domesticated dogs gone feral, but of an ancient lineage shaped by early divergence, possible maritime movement and a uniquely Australian evolutionary path.

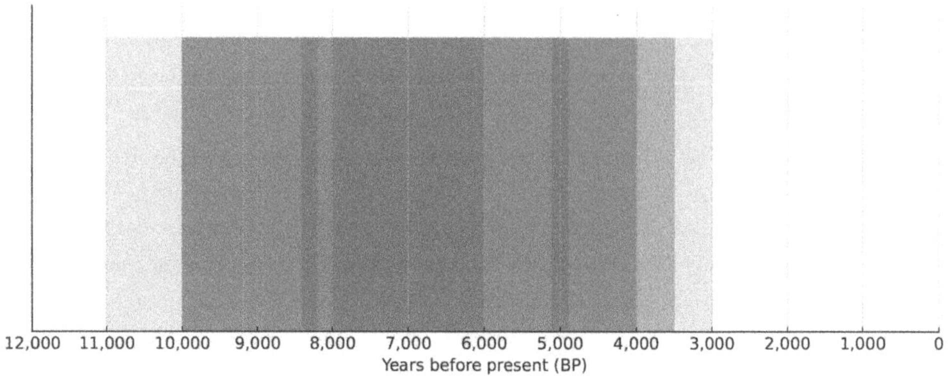

Fig. 2.2. Timeline of dingo arrival in Australia based on genetic and genomic research, alongside synthesis studies that also incorporate archaeological and biogeographical evidence. Each band represents the estimated arrival window suggested by a different study (Savolainen *et al.* 2004; Oskarsson *et al.* 2011; Cairns and Wilton 2016; Zhang *et al.* 2020; Kumar *et al.* 2023; Souilmi *et al.* 2024; Fillios and Taçon 2016; Smith *et al.* 2019; Cairns 2021), with darker shading indicating greater overlap. The fully shaded range spans from about 3,000 to 11,000 years before present, but two clusters of significant overlap emerge: a secondary one around 8,000 years ago and a stronger convergence around 6,000–5,000 years ago, which most likely marks the timeframe of dingo introduction and population expansion in Australia. Source: Bradley Smith.

DIVERGING PATHS: NEW GUINEA SINGING DOGS AND SOUTH-EAST ASIAN DOGS

Dingoes share ancestry with an ancient lineage of dog-like canids, and their closest genetic relatives are the NGSDs. These two canids are nearly indistinguishable at the genetic level, with mitochondrial DNA differing by only one or two mutations. Both descend from an ancient East Asian canid population that existed thousands of years before agriculture and domestication reshaped dogs into their modern forms. This shared ancestry links dingoes and NGSDs to a time when humans and canids were only beginning to interact. However, their evolutionary paths diverged: the NGSD adapted to the isolated highlands of New Guinea, while the dingo spread across Australia's diverse and challenging landscapes.

Comparative genomic studies confirm that dingoes and NGSDs form a distinct lineage, separate from modern domestic dogs and Asian village dogs. As illustrated in Fig. 2.3, dingoes and NGSDs diverged from an early Eastern dog lineage, distinct from Arctic and Western dog lineages. The Eastern lineage includes canids adapted to diverse environments, with dingoes and NGSDs remaining isolated from the genetic mixing seen in modern domestic dogs. This separation highlights their unique evolutionary path, shaped by geographical isolation and adaptation to their respective environments: Australia for dingoes and the rugged highlands of New Guinea for NGSDs.

Misinterpretations can arise from phylogenetic trees that sometimes position village dogs as basal to dingoes. This has led some to mistakenly assume that village dogs are the ancestors of dingoes. In reality, these analyses often fail to account for the extensive admixture and high genetic diversity found in village dog populations, factors that can pull their placement towards

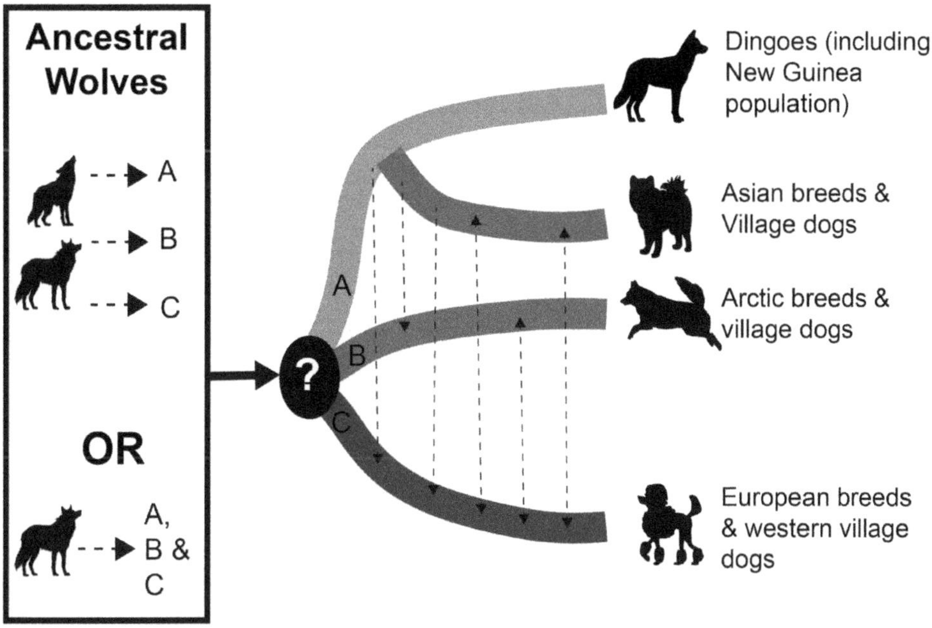

Fig. 2.3. Schematic representation of the hypothesised evolutionary origins of the three primary dog lineages – (A) the Eastern or Asian, (B) the Arctic and (C) the West Eurasian – and their relationships to each other and to dingoes. Wolves are the ancestors of all domestic dogs, though the specific populations involved remain uncertain (see inset). Dashed black lines indicate potential gene flow (genetic exchange) between lineages over time. Dingoes and New Guinea singing dogs (NGSDs) belong to the Eastern lineage and have remained largely isolated from the genetic mixing that characterises modern domestic dogs. Source: Reproduced from Cairns, K.M *et al.* (2025) with permission from CSIRO Publishing.

the base of the tree. But being 'basal' doesn't mean ancestral; it just means that a population appears closer to the root of the tree, often because of retained diversity or fewer recent changes. Whole-genome and genome-wide SNP studies provide a clearer picture, placing dingoes in a separate evolutionary lineage alongside NGSDs. As Cairns explains, 'Dingoes and singing dogs are nearly identical genetically, which suggests their descent from an ancient East Asian canid population before their divergence in Australia and New Guinea.'

These findings challenge earlier theories, such as Corbett's suggestion that dingoes descended directly from South-East Asian village dogs, by revealing a more complex evolutionary picture. Although both share a common ancestral population, their paths diverged thousands of years ago. Village dogs remained closely tied to human settlements, continually interbreeding with other domestic lineages. Dingoes, by contrast, became genetically isolated upon arrival in Australia. They lack many of the genetic markers associated with domestication and admixture, instead reflecting an evolutionary trajectory shaped by natural selection in the wild.

These genetic distinctions not only illuminate the dingo's deep evolutionary roots but also raise important questions about their identity, questions we will return to later when we explore what it means to be wild and how dingoes differ from dogs and their descendants.

In short, dingoes are not simply a branch of South-East Asian village dogs, but rather a preserved remnant of a much older canid lineage. Their close relationship with NGSDs points to an ancient connection, while their adaptation to the Australian landscape highlights their evolutionary independence. The dingo's story is one of divergence, resilience and redefinition, a powerful reminder that shared appearances do not always reflect shared histories.

PATHWAYS TO AUSTRALIA

Almost all proposed migration routes for dingoes trace their origins to Asia, most likely southern China or mainland South-East Asia, before passing through island South-East Asia, New Guinea and ultimately northern Australia (Fig. 2.4). Although the broad regional origin is well supported, scientists are still working to pinpoint the precise source population and pathway.

To understand how dingoes made this journey, it is essential to consider the geographical and geological barriers they faced. Between South-East Asia and Australia lies Wallacea, a region of scattered islands separated by deep-water channels that were never fully connected, even during the last Ice Age, when sea levels dropped by as much as 120 m. In the 19th century, British naturalist Alfred Russel Wallace identified this divide, now known as Wallace's Line, which separates the placental mammals of Asia from the marsupial- and monotreme-dominated fauna

Fig. 2.4. Proposed origins and migration routes of the dingo, based on genetic studies as summarised in 2016 by archaeologists Melanie Fillios and Paul Taçon. The exact pathways remain uncertain, but the general picture has held steady: dingoes most likely came from China, dispersing through South-East Asian islands before reaching Australia, in some areas with human assistance. Ongoing refinements, driven by advances in DNA and genomic research, have added detail, but the overarching model of origin and movement remains robust. Source: Fillios and Taçon (2016).

of Australasia. Crossing Wallacea would have required navigating sea gaps of up to 150 km, a formidable challenge for most terrestrial animals, outside of rare events such as tsunamis or storm-driven rafting.

During glacial periods, the Sahul Shelf connected Australia, New Guinea and Tasmania into a single landmass, while the Sunda Shelf united much of mainland South-East Asia with the islands of western Indonesia. Yet Wallacea remained a persistent biogeographical barrier, making the successful movement of species like the dingo especially remarkable.

Estimates for when the land bridge between New Guinea and Australia became submerged range from around 12,000 to as recently as 6,500 years ago. This variability reflects differences in dating methods and regional topography. Despite this, the broader pattern is well established: the final inundation of the Torres Strait marked a defining moment, permanently severing terrestrial connectivity between New Guinea and mainland Australia. The dingo's path into Australia may have involved more than one stage or more than one kind of journey.

The case for human-assisted migration

The challenges posed by Wallacea make it likely that humans played a key role in the dingo's journey to Australia. The deep-sea gaps between islands in this region, with some distances reaching up to 150 km, were significant barriers for most terrestrial animals. Dingoes, or more accurately, their ancestors, would not have been capable of swimming these distances unaided. It is theoretically possible that natural disasters, such as tsunamis, could have swept animals between island systems, but this form of dispersal is considered extremely unlikely. These geographical realities help explain why Wallace's Line remains a major faunal boundary to this day and why the presence of dingoes in Australia almost certainly required human assistance at some stage of the journey.

It is unlikely that Aboriginal Australians brought dingoes to the continent, given that they had been living in Australia for more than 50,000 years before dingoes appeared in the archaeological record, and there is no evidence of large-scale animal transport that far back. However, evidence does suggest multiple waves of human movement and contact through Torres Strait and Papua New Guinea as recently as 15,000 years ago. These later migration events open the door to human-assisted dispersal of animals such as dingoes.

The most plausible explanation is that dingoes were transported by seafaring cultures from regions to the north, such as New Guinea, the Torres Strait or Indonesia, areas that had long-standing maritime traditions and some degree of cultural exchange with Aboriginal Australians (Fig. 2.5). These communities used canoes and rafts to traverse open water long before the advent of modern navigation (the very process Macintosh attempted to replicate). Humans were responsible for moving a range of land mammals between islands in South-East Asia, many of which were not fully domesticated, and dingoes could well have been part of that broader pattern of animal mobility.

There is also archaeological evidence of cultural exchange between northern Australia and neighbouring regions. The arrival of new tools and technologies around 5,000 years ago, for example, may coincide with the introduction of dingoes or their ancestors. Perhaps even more

Fig. 2.5. Double canoes in the Society Islands, around the time of Captain Cook's first voyage (ca. 1769). A Polynesian dog sitting on the aft deck. This showcases what such craft might have looked like. Almost all sketches of such craft that exist include a dog on board. Inset: close-up of the dog. Source: Luomala (1990).

compelling is the archaeological and linguistic evidence of trade and potential human migration into northern Australia from the Torres Strait and Papua New Guinea, dating to around 10,000 years ago. If dingoes were introduced during this period of cultural contact, they may have come as hunting companions, camp animals or even spiritual entities.

Why would early humans bring dogs on such journeys? According to Cairns, the relationship between early humans and dingoes was likely symbiotic: 'Humans offered dingoes food and shelter, while dingoes contributed to hunting and protection. This bond may have been critical in their shared survival.' Some, such as Corbett, have suggested they may have been carried as a potential food source, but this seems unlikely given the abundance of fish and other protein-rich resources available in the waters north of Australia. More plausibly, dingoes played flexible roles, adapting to the needs and practices of the people they accompanied.

Still, one unanswered anomaly remains: if dingoes were introduced as part of wider trade or exchange networks, why do we not see the arrival of other commonly traded animals, such as pigs, chickens or additional dog lineages, that were moved around the Pacific? In fact, many classic elements of Austronesian cultural exchange, including specific pottery styles and domesticated species, never reached Australia. Nor do we see evidence of any new (multiple waves) dog arrivals between the introduction of dingoes and the arrival of Europeans. This conspicuous absence suggests that although the movement of dingoes may have occurred during a period of contact, it was limited in scope and duration. For reasons still unknown, the dingo arrived, and then the door closed.

The possibility of natural dispersal

The case for human-assisted migration is strong, but natural dispersal cannot be ruled out. If dingoes arrived before the land bridge connecting New Guinea and Cape York at the very northern tip of Australia was fully submerged, potentially as recently as 6,500 years ago, they could have walked into northern Australia via the Sahul Shelf. Even if some gaps in the land bridge remained, dingoes' ancestors may have been resourceful enough to cross shorter sea stretches during periods of low sea level. Natural rafts of vegetation, strong swimming ability or opportunistic movements across island chains may all have played a role.

The genetic similarities between dingoes and NGSDs support this idea, as both are considered part of the same early lineage. However, Wallacea remains the critical obstacle. Although natural dispersal mechanisms may explain movements *after* Wallacea, they cannot easily account for how dingoes, or any non-flying, non-marine mammal, crossed Wallacea in the first place.

A legacy of exploration and debate

The dingo's arrival in Australia likely unfolded over thousands of years and involved both human and natural forces. I propose that this journey occurred in two distinct phases. The first phase began around 10,000 years ago, when the ancestors of dingoes were transported across Wallacea by early seafaring peoples. These animals were likely not formal domesticates, but free-ranging village dogs, tolerated, semi-managed and moving with people through trade and migration routes that connected South-East Asia to New Guinea. Crossing Wallacea's deep-water gaps would have required human assistance, just as it did for other land mammals moved across the region during this period.

Once in New Guinea, some of these dogs likely lived in and around human settlements, while others roamed more freely. Over time, their descendants became more ecologically self-sufficient, adaptable, opportunistic and increasingly untethered from humans. Then, sometime around 6,000 years ago, a small number of these animals made the journey into northern Australia. This final leg did not require seafaring. During periods of low sea level, the Torres Strait narrowed to a land bridge or a series of short crossings, allowing mobile, free-ranging animals to disperse naturally. These were not leashed dogs in canoes, they were wild-living canids, moving through the landscape on their own terms.

This two-stage model helps makes sense of several longstanding questions. It explains why dingoes are genetically closest to NGSDs, yet appear to have arrived in Australia without the full suite of Austronesian domesticates. It also fits the genetic evidence showing a small founding population, consistent with natural dispersal rather than sustained human trade. And it captures what makes dingoes so unique: they are not domestic, but they are not wolves either. They emerged from the grey space in between, shaped by early human contact, but defined by their independence and ability to thrive in the wild.

THE FOUNDING POPULATION

The founding population of dingoes, however they arrived in Australia, likely consisted of a small group of animals that overcame remarkable odds to establish themselves.

Genetic studies reveal that dingoes have significantly lower genetic diversity than other canids, such as domestic dogs and wolves. This lack of diversity points to a phenomenon known as a genetic bottleneck, which is a sharp reduction in population size that limits the variety of genes passed to future generations. Evidence of this can be seen in the high number of 'runs of homozygosity' in dingo DNA, where long stretches of identical genetic material are inherited from both parents. In simple terms, this suggests the founding population size was small and relatively closely related.

Early mitochondrial research by Savolainen once proposed that all Australian dingoes may have descended from a single pregnant female. However, this idea has since been refuted by more comprehensive genetic studies, including recent studies led by Cairns, which reveal at least two distinct mitochondrial lineages and at least two Y-chromosome clades among dingoes. This indicates that both male and female founders contributed to the gene pool and that the founding population, despite being small, included multiple individuals.

These bottlenecks fit well with the challenges dingoes likely faced during their journey to Australia. Crossing the island chains of South-East Asia, whether through human transport or natural dispersal, would have been arduous. Long stretches of open water and limited resources would have restricted the number of animals that could successfully make the crossing, further reducing genetic diversity. Strikingly, the genetic evidence suggests that dingoes did not arrive as part of an ongoing trade network or repeated introductions. Instead, the data point to a short time frame, a small founding population and a long period of isolation from their source population.

But the story of dingo origins is more complex than a single migration event. Genetic research led by Cairns identifies two distinct dingo clades: north-western and south-eastern. This split may reflect either separate introduction events or early divergence following arrival. Intriguingly, south-eastern dingoes share closer genetic ties to NGSDs, raising the possibility that these populations arrived through different routes, or at different times. Whatever the case, the genetic structure we see today likely emerged early in their history on the continent.

DINGO POPULATION STRUCTURE: 'ECOTYPES'

Before the advent of genetic research, dingoes in Australia were broadly categorised into three main types based on their physical appearance and the environments they inhabited. This typology, proposed by Laurie Corbett, comprised the alpine dingo, found in the cooler south-eastern highlands; the tropical dingo, common in the hot, humid regions of northern Australia; and the desert dingo, adapted to the arid interior. These classifications were based primarily on morphology, coat colour and geographical range, and they offered a useful, if somewhat simplified, view of regional variation in dingo populations.

Recent genetic research, particularly work led by Cairns, has greatly expanded our understanding of dingo population structures. Rather than three broad types, genetic data now reveals distinct regional clusters, including populations from the West, East, South and the Big Desert region (a large stretch of Mallee country in north-west Victoria). These clusters suggest that Australia's dingoes are not just a single homogeneous population but may include emerging ecotypes, which are populations that are genetically and morphologically distinct, shaped by local environments and selective pressures.

Natural barriers such as the Great Dividing Range and the arid interior have likely limited gene flow, allowing regional populations to remain genetically distinct over time. Cold alpine temperatures may have further restricted movement into highland areas, while local conditions, such as prey availability and habitat type, appear to have influenced regional adaptations. Similar patterns are seen in other wild canids, such as wolves, where distinct ecotypes arise in response to prey specialisation and environmental variation.

Genetic data from Western Australia suggests northern entry routes, with these populations still carrying signatures of their initial migrations. Eastern dingoes show greater genetic diversity, likely reflecting overlapping dispersal events and broader migration corridors through more temperate landscapes.

Cairns describes the central deserts as 'genetic reservoirs', preserving lineages that highlight the dingo's remarkable ability to survive in some of the harshest environments on Earth. Despite growing contemporary pressures, particularly from targeted lethal control programs, many south-eastern populations remain genetically distinct, thanks to natural isolation provided by environmental barriers.

Further support for regional differences comes from morphological studies led by archaeologist Loukas Koungoulos, who found that dingoes from south-eastern regions, which experience cooler climates, exhibit distinct cranial features compared to their north-western counterparts. South-eastern dingoes also share morphological affinities with the smaller NGSDs, while north-western animals more closely resemble larger ancestral populations from the initial migrations into Australia. As Koungoulos explains:

> *The origins of this controversial Australian native animal have been heavily debated for more than a century. Our research has found the elusive first links between fossil material, suggesting dingoes evolved locally from an East Asian dog-like ancestor.*

This emerging picture of regional diversity in both form and genetics reflects the dingo's evolutionary journey, a blend of early ancestry, adaptation and the imprint of complex migration histories. Crucially, the dingo's shared genetic heritage with the NGSD reinforces the idea that these unique canids likely stem from a common East Asian ancestor. Together, the morphological and genetic evidence tells a story of a highly adaptable species that has diversified in response to Australia's varied and often challenging landscapes.

AN ONGOING MYSTERY

The origin of the dingo remains one of the most fascinating and evolving stories in Australian zoology. These canids diverged from the lineage of ancient domestic 'dogs' long before humans began farming or deliberately shaping dogs through selective breeding. Genetic evidence suggests that dingoes split from this dog lineage around 10,000–12,000 years ago, likely thousands of years before they ever set foot on Australian soil.

To put that in perspective: the Pyramids of Giza, monuments of an advanced civilisation, are less than 5,000 years old. The dingo's genetic story begins well before that, in a world where human societies in South-East Asia were still shaped by foraging, early cultivation and close contact with the natural world.

But it's critical to understand that genetic divergence does not equal arrival. Just because dingoes became genetically distinct from other dogs 10,000–12,000 years ago does not mean that was when they reached Australia. The archaeological record places their arrival much later, around 6,000 years ago, suggesting they spent millennia living in parts of island South-East Asia and New Guinea, likely as village dogs and/or free-ranging canids, before migrating to the Australian mainland.

This aligns with the theory proposed in this chapter: that humans helped dingo ancestors across the deep-sea barriers of Wallacea, but that the final leg of their journey, from New Guinea into northern Australia, may have been made without human assistance. As adaptable, resourceful animals, they may have walked themselves across the last remaining land bridges or narrow sea passages, eventually spreading across the continent and establishing themselves as wild-living canids.

Although the exact pathway remains unresolved, the evidence points to a complex journey, one shaped by both human influence and the dingo's remarkable ability to find its own way.

3

Dingo Dreaming

WHEN THE DINGO FIRST APPEARED ON AUSTRALIAN SHORES, IT QUICKLY became an integral part of Aboriginal Australian life. Known by many names across the country, including (but not limited to) Binure, Dapin, Gadli, Ganibarra, Ngugum, Papa-inura, Warrigal, Wartaji, Wilkerr, Wongari and Yirrangin, the dingo was a companion, a guardian and a spiritual presence.

To many Aboriginal Australians, the dingo's origins transcend Western timelines. In the Dreaming, the foundation of Aboriginal cosmology and spirituality, the dingo has always been here, its existence interwoven with the origins of humanity itself. The Dreaming encompasses the creation stories, ancestral beings and timeless connections that shape the land, its people and all living things. The dingo holds a place of immense significance, appearing as one of the most common and revered creatures in Dreaming narratives, second only to the legendary Rainbow Serpent and Bunjil.

For the Yarralin people of the Victoria River region in the Northern Territory, the dingo was an ancestor and a creator. They believe that humans have dingo origins, quite literally shaping us into who we are. The dingo gave humans their upright stance, distinctive heads and even their reproductive features, which were the only ones that closely resembled their own. Because until the dingo's arrival, they had only known marsupials, whose anatomy, including backward-facing pouches and genitals, was entirely different. In this way, the dingo was seen as a bridge, linking humanity to the broader animal world. In their Dreaming, humans and dingoes were once one and the same, walking side by side, their fates and lives intertwined. Death itself, also originates with the dingo, who turned down the moon's gift of eternal life, establishing the cycle of life and death for both species.

The dingo represents a 'cultural keystone species', a plant or animal so vital to a community that it shapes their identity and cultural practices. The dingo easily fulfils this role, serving as a spiritual guide, practical ally and enduring symbol of balance. This unique relationship reflects the deep interconnection between people, animals, and the land.

Well-respected anthropologist Deborah Bird Rose, who worked closely with Aboriginal communities, captures the uniqueness of this relationship:

> *Dingoes provided a companionship that had never before existed in Australia. These creatures were the first non-humans who answered back, came when called, helped in the hunt, slept with people and learned to understand some of the vocabulary of human languages. People gave them names, fitted them into the wider kinship structure and took care of dead dingoes in the same way they took care of dead people. Dingoes have been fitted into the sacred geography as extremely powerful Dreamings and they now figure prominently in ritual, songlines and stories.*

But before we continue, it is important to acknowledge that these relationships vary greatly across the country, shaped by local customs and beliefs. Much of what is recorded reflects Western interpretations, which may not always capture the nuance of Aboriginal worldviews. As such, this chapter reflects available sources and does not claim to be definitive.

SYMBOLISM AND SPIRITUAL ROLES

In Aboriginal cultures, the dingo is a spiritual being, a guide and a symbol of profound connection to the land. Through Dreamtime stories, ceremonies and sacred artefacts, the dingo emerges as a figure of creation, transformation and balance. Its intelligence and independence reflect the values of resilience and adaptability, making it both a vital presence in the ecosystem and a bridge between the human and spiritual worlds.

The Dreamtime contains many stories in which dingoes shape both the land and the lives of people. In these traditions, dingoes are often associated with ancestral beings and depicted as companions to humans, offering protection and imparting knowledge essential for hunting, movement and survival across the land.

A story from the Yamatji (Tjupan) nation in Western Australia recounts the 'birth of the dingo'. In this tale, a clever man creates a guardian creature from sacred mulga wood to protect his people from a monstrous kangaroo. After secret rituals and chanting, the wooden figure is brought to life as a great dingo, which successfully defeats the kangaroo, allowing the people to live without fear.

From the lower Cape York Peninsula, as retold in Dick Roughsey's *The Giant Devil Dingo*, the story of Gaiya reflects a similar theme of protection. Gaiya, initially a fearsome devil-dingo, is transformed and reborn as a companion to humans. In this new role, Gaiya becomes a protector and helper, symbolising the dingo's capacity to safeguard and guide people.

In some Central Australian desert stories, ancestral dingoes are said to have shaped the land, digging out waterholes as they travelled. These waterholes provided life-sustaining resources for humans and animals alike, making survival possible across arid country. For many Aboriginal peoples in desert Australia, water sources are also places of gathering, ceremony and deep cultural meaning, tied to ancestral journeys and songlines. In the Kimberley region, stories tell of a dingo that became the guardian of a sacred mountain. Its howl echoed through the valleys, warning of impending danger and reminding people of their duties to protect the land.

Across Australia, ancient rock art depicts dingoes in scenes of daily and spiritual life (Fig. 3.1). These artworks often show dingoes alongside human figures, ancestors or sacred symbols, highlighting their integral role in Aboriginal existence. In the Kakadu region, one striking painting shows a dingo flanked by hunters, its elongated and stylised body reflecting its spiritual essence. Such depictions suggest that dingoes were seen not only as earthly companions, but also as conduits to the spiritual world, their presence bridging the gap between the physical and the divine.

Fig. 3.1. (Top) The Bunjil Rockshelter in Victoria (undated), illustrating Bunjil, the creator spirit in Koori culture, accompanied by his two dingo (wirringan) helpers. According to legend, once Bunjil completed his creation of the land, people and laws, he transformed into an eagle and ascended to the sky, where he remains today. (Bottom) Rock art in Cape York, Queensland (undated), depicting Quinkans (evil spirits) and two dingoes. Source: (Top) Eric Smith; (Bottom) Bradley Smith.

Dingo burial practices in Australia highlight their deep and enduring connections with Aboriginal people. Archaeological evidence shows that dingoes were often buried alongside humans, suggesting their status as cherished companions rather than mere camp animals. Koungoulos has documented over 50 archaeological dingo burials across the country, with a notable concentration in south-eastern and northern regions. At the Curracurrang Rockshelter in New South Wales, for instance, burials dating back over 2,000 years reveal dingoes that lived into old age and received clear signs of care. Some had healed injuries, while others showed unusual tooth wear or fish remains in their stomachs – clues that they were regularly fed and lived closely with people. As Koungoulos explains, 'dingoes were usually buried in the same manner as Aboriginal community members and often in areas used for human burial, sometimes alongside humans.' These practices offer striking archaeological evidence of the dingo's profound place in Aboriginal society: as kin, companion and sacred presence.

The dingo's symbolic importance is also evident in Aboriginal ceremonies and artefacts. Fur and teeth from dingoes are often used to adorn ceremonial objects like headpieces and necklaces, symbolising strength and protection. In some communities, dances and rituals mimic the movements of the dingo, celebrating its agility and independence. These ceremonies often invoke the spirit of the dingo to bless hunts or protect the community, reinforcing its role as a sacred being and a spiritual ally.

PRACTICAL CONTRIBUTIONS TO DAILY LIFE

As well as companions, dingoes were indispensable partners in daily life, where survival often depended on using every available resource wisely. They forged their unique bond with humans early, through the careful collection of pups from dens during the whelping season, which begins around July. This practice, documented across various communities, reflects both opportunistic encounters during hunting activities and deliberate efforts to integrate these animals into daily life. It is likely that celestial cues, such as the seasonal prominence of certain stars, informed the timing of these activities. Healthy pups were hand-reared within the community, ensuring they formed close bonds with humans.

Camp life

At night, dingoes became vital companions for warmth during the desert's frigid temperatures (Fig. 3.2). The extreme temperatures made mutual warming essential for survival. Dingoes would lie close to their human companions, sharing body heat and staving off the chill. This practice gave rise to the Australian slang term 'three-dog night', used to describe an especially cold night requiring the warmth of three dogs to stay comfortable. The colder the temperature, the more dogs you would need. Early explorers often observed this practice among Aboriginal groups, noting how people and dingoes huddled together around fires to keep warm. However, this required dingoes to be carefully raised from puppyhood to ensure trust and reduce fear or aggression. Dingoes unsuitable for this role were often discarded, reflecting the pragmatic nature of these relationships.

Fig. 3.2. A traditional method of carrying a dog while walking, where the animal rests across the small of the back, supported by the woman's bent arms, leaving her hands free and providing warmth. The photograph on the right, taken by Charles Mountford in the 1940s in Central Australia, depicts this practice. It was later published in his 1976 book *Nomads of the Australian Desert*. Source: National Library of Australia (Trove).

Dingoes also played a role in camp sanitation. By scavenging food scraps and even human waste, they helped reduce refuse around living areas, which may have limited the spread of some diseases. However, this role was not without risk. Dingoes also contributed waste of their own and could serve as vectors for disease, raising debate over whether their presence ultimately improved or worsened health conditions.

Guardianship and warning roles

Distinctive dingo barks communicated specific warnings – one for snakes and scorpions, another for strangers and yet another for spiritual threats. As one Yarralin elder, Peter, explained in conversation with Deborah Bird Rose:

> *Dogs help people from jarnpa, they can see and smell them, they bark, or push you away from that jarnpa, try and take you another way. It's a different bark from their snake warning. And different again when ordinary person, [a] stranger comes.*

This physical guardianship extended to spiritual realms, where dingoes were believed to ward off evil spirits (or mamu). Women often relied on dingoes for protection during hunting or other tasks. In some Dreaming narratives, dingoes were seen as spiritual sentinels, their keen senses capable of detecting threats invisible to humans. This dual role as protector and spiritual guardian reinforced their value within the community.

The significance of this protective role also shaped how dingoes were integrated into Aboriginal kinship systems, which relates to the complex networks of social relationships and obligations that structure Aboriginal life. In some communities, dingoes were given skin names, titles that define relationships and responsibilities in Aboriginal kinship systems, typically assigned to parental figures. Elders have explained that dingoes were given the same skin names as parents because dogs look after people. It symbolised their role as guardians rather than dependents, a contrast to Western notions of pets as subordinate animals.

Ignoring a dingo's warning could have dire consequences, as preserved in traditional stories. One Warlpiri tale recounts two people dismissing their dogs' barks about approaching danger. Irritated by what they saw as needless barking, one man cut off the dogs' ears. Without their guardians' acute hearing, the camp was left vulnerable, leading to tragedy, a powerful lesson in respecting dingo instincts.

Resource conflicts and management

Living with dingoes and camp dogs was not without challenges. They often competed for the same resources as humans, targeting prey such as young or weak animals and scavenging scraps around camps. As Annette Hamilton observed in her discussion of Aboriginal relationships with dogs, these animals could be significant competitors for food, both during hunts and around camps. Some estimates suggest that dingoes and camp dogs consumed a notable portion of the available food, although their contributions to the community's diet often fell short of their intake. This imbalance placed pressure on resources and created practical difficulties. Dingoes were often described as 'camp hanger-ons' or scavengers. That is, disruptive presences that brawled, stole food, barked through the night and even tripped up dancers during ceremonies.

To protect their food, people employed creative solutions to outsmart the cunning animals. Six-foot-high shelves or tables built across branches served as makeshift pantries and children were often forced to eat standing upright to keep their meals out of reach of hungry dingoes. Similarly, burial practices were adapted to thwart dingoes' scavenging. Rocks or cairns covered terrestrial graves and aerial tombs were constructed in trees to keep the deceased safe from unwelcome attention (Fig. 3.3).

Fig. 3.3. Aboriginal burial platform (sepulchre) from an engraving by Samuel Thomas Gill in his *Australian Sketchbook* (1864). Such elevated structures were used in a two-stage mortuary process: the body was placed above ground to decompose safely before bones were collected for secondary burial, reflecting both practical concerns and cultural beliefs. The image represents the structure rather than the broader practices. Source: Gill, Samuel Thomas & Hamel & Ferguson. 1864, *Native sepulchre* Printed in colors by Hamel & Ferguson, Melb. [i.e. Melbourne] viewed 11 December 2025 http://nla.gov.au/nla.obj-139539176).

Working partners

In some communities, dingoes were even trained to assist in carrying small loads. Hunters fastened tools or firewood onto the dingo's back, allowing the animal to transport supplies over short distances. This practical relationship was balanced with acts of care and respect, dingoes were often rewarded with portions of the hunt, recognising their vital contributions.

The behaviours of dingoes also inspired and educated young hunters. Boys learned stealth and strategy by mimicking dingoes' hunting techniques, while girls observed their nurturing instincts, which mirrored their future caregiving roles within the community. Stories and songs passed through generations celebrated the dingo's resourcefulness, teaching lessons about survival and the importance of living in harmony with nature.

CONNECTION TO WOMEN AND CHILDREN

For Aboriginal women, the bond with dingoes often mirrored the care and nurturing they provided to their own children. Orphaned dingo pups, rescued from the wild or found abandoned, were treated with tender affection. On Mornington Island, elders recount how women would breastfeed these tiny pups, an act that ensured their survival during critical early stages of development. This practice went beyond necessity, it reflected a profound recognition of the dingo's place within the community, where animals were seen not as separate from human life but as interconnected beings. These hand-raised dingoes grew with an unshakable loyalty to the women who nurtured them, serving as companions and protectors in return.

Dingoes watched over children with care, stories tell of them trailing wandering toddlers or standing guard at the edge of play areas, alert to unseen threats. Stories from Arnhem Land describe dingoes encircling sleeping infants, forming a protective barrier against snakes and other dangers. If a child wandered too far from the camp, the family dingo would often accompany them, a silent chaperone ensuring their safety. These behaviours were not trained, but arose naturally from the deep bond formed between the animal and its human family. In many ways, dingoes were seen as extensions of the community's caregiving network.

Dingoes also played a role in teaching children about the natural world. Young boys, preparing for the responsibilities of hunting, would watch dingoes in action, learning stealth, strategy and timing from their movements. This symbiotic relationship extended beyond the practical, reaching into the spiritual realm. Dreamtime stories often depict dingoes as spiritual guides for children, leading them safely through both physical landscapes and metaphorical journeys of growth and learning.

In some communities, the connection between children and dingoes was immortalised in song and story. One tale from the Western Desert speaks of a dingo that befriended a group of children, teaching them the locations of safe shelters and vital food sources. These lessons ensured their survival during hard times. When the children grew up, they passed the story on to future generations, preserving the dingo's wisdom and their own gratitude for its guidance.

MYTHOLOGY AND CAUTIONARY TALES

Storytelling traditions in Aboriginal cultures intertwine lessons about the natural world with deeply held cultural values. Mythological and cautionary tales serve as vehicles for passing down knowledge, teaching respect for the land and its creatures, and instilling an ethical framework for coexistence with nature. The dingo's dual nature – wild yet cooperative, fiercely independent yet loyal – makes it a central figure in these narratives, which highlight its role in the ecosystem and the responsibilities it demands of humans.

In northern Australia, stories and teachings emphasise the responsibility humans hold toward animals that share their camps and landscapes. Spiritual beings such as the Mimih are said to enforce moral conduct, while dingoes are remembered as vigilant protectors capable of warning against danger. At the same time, these traditions caution against over-reliance, reminding people that even helpful wild animals require clear boundaries. Together, such stories reinforce a careful balance between respect, gratitude and restraint.

In some Dreaming traditions, dingoes take on roles that caution against greed, arrogance and neglect, reinforcing lessons about vigilance, restraint and care for the land. They may appear as tricksters, moral enforcers or resilient survivors, using their actions to teach respect for shared resources, purposeful hunting and the balance that sustains life.

One well-known example appears in *The Bora of Baiame,* where the Madhi, or Dreaming dingoes, disrupt a sacred gathering by refusing to listen and laughing through ceremonial business. Despite repeated warnings, they persist in their behaviour and are ultimately punished by losing human speech, left only with the ability to howl. The story serves as a cautionary lesson about heedlessness, respect and the consequences of ignoring cultural law.

HUNTING ASSISTANCE

The role of dingoes in Aboriginal hunting practices was multifaceted and varied significantly depending on prey type, environmental conditions and the structure of hunting groups. Their sharp senses, natural agility and attentiveness to human activity made them valuable in many hunting scenarios, though not universally relied upon or without limitations.

For many Aboriginal groups, dingoes were considered useful in the pursuit of large game, including kangaroos, emus and wallabies (Fig. 3.4). One notable method, described by early ethnographers and supported by archaeological interpretation, involved large-scale game drives. These coordinated events, often described in early accounts, brought together entire communities, with men forming ambush lines armed with spears while women, children and dingoes acted as beaters. The dingoes reportedly flanked and helped flush out prey, encouraging movement towards hunters or into natural traps. In south-eastern Australia, British naturalist Herbert Ward recorded an account from 1907 in which a group successfully drove kangaroos into pits, killing 8–10 animals in a single event. In such contexts, dingoes may have acted as force multipliers, disrupting or redirecting prey, adding pressure to drives and amplifying human efforts.

That said, I think it's worth exercising a little caution in interpreting these accounts. Although I don't doubt that dingoes contributed meaningfully to some of these hunting activities, it's less clear how much of their behaviour was intentional or coordinated in the way we might associate with trained dogs. As wild animals with an independent streak, dingoes likely aligned their actions with human hunters opportunistically, drawn by movement, noise or the prospect of food, rather than operating in planned cooperation. Some descriptions may also reflect a mix of dingo and camp dog behaviours, or reveal European assumptions about how humans and canines typically interact. Still, their repeated presence in these stories and scenes points to a complex, adaptive partnership worth taking seriously.

In contrast, dingoes were equally effective in aiding the capture of smaller prey, such as bilbies, goannas and bandicoots, which are critical resources in arid regions where food was often scarce. Women and children, often accompanied by dingoes during foraging trips, relied on the animals' acute sense of smell and powerful digging abilities to locate hidden prey. For example, dingoes have been observed sniffing out burrows and using their forepaws to dig. Hunters would stand ready to catch or dispatch the prey as it emerged. This opportunistic method likely emerged from observation and experience rather than formal training and it

Fig. 3.4. Pintubi (NT) men with young dingoes used for hunting. Source: National Library of Australia.

blurred the line between foraging and hunting, demonstrating the practical value of dingoes in supplementing daily subsistence efforts.

Despite their many advantages, dingoes' involvement in hunting was not without challenges. Their strong prey drive sometimes led them to disrupt hunts by barking prematurely or consuming captured prey before hunters could intervene. To manage this, some groups selected certain pups, those showing promise or a calm temperament, and reared them closely to ensure they bonded with humans and followed cues during hunts. In some cases, unwanted behaviours were curbed through training or physical interventions, such as removing teeth to prevent independent kills. This was not selective breeding in the modern sense, but rather a practice of investing time and trust in individuals who showed potential.

The relationship between dingoes and gendered hunting roles offers additional insights. Men typically engaged in structured hunts for large game, while women often relied on dingoes for assistance in small game hunting and foraging (Fig. 3.5). These activities played a significant role in expanding dietary breadth during the Holocene. Animal remains discovered at ancient sites suggest that dingoes' assistance in targeting small prey may have contributed to shifts in hunting and subsistence strategies over time.

Overall, the roles of dingoes in Aboriginal hunting practices demonstrate the adaptability and ingenuity of these partnerships. From aiding in large-scale hunts to supplementing daily

Fig. 3.5. Women and children gathering murrnong (daisy yam) and other food in the Goulbourn River area, as sketched by squatter Henry Godfrey, November 1, 1843. Source: the artist's sketchbook, La Trobe Picture Collection State Library of Victoria.

foraging efforts, dingoes were integrated into subsistence strategies in diverse and innovative ways, shaped by the ecological and cultural contexts of each community.

DINGOES AS FOOD

Dingoes held a complex position in Aboriginal diets. Some groups actively hunted and consumed wild dingoes, while others viewed their meat with disdain or adhered to cultural taboos that prohibited their consumption altogether. For those who did consume dingoes, the meat was prepared by roasting in earth ovens (traditional cooking pits dug into the ground and lined with hot stones), much like other game. Hunting efforts often focused on dens during whelping season, where both adults and pups could be captured. However, the archaeological record suggests dingoes were never a staple food source, as their remains are relatively rare in faunal assemblages. That is, there were very low frequencies of dingo bones in middens and camps compared with kangaroo, wallaby, emu or reptiles. Among groups that did not consume dingoes, taboos often stemmed from their symbolic significance or totemic restrictions. Some Central Australian groups placed restrictions on harming female dingoes and pups, with killing more likely to be directed toward adult males.

ECOLOGICAL AND NAVIGATIONAL ROLES

In the Australian outback, where water often meant the difference between life and death, dingoes were not just companions, they were essential guides to survival. Their sharp senses allowed them to detect water sources hidden from human perception, a skill honed through millennia of adaptation to arid environments. Aboriginal trackers observed how dingoes would pause at seemingly barren stretches of desert, their noses twitching as they caught the faintest scent of moisture. Moving deliberately, they might circle an area before beginning to dig, their claws revealing shallow waterholes or water just beneath the surface.

Accounts from desert Aboriginal groups describe moments of survival in which dingoes played a crucial role in locating water. During periods of intense heat and dwindling supplies, people observed their dingo companions behaving unusually, sniffing and scratching at areas of sand that appeared no different from the surrounding landscape. Trusting the animal's instincts, they dug alongside it, uncovering hidden springs that sustained them. Similar accounts are shared across many Aboriginal groups, illustrating the deep trust placed in the dingo's extraordinary abilities, particularly during the dry season when even experienced human trackers struggled to locate water.

Dingoes' water-finding abilities extended beyond immediate survival to the long-term ecological knowledge of Aboriginal communities. Knowledge of springs and soaks uncovered by dingoes was passed down through generations, and these sites often became culturally significant. They were seen not only as practical resources but as places imbued with spiritual meaning, gifts from the Dreamtime. In some traditions, dingoes were believed to be spirit beings that had shaped the land itself. Across several desert traditions, ancestral dingoes are associated with the creation of waterholes, their movements across the land leaving behind pools that sustained future generations.

Beyond their role as water seekers, dingoes were integral to the practice of fire management, a cornerstone of Aboriginal land stewardship. Controlled burns, carefully planned and executed, were used to clear old vegetation, promote regrowth and encourage biodiversity. These burns also created opportunities for hunting. Dingoes, ever alert, would patrol the fire's edges, watching for animals fleeing the advancing flames. Bandicoots, lizards and even birds, disoriented by the smoke and heat, became easy prey for both dingoes and human hunters. This collaboration turned the fire into a tool not just for managing the landscape but for sustaining life.

During controlled burns, dingoes were often observed moving along the fire line, their movements fluid and purposeful. As flames advanced through the underbrush, dingoes would flush out animals seeking to escape the smoke and heat, creating hunting opportunities for both themselves and nearby people.

The relationship between dingoes and fire extended beyond hunting. Aboriginal people often observed dingoes resting near freshly burned ground, where the heat would drive out ticks, fleas, and other pests, as well as clearing away dense vegetation. This behaviour not only offered relief from parasites, but also gave dingoes a clearer view to detect potential predators or prey, showing their ability to adapt to a transformed environment. Some communities believed that the dingo's instinctive affinity for fire was tied to its role in the Dreamtime, where it was said to have walked alongside the first people, teaching them how to use fire as a tool.

Through their interactions with water and fire, dingoes embodied a balance between survival and stewardship. Their behaviour not only supported human life, but also sharpened understanding of how the land responds to heat, drought and renewal. Aboriginal communities drew these observations into cultural practice and oral knowledge, recognising dingoes not simply as animals, but as co-navigators of Country whose actions revealed the deep interconnectedness of life in the Australian outback.

INTANGIBLE HERITAGE

The profound relationship between Aboriginal Australians and dingoes continues to resonate today. Stories of partnership, survival and spirit are preserved in oral traditions, songs and dances, connecting present-day communities with an ancient bond. For many, the dingo remains a living symbol of the deep connection between people and Country.

For many Aboriginal people, the dingo represents resilience and harmony. Their relationship with the dingo is not one of domination or domestication; such acts would conflict with their deep respect for the natural law that guides life on Country. Instead, the dingo is recognised as an essential part of maintaining balance, embodying the enduring responsibility to care for Country. This responsibility includes ensuring the continued presence of dingoes in the landscape, reflecting a worldview profoundly rooted in coexistence with nature. This deep understanding of coexistence and balance offers important perspectives for contemporary wildlife conservation and management.

4
The case of the vanishing tiger

THE STORY OF THE THYLACINE AND THE DINGO IS ONE OF DRAMATIC contrasts. These two predators, one an ancient marsupial carnivore, the other a relative newcomer (Fig. 4.1), found themselves sharing the Australian landscape. Their interactions would reshape Australia's ecological communities and offer insights into how the arrival of new predators can transform environments.

To understand one of Australia's most intriguing extinction mysteries, we first need to meet the protagonist: the thylacine. Though nicknamed the Tasmanian tiger, this was no big cat. This was a dog-like marsupial predator with a distinctive stiff-backed gait, a trait typical of marsupials. Its most striking feature was a series of dark stripes that ran across its tawny fur from shoulders to tail – a distinctive pattern that inspired the 'Tasmanian Tiger' moniker. But it was the thylacine's jaw that set it apart. Those powerful mandibles could stretch to an almost impossible 120-degree angle. As a marsupial, the thylacine female raised her young in a rear-facing pouch, where tiny, jellybean-sized young developed into future predators. The fact that it looked a bit like a dog, and had a pouch, naturally earned it the scientific name *Thylacinus cynocephalus* – the 'dog-headed pouched one'.

The thylacine was a uniquely Australian predator, one of several marsupial carnivores that shaped the continent's ecosystems. The largest of these predators, it shared its hunting grounds with other significant hunters such as the Tasmanian devil (*Sarcophilus harrisii*) and various species of quolls (*Dasyurus* spp.). Together, these marsupial predators influenced prey populations and maintained ecological balance across diverse habitats from open plains to dense forests for tens of thousands of years. Its story, however, is often overshadowed by its final chapter: extinction. By the 1930s, the thylacine was no more, the last known individual dying alone in a Hobart Zoo cage in 1936.

Despite its name, the 'Tasmanian tiger' wasn't always confined to Tasmania. For most of its existence, the thylacine thrived on mainland Australia, where it roamed long before Tasmania became an island. The species disappeared from the mainland around 3,000 years ago, with the dingo's arrival typically cited as a key factor in its local extinction.

Because thylacines survived in Tasmania – the one place dingoes never reached, it seemed reasonable to assume they had been outcompeted by a more advanced placental predator. It was a simple story, and for a time, a satisfying one. However, recent research suggests dingoes likely arrived much earlier than once thought, perhaps as far back as 4,000–8,000 years ago. This would mean that the two predators coexisted for thousands of years before thylacines disappeared from the mainland.

To make sense of this long overlap, we need to consider not just the arrival of the dingo, but the broader changes taking place across the continent. As discussed in the previous chapter, a wave of

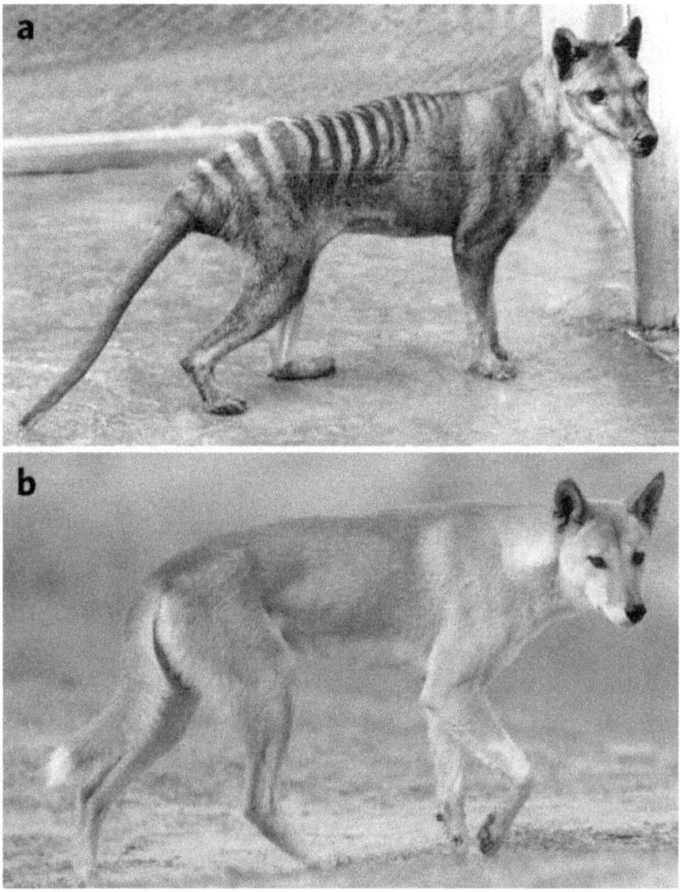

Fig. 4.1. A comparison of the thylacine (a) and the dingo (b), illustrating their convergent body shapes despite very different evolutionary lineages. The thylacine was a marsupial and the dingo is a placental mammal, but both evolved similar forms suited to their role as mid-sized predators. Subtle differences in posture, skull shape and tail structure reflect deeper distinctions in behaviour, hunting style and physiology. Source: recreated from Feigin *et al.* (2018).

cultural and technological transformation, (or intensification as mentioned in Chapter 2), swept across mainland Australia beginning around 4,000 years ago. This period brought new hunting tools, population growth and shifting relationships between humans, animals and ecosystems. Add to that a changing climate and we begin to see how the thylacine may have been gradually squeezed out not just from a single new competitor, but from a rapidly transforming world.

NEW RIVALS: AN EVOLUTIONARY ARMS RACE

Although these two species superficially resembled each other, their differences in physical capability, social behaviour and metabolic needs would profoundly influence their coexistence – and, ultimately, their fate.

In 2007, palaeontologist Stephen Wroe and his colleagues brought modern engineering tools to bear on the ancient mystery. Their biomechanical analysis of thylacine and dingo skulls – essentially, a scientific study of how their skulls handled various forces – revealed significant differences in how each predator hunted prey. The thylacine could generate a powerful bite, but its skull was poorly adapted to withstand the stresses of wrestling with larger, struggling prey such as kangaroos. This limitation likely confined the thylacine to hunting smaller, more agile animals such as wallabies and bandicoots.

This specialisation, however, imposed a vulnerability, noted by Wroe *et al.* 'Especially among large predators, the more specialised a species becomes, the more vulnerable it is to extinction', emphasising the delicate ecological balance the thylacine relied upon. Even small disturbances to its environment could have tipped this finely tuned predator towards decline.

By contrast, the dingo, with its robust skull and versatile hunting strategies, thrived. Biomechanical analysis shows that dingo skulls are not only stronger overall, but better suited to absorbing the forces involved in subduing large, struggling prey. This structural resilience allowed dingoes to tackle a broader range of prey, from small marsupials to large kangaroos (Fig. 4.2). They could hunt alone, in pairs or in packs. This flexibility further enhanced their predatory reach. Colin McHenry, a co-author of Wroe's study, explained, 'The introduction of the dingo meant the thylacine was suddenly competing with a more sophisticated predator that also hunted in packs for the same food. History would support the view that there wasn't room for both types of predators.'

The differences were also metabolic. Dingoes had a significantly higher energy requirement than thylacines, with estimates suggesting they needed up to 40% more food than a thylacine of the same size. This made them more active and persistent hunters, capable of covering long distances in search of prey. A thylacine might abandon a chase after a short burst, whereas dingoes could afford to keep going, supported by their stamina and metabolic drive. But this

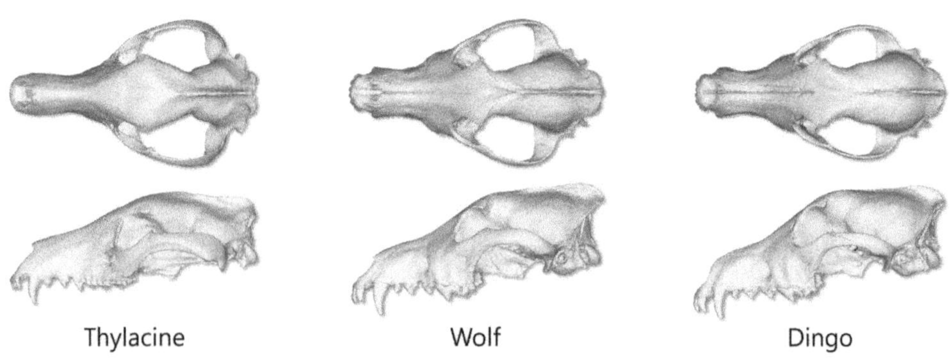

Fig. 4.2. Skull comparisons of the thylacine, grey wolf and dingo in both dorsal and lateral views modified from Rovinsky *et al.* (2021). These images, scaled to the same length for comparison, highlight that dingoes and wolves are far more robust in the snout, 'face' and jaw structure. The dingo's cranium is broader and more reinforced, supporting stronger bite forces and better shock absorption – traits useful for subduing large, active prey. In contrast, the thylacine's skull, while capable of generating a powerful bite, was more fragile under stress. Source: Douglass Rovinsky.

came at a cost: they had to hunt more frequently and were more vulnerable during periods of prey scarcity. In contrast, the thylacine's lower energy needs may have been an advantage in lean times, allowing it to survive on less. It's a classic evolutionary trade-off. Endurance and flexibility versus efficiency and restraint, a pattern still reflected today in the contrast between the dingo's wide-ranging hunting style and the more conservative strategies of native predators like quolls.

Adding to the complexity was the dramatic size difference between male and female thylacines – a feature biologists call sexual dimorphism. Building on Wroe's study, research by Mike Letnic and colleagues in 2012 revealed that female thylacines were significantly smaller than their male counterparts, often less than half the size of a dingo. Research on specimens from Western Australia revealed just how pronounced this size difference was. In the Nullarbor region, the smallest thylacines were ~20% smaller than the smallest dingoes, while in the south-west, this difference increased to 30%. This size disparity would have made female thylacines particularly vulnerable in direct confrontations with dingoes, undermining their ability to reproduce and maintain population numbers.

Larger predators often kill smaller ones – a behaviour ecologists call interspecific killing – but this rarely leads to complete extinction. Today's African lions may kill leopards and wolves may kill coyotes, but these smaller predators persist despite the pressure. Even more tellingly, foxes and feral cats survive in modern Australia, despite frequent killing by dingoes. Simple predator competition seldom tells the whole extinction story.

Evidence from Aboriginal rock art in the Kimberley region provides additional clues about the timing of these species' interactions. Thylacine images disappear from the art record well before dingoes appear in more recent paintings, hinting at the possibility that thylacines may have already been declining in some regions before dingoes became firmly established.

In 2013, ecological modeller Thomas Prowse and his colleagues at the University of Adelaide challenged the prevailing extinction narrative. 'Perhaps because the public perception of dingoes as "sheep-killers" is so firmly entrenched, it has been commonly assumed that dingoes killed off the thylacines and devils on mainland Australia", Prowse observed.

If dingo competition alone didn't drive thylacines extinct, however, then what did? To solve this mystery, we need to examine what else was happening in Australia during this pivotal period.

THE HUMAN FACTOR

By 4,000 years ago, we know that dingoes had well and truly spread across the continent. During this same period, an equally dramatic but less visible transformation was taking place (mentioned in Chapter 2 as a period of 'intensification'). Human populations surged and new hunting tools like spear throwers and barbed spears significantly increased the effectiveness and reach of hunters. These innovations reshaped the predator–prey dynamic, allowing people to hunt with greater precision and at safer distances. Rock shelters across mainland Australia show a dramatic increase in occupied sites, artefact densities, and revolutionary hunting technologies. Between 4,000 and 3,000 years ago, sophisticated stone tools emerged, including new spear points and small, sharp blades that, when combined with spear throwers, dramatically extended hunting capabilities.

'Careful analysis of radiocarbon dates suggests that the Aboriginal population may have tripled during this period', notes Australian extinction expert Chris Johnson. This represented a fundamental shift in how people lived. Settlements became more permanent, people expanded into new territories and hunter–gatherer strategies became more elaborate and efficient.

Yet Tasmania tells a different story. The island's human population remained relatively stable, with hunting technology evolving more gradually than the rapid innovations seen on the mainland. This striking contrast in human development, according to Johnson and Barry Brooks's interpretation, might explain the thylacine's survival in Tasmania better than the simple absence of dingoes.

CHANGING HUNTING PATTERNS

Human hunting practices transformed across mainland Australia, influenced by both population growth and the presence of dingoes. Archaeologist Loukas Koungoulos's work shows how, rather than being simple competitors, dingoes shaped hunting in multiple, sometimes unexpected ways over thousands of years.

Historical records and oral accounts suggest dingoes were sometimes used as companions in hunting small game like possums and bandicoots, though they were considered less reliable when pursuing larger prey such as kangaroos and emus. Although this partnership may not have been widespread or formalised, it likely shaped local hunting strategies in some regions.

Archaeological remains reinforce the broader picture: over time, hunting practices gradually shifted away from large prey like kangaroos and emus towards a more varied array of smaller animals. These included possums, bandicoots and rodents, but also echidnas, reptiles and ground-dwelling birds. 'We see a clear transition from predominantly large prey to a more diverse range of smaller animals,' notes archaeologist Melanie Fillios, 'suggesting human hunters were adapting to new ecological realities.'

Koungoulos identifies specific mechanisms through which dingoes may have contributed to this shift. Their predation on large herbivores like kangaroos reduced grazing pressure, allowing more vegetation to flourish. This improved habitat supported larger populations of small animals, while simultaneously making the pursuit of large game more unpredictable and riskier, in the sense that large prey became harder to find and more energetically expensive to pursue, often with no guarantee of a successful return – for both humans and thylacines.

Their theory is supported by several lines of evidence. At Tunnel Cave in south-western Australia, an Aboriginal occupation site, researchers found a steady decline in the proportion of large mammals (over 5 kg) in deposits spanning thousands of years. The percentage drops from 15% in layers dated 6,780–10,200 years ago to just 3% in layers from 1,290–1,450 years ago. Similar patterns emerge at other Aboriginal sites, such as Graman (GB4), a rock shelter in northern New South Wales, where the proportion of large macropod remains declines dramatically over 4,000 years, with no identifiable macropod bones in the uppermost layers. Instead, smaller animals, particularly possums, become the dominant prey species.

Importantly, this pattern doesn't appear in naturally accumulated bone deposits. At Skull Cave, a pit trap in Western Australia that was never occupied by people, the proportion of large animals like western grey kangaroos remains stable at ~2% throughout the deposit. Unlike Aboriginal sites where bones accumulated through human hunting and discard, natural sites like this one reflect the composition of local fauna unaffected by human foraging. This contrast suggests that the changes seen at sites like Tunnel Cave reflect genuine shifts in human hunting behaviour, not broader environmental changes.

'What's particularly striking,' notes Johnson, 'is that this shift towards smaller prey isn't just a matter of changing preferences. The evidence suggests a fundamental reorganisation of the entire predator–prey system, affecting everything from hunting technologies to settlement patterns.'

For thylacines, already specialised in hunting medium-sized prey, these changes created an ecological nightmare. They now faced unprecedented competition from two fronts: dingoes and human hunters both increasingly targeting similar quarry. This pressure would have been particularly intense during periods of environmental stress, when prey became scarce, squeezing the thylacines into an ever-narrowing survival corridor

A CHANGING CLIMATE

The human transformation of mainland Australia wasn't happening in isolation. The climate itself was shifting in ways that would have stressed both predators and their prey. During the Holocene period, after relatively wet and stable conditions between 8,000 and 5,000 years ago, Australia entered a period of increasing environmental instability.

The key driver of this change was an intensification of the El Niño-Southern Oscillation (ENSO) system. Around 5,000 years ago, El Niño events became more frequent and more intense, creating a climate marked by dramatic swings between drought and abundance. For predators dependent on stable prey populations, this new climatic regime would have been particularly challenging. As researcher Owen Brown demonstrated in 2006, these heightened ENSO cycles created boom-and-bust dynamics that would have severely stressed both predators and their prey.

Modern studies of predator–prey relationships during drought cycles give us insight into how these changes might have affected thylacines. Today, when El Niño brings drought to Australia's arid regions, kangaroo populations can crash dramatically. Large herbivores often congregate around remaining water sources, creating temporary concentrations of prey, but also increasing competition between predators. Even contemporary dingoes struggle during these periods, sometimes forced to shift from hunting kangaroos to smaller prey or scavenging.

For thylacines, these boom-and-bust cycles in prey availability would have been even more of a problem. 'The fossil record shows that different species responded to climate stress in different ways', explains palaeontologist Tyler Faith.

'Some adapted by changing their behaviour or diet, while others, like the thylacine, found their specialised hunting strategies increasingly challenging to maintain.' Unlike the dingo, thylacines were specialised carnivores with less ability to fall back on alternative food sources.

When preferred prey became scarce, they had fewer options. The combination of unpredictable rainfall and more variable prey abundance would have made it increasingly difficult for thylacines to maintain stable populations.

Once again, Tasmania tells a different story. The island's position in the path of prevailing westerly winds meant it was less affected by ENSO cycles. The climate remained relatively stable and predictable, offering yet another reason why thylacines persisted there while disappearing from the mainland.

A FATAL COMBINATION

Over the past decade, research, particularly the modelling work by Thomas Prowse and colleagues, has revealed how multiple pressures on thylacine populations combined in devastating ways. Environmental stressors didn't act in isolation. During droughts, predators and prey were pushed into the same shrinking pockets of habitat, increasing encounters with both dingoes and human hunters. In these so-called 'compression points,' competition intensified, and survival strategies unravelled.

As climate variability increased, prey became more unpredictable. Dingoes, with their flexible diet and hunting strategies, adapted. Thylacines, more specialised and less adaptable, struggled. Meanwhile, human hunters, now armed with advanced tools and sometimes accompanied by dingoes, could efficiently target the same medium-sized prey thylacines relied on.

Tasmania provides a striking natural experiment that helps us understand how these factors interacted. There, the absence of dingoes was just one part of a broader pattern – the island experienced more stable climate patterns, had a more gradually evolving human population and maintained consistent prey availability. Without these compounding pressures, thylacine populations remained viable for thousands of years longer than on the mainland. This contrast reinforces how the combination of pressures, rather than any single factor, drove mainland populations to extinction.

The implications of this more sophisticated understanding extend beyond the thylacine's story. For modern conservation efforts, it highlights how species extinctions often result from complex cascades of interacting pressures rather than single causes. Understanding these dynamics helps us recognise how different threats might combine and amplify each other – a crucial consideration for preventing similar losses in our own time.

THE DINGO'S TRUE IMPACT

Although the dingo has often been blamed for multiple extinctions on mainland Australia, the latest research suggests its ecological impact after it arrived was largely neutral or even positive, with only one species – the Tasmanian native hen (*Tribonyx mortierii*) – being clearly driven to extinction by dingo predation. As a flightless bird weighing ~2 kg that lived close to water sources, the native hen would have been particularly vulnerable to this novel predator. As argued by Johnson and Letnic, rather than causing widespread extinctions, the dingo appears to have effectively stepped into the ecological roles previously filled by the thylacine and Tasmanian devil after their disappearance. This more nuanced understanding shows that while the dingo's arrival certainly impacted Australian ecosystems, its role was primarily one of ecological substitution rather than destruction, with the native hen being the sole exception.

LEARNING FROM THE PAST

History has a way of repeating itself – sometimes in painfully ironic ways. Our fascination with the thylacine hasn't dimmed with time. Each new discovery reignites hope and debate about this lost predator. When scientists in late 2024 uncovered a perfectly preserved thylacine head in a dusty cupboard at the Melbourne Museum – a specimen that had quietly waited in ethanol for over a century – it seemed like the species was giving us one final gift. This remarkable find provided the most complete genetic blueprint yet of the thylacine, with scientists successfully sequencing more than 99.9% of its genome.

In partnership with the University of Melbourne, the American company Colossal Biosciences stands poised to attempt something that sounds like science fiction: bringing the thylacine back from extinction. Their plan involves modifying the DNA of a living marsupial, the fat-tailed dunnart (*Sminthopsis crassicaudata*), until it matches that of the thylacine. The company speaks optimistically about restoring an apex predator to its native ecosystem and helping to 'balance the ecological scales'.

Yet, this dream of resurrection reveals a profound contradiction in our relationship with predators. Millions of dollars are poured into bringing back an extinct predator, while Australia's current apex predator, the dingo, faces mounting threats to its survival. Across vast swaths of the continent, dingoes are systematically poisoned to protect livestock. Barrier fences stretch thousands of kilometres to exclude them from pastoral lands. Their genetic identity is being diluted through isolation, fracturing packs due to lethal control and in some places even hybridisation with domestic dogs.

The parallels are striking and sobering. The remaining thylacine in Tasmania's final extinction in the 30s came at the hands of farmers and government bounty hunters who saw it as a threat to sheep (Fig. 4.3). Even if scientists succeed in creating a living thylacine, what future would await it? The same conflicts that drove it extinct still exist. We haven't learned to coexist with the dingo, so why would a resurrected thylacine fare any better?

More ironically still, even Tasmania's current top predator, the Tasmanian devil, faces severe population declines due to a devastating transmissible cancer, which is a devil facial tumour disease, a uniquely aggressive form of cancer that spreads directly between animals through physical transmission. So rather than resurrecting extinct predators, perhaps our focus should be on protecting the ones we still have.

The bitter truth, as we will discover in chapters to come, is that we now understand dingoes play a crucial role in maintaining healthy ecosystems. The very role once filled by thylacines. The loss of the thylacine created an ecological vacancy that dingoes stepped into, becoming Australia's new apex predator. To lose the dingo would be to repeat a profound ecological disruption, one that we now know was more complex and preventable than previously understood.

Fig. 4.3. A hunter poses with a dead thylacine in 1869. Once widespread across mainland Australia, thylacines disappeared from the mainland around 3,000 years ago, likely due to a combination of human pressures, dingo competition and environmental change. The species survived in Tasmania until the 20th century, with the last known individual dying in Hobart Zoo in 1936. The thylacine was officially declared extinct in 1986. Source: Tasmanian Museum and Art Gallery.

The thylacine's story offers a clear warning. While the dream of resurrecting extinct species captures our imagination, the more urgent task is preventing extinctions in the first place. The complex interplay of factors that drove the thylacine extinction, namely human pressures, environmental change and competition, mirrors the challenges facing many of Australia's predators today. Perhaps the most fitting memorial to the thylacine would be ensuring its successor doesn't meet the same fate.

Part 2

The dingo

5

Secrets to survival

SINCE THE DINGO FIRST SET PAW ON AUSTRALIAN SOIL, IT HAS REIGNED AS the 'king' of the Australian bush, a title often used by bushmen, naturalists and locals to describe its dominance and adaptability. As the continent's largest land-based predator, it has claimed nearly every habitat, from arid deserts and subalpine forests to snowy alps, shaping prey populations and influencing the behaviour of smaller predators.

Every feature of its body and behaviour tells the story of a predator perfectly moulded by this land, for this land (Fig. 5.1). From its keen senses to its versatile physique, each adaptation serves a purpose.

But for Australia's largest apex predator, the dingo is deceptively small, standing around knee height to an adult human (50–60 cm at the shoulder) and weighing just 13–20 kg. This compact form allows it to slip through dense scrub and outmanoeuvre prey, making it a creature of balance: muscular, but not bulky, designed for endurance over power.

Small, yes, but effective. Unlike larger predators from other continents, like wolves, which evolved to hunt large-bodied, hoofed animals in abundant grasslands and forests, the dingo thrives in a land where prey hops rather than gallops. Australia offers no gazelles or deer; instead, the dingo's menu is made up of kangaroos, wallabies, rabbits, reptiles and rodents. The dingo's moderate size and flexible social structure make it uniquely suited to this environment: capable of bringing down kangaroos in pairs or small groups during times of abundance and equally adept at surviving solo on smaller prey when food is scarce.

A wolf or jackal dropped onto the Australian continent would be out of place. Built for cooler climates and larger energy-dense prey, wolves require high-calorie food sources that the Australian bush rarely provides. Their large bodies and pack-based strategies are ill-suited to Australia's leaner, less predictable conditions. They would overheat, burn too much energy chasing nimble marsupials and likely starve. Likewise, domestic dogs lack the independence and adaptability needed to navigate the boom-and-bust cycles of this continent. The dingo alone is built for this balancing act: lean, efficient and endlessly resourceful.

Visually, the dingo embodies Australia. Its coat, typically golden or sandy, blends seamlessly with dry grasses and red sands (Fig. 5.1). Yet the dingo is not a one-size-fits-all predator. Desert-dwelling dingoes are red or sandy coloured; those living in forests often sport darker hues; tropical dingos' coats are thin and wiry, while alpine dingos grow thicker and softer double coats in winter, their fur taking on a luxurious density. Whatever the climate, the coat is short and coarse, practical, durable and low-maintenance. This unassuming pelt serves as the perfect armour.

Table 5.1 provides a quick reference to some of the key traits, behaviours and ecological roles of the dingo. It highlights the remarkable adaptability and importance of this apex predator across Australia's diverse landscapes.

Fig. 5.1. Free-ranging dingoes photographed in the alpine region of south-eastern New South Wales. The images reveal the striking variety of coat colours and patterns found within a single population, from pale sandy tones to rich gingers and darker saddled forms. Sources: Michell J Photography, Cooma, NSW.

Table 5.1. Key characteristics of the Australian dingo, highlighting its physical features, behavioural traits and ecological adaptations

Characteristic	Detail
General appearance	A lean, medium-sized canid with a narrow chest, broad wedge-shaped head, strong neck, erect ears, long muzzle and a bushy, low-set tail. Built for endurance and agility, with long legs, compact muscles and an efficient, ground-covering gait. Always alert, with a poised, athletic stance and a highly adaptive physique
Size	Wild adults typically weigh 13–20 kg. Shoulder height ~48–60 cm. Total length including tail ~117–124 cm. Captive dingoes may be heavier
Eyes and ears	Almond-shaped with dark rims, eyes ranging from dark brown to dark amber in lighter coloured dingoes. Upright, pointed ears, small to medium in size, with a clean, fur-lined edge
Tail	Bushy and bottle-shaped, usually carried low with a white tip; thickness and carriage variables. Features a scent gland near the base (precaudal gland)
Coat	Short to medium coat; coarse outer layer with a seasonal undercoat. Varies with climate – thicker double coat in alpine regions, sparse in tropics. No strong body odour
Coat colour	Highly variable – most are ginger, but also black-and-tan, white, and solid black. Hybrids may show brindle, patchiness, or sable, however coat colour is not a reliable indicator of ancestry
Gait and movement	Moves lightly, efficiently, and in straight lines. Hindfoot typically lands in the print of the forefoot. Capable of trotting, cantering, and suspended gallop

continued

Table 5.1. (Continued)

Characteristic	Detail
Speed	Up to 48 km/h (30 mph); valued more for stamina and endurance than top speed
Bite force	Estimated at around 320 psi – comparable to similarly sized canids
Eye colour	Typically amber to golden-yellow, with a dark eye rim
Habitat	Occurs across mainland Australia in diverse environments: deserts, woodlands, grasslands, tropical forests, and alpine regions. Absent from Tasmania
Home range and territory	Highly variable: typically 10–300 km^2, influenced by habitat, prey density, and social structure
Diet	Primarily carnivorous. Diet includes small to medium-sized mammals (e.g. rabbits, wallabies), birds, reptiles and carrion. Also consumes insects, fruits and other plant material when needed. Cooperative hunting of large prey (e.g. kangaroos) occurs in some regions
Hunting strategies	Opportunistic and adaptive predators; use stealth, ambush and cooperative pack tactics when targeting large prey
Activity	Mostly crepuscular (active at dawn and dusk), but may be nocturnal in hot climates or where humans are active
Social organisation	Typically live in stable family packs with defined territories. Some individuals may disperse and roam alone temporarily

To truly understand these remarkable adaptations requires closer observation. In the wild, dingoes offer only fleeting glimpses: a silhouette against red sand, a shadow in the scrub. To uncover the details hidden beneath fur and bone, I am going to turn to the Dingo Discovery Centre and Research Sanctuary located in Toolern Vale, Victoria.

ANATOMY OF SURVIVAL

I turn off the main road and guide my car along the gravel driveway. Lyn Watson's house, nestled in the Macedon Ranges, comes into view. The house is as much a part of the sanctuary as the pens and paddocks that it overlooks. From its rear deck, you can see it all: clusters of enclosures terraced into the slope, spacious paddocks beyond, and beyond that, the endless horizon and Port Phillip Bay.

I park and make my way towards the house. Lyn appears at the porch, waving as I approach. 'Good to see you again!' she calls, her voice full of warmth. At eighty-something, she still moves with the energy of someone decades younger, her passion for dingoes as untamed as the dingoes she cares for, and the same as when I first met her almost 20 years ago when I started coming here to conduct behavioural experiments for my PhD.

Here, among the sanctuary's enclosures, we can examine the dingo's anatomy with an intimacy impossible in the wild, guided by Lyn's remarkable expertise and background as a respected all-breeds dog judge. Her knowledge, honed over years of living alongside these animals, goes far beyond what you'd find in scientific journals or textbooks. While modern science often relies on detached methodologies, Lyn's hands-on approach, built on touching, observing and even holding a dingo upside down to study its reactions, offers insights that are both practical and profound.

Today, I arrived just in time for one of Lyn's presentations, a chance for visitors to see a living, breathing dingo up close, learn about their biology and plight, and even interact with them.

A volunteer emerges from the enclosures down the hill, leading a striking golden dingo named Pumbah, a name humorously earned during his chubby puppy days. Now a fully grown adult male, Pumbah serves as one of the sanctuary's invaluable ambassador dingoes. Known for his calm demeanour and confidence, he stands out in a species often wary of people. Many dingoes at the sanctuary prefer to keep their distance, whereas Pumbah takes it all in stride, embodying the role of his species' representative with ease.

Under a shaded verandah that serves as an impromptu education space, a small group of people gathers. The back wall is painted with a mural depicting the three primary coat colours: ginger, black and tan, and white. Lyn gestures for everyone to take a seat as Pumbah springs lightly onto a wooden table, standing tall and composed. He knows the routine: treats and admiration are part of the deal.

'Everyone, meet Pumbah', Lyn begins, resting a hand on his back. 'Let's talk about what makes dingoes so remarkable.' Her voice is warm yet authoritative. She pats his lean, muscular flank. 'Dingoes are built for survival. Unlike domestic dogs, they're stripped down to essentials – no wasted energy, no unnecessary frills.'

The comparison to domestic dogs resonates with the audience, many of whom likely have pets at home. Nods ripple through the group as they mentally contrast their own dogs with the dingo standing before them.

Lyn gestures towards Pumbah, highlighting one of his most striking features. 'Notice how the head is the widest part of his body?' Viewed head-on, this is unmistakable. In most domestic dog breeds, the chest is broader than the head – take a bulldog, for an extreme example. Bulldogs are heavily built, with wide, muscular chests and compact bodies that give them a stocky, almost barrel-like appearance.

'Look at his body proportions', Lyn continues, stepping back to give visitors a full view. 'Unlike most domestic dogs, which are often taller than they are long, dingoes maintain equal height and length, what we call a square body index – essentially forming a boxy, balanced silhouette.' She runs her hand along Pumbah's lean frame. 'This balanced proportion, shared with wolves and other wild canids, creates a long, slender silhouette perfect for endurance and agility.'

Standing beside Pumbah, Lyn places her hand on his ribcage. 'Feel how the ribs extend all the way back, protecting the entire torso.' She traces the length of his body. 'While domestic dogs and dingoes have the same number of ribs, a dingo's ribcage extends further back, protecting their vital organs. In domestic dogs, the shorter ribcage leaves the belly exposed and vulnerable.' She pauses, her hand hovering thoughtfully over the ribcage. 'This extended protection helps them survive against threats like a kangaroo's powerful kicks or the defensive strikes of large prey.'

'Their narrow chest and streamlined build create perfect proportions for efficient movement', Lyn explains, drawing attention to Pumbah's legs. 'Watch how their front and back legs align perfectly – they're single-tracking, a trait shared with wolves and other wild canids.' She

Fig. 5.2. (Left) A front-on view of a wild dingo from K'gari, illustrating how its head is the broadest part of its body. (Centre and Right) The dingo's streamlined body shape results in a single line of footprints when it moves. Source: Bradley Smith.

demonstrates by having Pumbah walk slowly across the deck. 'See how their back feet step precisely into the prints left by their front feet? It's an adaptation that minimises noise and conserves energy during long-distance travel and hunting.' (Fig. 5.2). By contrast, she explains, most domestic dogs leave a staggered pattern of four prints, reflecting their modified body shapes bred for different purposes.

Flexibility is another marvel of dingo anatomy. 'Their skeletal structure is almost cat-like', Lyn says, demonstrating by gently twisting Pumbah's wrist. The motion, seeming uncanny for a canid, reveals the extraordinary mobility of their joints. 'This flexibility', she explains, 'combined with their lean build and narrow chest, allows them to squeeze through any space their head can fit – another advantage of having the head as the widest part of their body.' The agility proves invaluable when pursuing prey into burrows or navigating dense underbrush.

The neck showcases perhaps the most remarkable example of dingo flexibility. 'Like all canids, dingoes have seven cervical vertebrae', Lyn explains, 'but their joint structure allows for an exceptional range of motion.' She demonstrates by gently rotating Pumbah's head until he looks almost directly behind himself, a feat that brings to mind *The Exorcist*, though far less unsettling when performed by this calm, cooperative dingo.

Lyn then demonstrates another impressive capability, tilting Pumbah's head back until he gazes skyward while his body remains forward-facing. 'This adaptation lets them spot prey in trees', she explains. 'And yes, dingoes can actually climb trees, combining their flexible joints with those agile wrists we saw earlier.'

Pumbah's ears twitch at a distant birdcall, catching sounds too faint for human hearing. Lyn seizes the moment. 'Their ears are shaped like the Sydney Opera House', Lyn notes, smiling as the crowd visualises the comparison. 'They're so sensitive that a dingo might even hear your heartbeat.' These remarkable ears can swivel backward independently, ensuring no sound escapes notice, whether it comes from prey, predator, or pack member (Fig. 5.3).

The sensory sophistication extends to their eyes. Lyn points out Pumbah's deep amber gaze. 'Notice how their eyes are angled, unlike the front-facing eyes we've bred into domestic dogs?' This positioning provides a wider field of vision, crucial for scanning their environment. The distinctive dark rims (the eyelid margin) aren't just for show; they reduce glare and enhance focus in bright conditions, much like natural sunglasses.

Lyn leans in and takes a deliberate sniff of Pumbah's coat. 'Notice something missing? That typical 'doggy' smell is absent.' She explains how this neutral scent benefits a predator, likely resulting from differences in their sebaceous glands. However, dingoes aren't completely scentless (all animals have a smell) and they possess a supracaudal gland (a small black patch of hair on their tail) and specialised paw glands, used primarily for communication with other dingoes.

Lyn saves her most dramatic demonstration for last. With a subtle nod to her volunteer, she signals what's to come. The volunteer gently lifts Pumbah by his hips and, in one fluid motion, turns him upside down. What happens next astounds the audience: instead of struggling, Pumbah immediately goes limp, hanging serenely like a relaxed ragdoll. 'If you try this with your dog at home', Lyn cautions with a knowing smile, 'you'll quickly understand how extraordinary this is.' I can attest to this, so you don't have to. My own attempt to replicate this with my pet dog resulted in a panic of flailing limbs and desperate scrambling.

Not all dingoes will let you do it, but this peculiar behaviour highlights a fundamental difference between dingoes and domestic dogs, one that raises intriguing questions about their evolutionary history. Lyn explains that this isn't a trained trick, it's an innate response shared by all dingoes, from adults to juveniles, even those with no previous handling experience. The purpose of this behaviour remains one of dingo science's most intriguing mysteries.

Fig. 5.3. A dingo running with its ears angled backward, alert to the pursuit of another dingo behind it. This dynamic moment captures the dingo's agility and acute awareness of its surroundings. Source: Lyn Watson/ Australian Dingo Foundation.

'We've seen this response consistently across many of dingoes', Lyn notes, 'but we're still debating why they do it.' She outlines several hypotheses: it might be an energy-conserving response when caught by larger predators, a protective reflex to prevent injury during falls or even a specialised submission behaviour that helped their ancestors navigate complex social hierarchies. As Pumbah is gently returned to his feet, shaking himself off as if nothing unusual happened, I'm struck again by how much we still have to learn about these remarkable animals.

The presentation ends with a visible impact. Curiosity transformed into awe on every face. As Pumbah gracefully leaps down and is led away, the visitors follow Lyn down the slope to meet this season's pups, their enthusiasm a testament to the power of understanding through connection.

Back on the balcony after everyone has left, Lyn and I watch the dingoes below, some basking in afternoon light, others playing with timeless grace. 'Every person who leaves here is inspired to save dingoes', she says softly, her gaze on the horizon. 'I might not have enough time to reach everyone, but that's what keeps me going (Fig. 5.4). Every visitor is another chance to help people understand what's at stake.'

Fig. 5.4. Lyn Watson, one of Australia's most passionate dingo advocates, photographed in a powerful portrait titled 'On Their Side'. The image was a finalist in the 'Human/Nature' category of the prestigious 2025 BigPicture: Natural World Photography Competition. It symbolises Lyn's lifelong commitment to protecting dingoes and challenging the fences, both literal and political, that confine them. Source: Doug Gimesy Photography.

The weight of her words settles as I take in the sanctuary that she and her late husband Peter built. She's right. Everyone who leaves this place carries a piece of it with them: more love, more awe and perhaps most importantly, more resolve to protect what matters.

THE ART OF VIGILANCE

Encountering a dingo in the wild reveals an animal that navigates its world on multiple sensory channels, each one finely tuned. Unlike specialised breeds, the beagle's singular focus on scent or the greyhound's fixation on visual pursuit, a dingo maintains a perfect balance of senses. It sees, hears and smells with equal attentiveness, weaving these streams of information into a nuanced picture of its environment. This is no domesticated hound waiting for a command; a dingo is an autonomous operator, selected by nature to serve no one's purpose but its own survival.

Where a domestic dog might become so absorbed in one stimulus, like a moving object or sudden sound, that it loses track of its surroundings, the dingo remains fully aware. Many dogs are nimble and agile, but their attention is often narrowly focused. They tend to follow one track at a time (see, then react), sometimes to the point of bumping into things if they try to do both at once. A dingo, in contrast, seems to operate on multiple channels simultaneously. It can move confidently in one direction while scanning another, seamlessly blending motion with perception. It's as if every sense (sight, smell, and hearing) is independently active but fully integrated, allowing the dingo to build a rich, three-dimensional awareness of its environment in real time.

Often, when dingoes sense my presence, especially the first time, they respond with what is often referred to as a snuff-bark. To call it a 'bark' is a stretch; it's more like a sharp, nasal exhalation that barely breaks the hush. In the world of the dingo, it's a subtle alarm, warning kin that something's up. There's a clever dual purpose here, too. By pushing air out through the nose and then inhaling deeply, the dingo draws in the scents stirred by its own breath. It's as though they've devised a system for both announcing a possible threat and taking a long, mindful sniff of it at the same time.

They are keen judges of strangers too. After this initial alert, I've noticed a pattern while standing in a sanctuary enclosure or encountering a dingo in the wild: they will slip downwind to take my measure, ensuring my scent reveals what my posture might hide. Later, when I've moved from that spot, the dingo returns to where I stood. Nose to the ground, it inhales the essence I've left behind, dissecting every chemical clue to learn what sort of creature I might be.

If there's one behaviour that stands out above all else in a dingo and ties together their keen senses, it is their ability to notice every small detail, react with extreme caution and remember everything they observe. At first glance, this behaviour might be mistaken for neophobia, which is the psychological term referring to the fear of anything or anyone new. The dingo isn't cowering in permanent distrust, however, it's simply taking the time to gather intel. When they first encounter something unfamiliar, their approach is cautious: tails tucked, bodies stretched taut to keep maximum distance from the object while still getting close enough to sniff or watch. They tremble on that threshold between curiosity and caution, looking for clues about whether this new thing might harm them. And once they decide it's not a threat, they shift gears and carry on, minding their own business as though nothing had changed. It's not fear so much as a hardwired vigilance. An evolutionary insurance policy in a land where surprises can often be fatal.

I recall a story from the sanctuary, courtesy of Jenny Morton, a Professor of Neurology at the University of Cambridge who had come to conduct cognitive studies. Her plan was straightforward: acclimate the animals to a testing enclosure for brief daily sessions and then conduct the experiments. Yet in the initial few days, when they first entered the test area, the dingoes were particularly vigilant, sniffing, checking and double-checking part of the space. Puzzled by their persistent wariness, Morton reviewed the video footage and discovered why: each day there were small subtle changes in the apparatus: a different coloured bowl, an uncoiled hose, an open camera. She began deliberately introducing minor alterations to test their awareness. Without fail, the dingoes spotted every change, as if comparing the space against a perfect mental snapshot of what they had seen the day before.

To highlight this extraordinary vigilance, Morton tested human volunteers, asking them to observe the enclosure's layout and later identify changes. Despite their best efforts, they spotted only about two out of ten alterations, and these weren't inattentive people. When given traditional spot-the-difference puzzles, they excelled. In a *real* environment, their awareness paled against the dingo's comprehensive vigilance.

It's a humbling thought. We like to imagine ourselves as observant, perceptive beings, yet we barely scratch the surface compared to an animal that treats each setting like a high-stakes scenario. Dingoes enter, be it desert scrub, eucalyptus woodland or a simple testing enclosure, and mentally map every exit and shadow, reading it all like a secret code.

In the wild, this level of environmental awareness is the difference between living to see another dawn and fading into the dust. It's a skill set sharpened by many thousands of years of survival, never dulled by complacency. The dingo's world demands attentiveness because, out there, a missed detail isn't just a puzzle piece left unturned. It's the predator or poison plant you never saw coming.

A MIND AS SHARP AS ITS TEETH

When we think of intelligent animals, we often picture great apes using tools, dolphins recognising themselves in mirrors or ravens solving multistep puzzles. Yet in Australia's wilderness, another animal's cognitive abilities rival these celebrated minds: the dingo. Far from being simple opportunistic predators, dingoes display problem-solving abilities that place them among the most intellectually sophisticated wild canids in the world.

Dingoes don't just react to the world, they think about it methodically. They're problem-solvers who can figure out complex puzzles, manipulate latches and mechanisms, anticipate human behaviours and devise hunting strategies that require multiple steps of planning. In captivity, they've been observed deliberately deceiving both humans and other dingoes to achieve their goals. In the wild, they demonstrate remarkable spatial memory, remembering precise locations of resources and threats across vast territories.

I've witnessed their cognitive prowess firsthand through years of research and observation. Two extraordinary cases, both of which I had the privilege to document scientifically for the first time in any canid, have fundamentally challenged our understanding of what these animals can do. The first involves their remarkable use of tools; a behaviour once thought exclusive to primates and certain birds. The second reveals their strikingly complex reactions to the death of another dingo, offering unprecedented insights into their emotional and social capabilities.

Tool use: Sterling the problem solver

One sunny afternoon in 2010 I was at the dingo sanctuary conducting research. It had been a few months since my last visit. During a break in testing, Lyn waved me over with a knowing smile. 'You've got to see what Sterling does', she said, pointing to the handsome young dingo lounging in his enclosure. Sterling had a reputation among the staff as a master manipulator, not of people, but of his environment. 'He's been dragging tables around his enclosure to reach things', Lyn explained. 'He'll clamp onto a table leg with his teeth and pull it across the enclosure to grab the name tags or anything else that catches his interest.'

It reminded me of Köhler's famous chimpanzees from undergraduate psychology class. Those clever primates were observed stacking crates to retrieve bananas dangling tantalisingly overhead. Tool use represents the manipulation of an object to achieve a goal, specifically when that object is employed in a way that differs from its original or typical purpose. This ability to repurpose objects demonstrates sophisticated cognitive processing, requiring both abstract thinking and the capacity to mentally transform an object's potential function. What made Sterling's case particularly fascinating was that, unlike Köhler's trained chimps, this was purely instinctual behaviour. Problem-solving that emerged without human coaching.

Eager to document this remarkable behaviour, we set up an experiment, hanging a small plastic envelope containing food on the steel mesh just out of Sterling's reach. Unlike the chimps, Sterling wouldn't perform on command. We waited on a bench near his enclosure, but he simply ignored it. It wasn't until we climbed the hill to the house balcony that his genius revealed itself. By the time we returned, the table had been moved and the prize was his.

When we reviewed the footage, we saw Sterling's problem-solving in action. He first attempted to leap for the food from underneath the envelope, but quickly abandoned that strategy as it was too high. So, he immediately turned his attention to the trestle table within his enclosure. And sure enough, he took hold of the table strut with his teeth, dragging the table backward until his back feet hit the mesh wall of the target. He leapt onto the table and, with precise movements, extended his body to swipe the prize (Fig. 5.5).

Sterling's behaviour didn't stop at tables. He also moved a kennel across his enclosure, positioning it under a blind wall so he could peer over at neighbouring dingoes. Sterling

Fig. 5.5. Sterling reaches for a small parcel of food after skilfully moving the table into position. This behaviour demonstrates problem-solving and tool-use capabilities. Source: Bradley Smith.

was deliberately transforming his surroundings, crafting opportunities where none existed before.

The discovery was published in the scientific journal *Behavioural Processes* in 2012. It marked a major turning point – not just in our understanding of dingoes, but in how we define cognitive complexity across the canid family.

Sterling's innovative problem-solving demonstrates just one facet of dingo intelligence. Although tool use reveals their capacity for abstract problem-solving and causal reasoning, dingoes also display remarkable social and emotional intelligence. This complexity became particularly evident during observations of wild dingoes on K'gari, where their response to death challenges our understanding of canid consciousness.

Grief and vigilance: a dingo's reaction to death

K'gari, with its sweeping sand dunes and pristine wilderness, feels untouched by time. A dingo paradise with freshwater, beaches, sand dunes and thick forests home to giant satinay (*Syncarpia hillii*) and brush box (*Lophostemon confertus*) trees. For wildlife ecologist Rob Appleby, who I began working with while he collected data for his PhD, the island was more than a study site, it was a living laboratory. Day after day, he trekked its sandy trails on foot, tracking dingoes to film their interactions with each other and with the few humans who ventured into their world. His work demanded patience, persistence and the occasional stroke of luck.

On this particular day, the trek was routine. Rob had found a group of pups at a rendezvous site, just as he had hoped, and settled in to observe them. The pups played together in a flurry of movement, their antics a perfect snapshot of youthful energy. The mother lingered nearby, resting in the shade, her golden coat blending seamlessly with the surrounding landscape.

Then, everything changed.

Rob's attention was drawn to one pup lying apart from the others. It was laying on the ground, its small body shivering with convulsions. The scene shifted from playful to sombre in an instant. Over the next 30 minutes, he watched as the pup's movements slowed, its breathing faded and the unmistakable stillness of death settled over the tiny figure. He suspected the pup had been struck by a snake, its symptoms consistent with envenomation, though it was hard to know for certain.

The mother approached, sniffing her pup's body with care before stepping back to rest nearby. The remaining pups circled their sibling, their rough-and-tumble play giving way to subdued nudges and hesitant sniffs (Fig. 5.6). The usual chaos of the rendezvous site fell silent, replaced by a heavy stillness.

Over the following days, the mother returned to the dead pup again and again, carrying its body to different locations. Rob observed her lift the small, lifeless form delicately in her mouth and move it across the landscape. On one occasion, she placed the pup near a shallow depression, a site that resembled the sleeping areas she had prepared for her living young. Even as the rest of the litter played nearby, the mother stayed close, often resting near the pup's body, her actions hinting at more than instinct.

Was she grieving? Protecting her offspring from predators? Or could her behaviour be explained by a practical strategy, a cautious refusal to give up? In the animal world, where life and death are often a matter of timing, it might be adaptive to wait and see. If the pup were incapacitated rather than dead, perhaps from injury or shock, delaying abandonment could mean the difference between survival and loss. This 'wait and see' approach, observed in other social species like chimpanzees, offers one lens through which to interpret her actions.

Yet there was also something deeply emotive about the scene. The mother's repeated returns to the pup and the care with which she moved it suggested an enduring bond. Dingoes, like other intelligent and social animals, form strong relationships, particularly between mothers and their young. Her actions hinted at both hope and reluctance, as if the act of leaving the pup behind would make the loss irrevocable.

Rob's observations, which we published in 2013, the first documented instance of such behaviour in a wild canid, blur the line between practical survival strategies and emotional complexity. They evoke parallels with behaviours seen in primates, elephants and dolphins, species known for their responses to death. Was this dingo mother holding onto the faintest possibility of recovery? Or was she grieving in her own way, her actions shaped by the same maternal bond that compels mothers across the animal kingdom to protect their young?

Fig. 5.6. Dingo pups on K'gari responding to the death of a sibling. Their shift from playful activity to cautious investigation, along with the mother's repeated return to the body, reflects the strong social bonds within dingo families and raises questions about grief, vigilance and emotional complexity in wild canids. Source: photograph by Rob Appleby, reproduced from Appleby *et al.* (2013).

This observation helped open up new conversations about emotion and social bonding in wild canids, challenging long-held assumptions.

Anecdotally, this aligns with zookeepers I have spoken with, who often talk about how difficult it is for dingos to deal with the death of one of their mates. Perhaps the truth lies somewhere in between, a delicate interplay of instinct, strategy and emotion. Whatever her motivations, her actions challenge our assumptions about how animals perceive and respond to death.

MASTERS OF ADAPTATION

To witness a dingo, whether stalking prey across desert plains or demonstrating remarkable problem-solving in sanctuaries, is to observe an evolutionary masterpiece. As Lyn Watson often observes, 'If a dingo is too big, too fat, too lazy, or too dumb, it will never make it in the wild.' This simple truth encapsulates thousands of years of evolutionary refinement. Every feature we have looked at, from their finely balanced bodies to their sharp environmental awareness and problem-solving minds, reflects traits shaped for survival.

Dingoes continue to adapt with remarkable resilience, meeting new challenges that require both physical and behavioural flexibility. This capacity to adjust and endure is why the dingo survives today: silent, watchful and always one step ahead.

6
Following the sun

THE SUN WRITES THE TRUE CALENDAR OF THE DINGO, ITS ARC ACROSS THE endless Australian sky marking the rhythms of survival more precisely than any human-made measure.

These predators are shaped by a landscape where seasonal shifts dictate the most intimate rhythms of their existence, the boom-and-bust cycles of drought and abundance directly determining when they breed, raise their young, hunt and disperse. Like their wild relatives across the globe, wolves threading through Minnesota's forests or packs ranging Italian valleys, dingoes form tight-knit families led by dominant breeding pairs, their lives a delicate choreography of cooperative hunting and intricate social bonds. Yet here, survival requires a constant negotiation with an environment that speaks in subtle, unforgiving dialects. Understanding their seasonal cycle allows us to see the wilderness through their eyes.

BREEDING SEASON (MARCH–MAY)

The first season, perhaps the most important as it sets in motion all others, is when dingoes enter their breeding period. It is also likely the busiest time in their diary. This is ultimately governed by the sun's annual cycle, its position in the sky altering the length of daylight hours (Fig. 6.1). As summer wanes and days begin to shorten, the subtle shift in light triggers a cascade of hormonal changes within the dingo, marking the start of its singular, meticulously timed breeding season.

Dingoes breed just once per year. The shortening photoperiod, the daily dwindling of sunlight, acts as nature's clock, triggering oestrus in females and preparing males as their testes swell in anticipation. These changes are not uniform across the continent; geographical latitude plays a role, with northern populations breeding slightly earlier than those in the south. Despite this variation, the overall timing remains tightly linked to environmental conditions.

This precision is critical. Breeding too early or too late could mean a litter born into drought or prey scarcity, conditions that might doom their survival. Synchronising their cycles with the declining sun, dingoes align their pups' births with the seasonal abundance of prey. Environmental triggers such as food availability or drought can further fine-tune the timing of reproduction. During lean years, male readiness may be delayed, with testes responding months later than usual. Female oestrus, too, may wait until signs of environmental renewal emerge, demonstrating the species' remarkable adaptability.

Within the pack, reproduction is hierarchical. Typically, only one male and female pair, traditionally referred to as the alpha, but more appropriately the dominant pair, will breed.

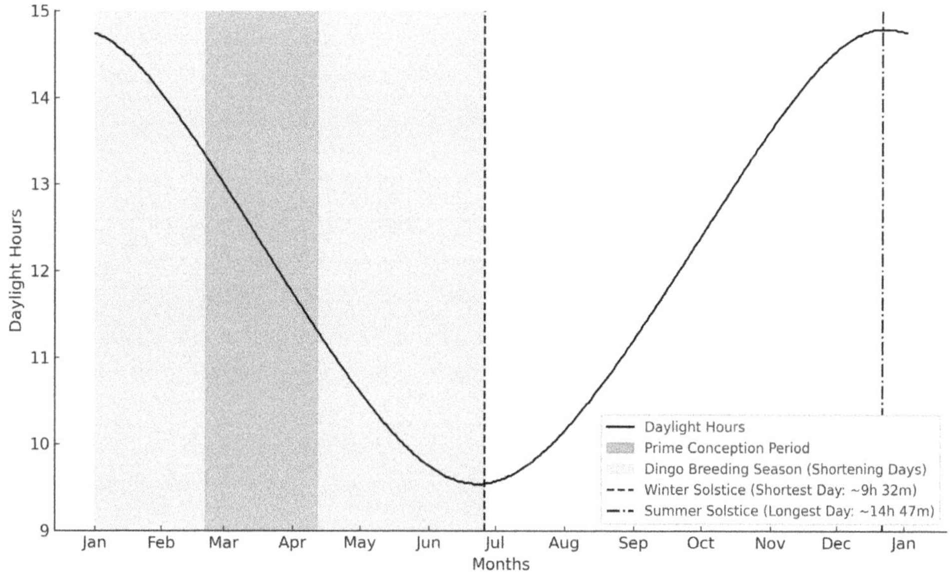

Fig. 6.1. The relationship between daylight hours and dingo breeding seasons in Australia. The graph shows how dingo breeding is synchronised with decreasing daylight hours preceding the winter solstice (dashed line). The dark-shaded region indicates the peak conception period and the lighter shading shows the broader breeding season window. This timing ensures pups are born when environmental conditions are most favourable for survival. Source: Bradley Smith, from data published in Lord *et al.* (2014).

Subordinate females are commonly prevented from breeding through a combination of social hierarchy and hormone-mediated suppression, ensuring the pack invests its energy in a single dominant litter. This suppression ensures better survival rates for the pups, which will benefit from the focused attention and care of the entire pack.

Courtship is as much a social process as a reproductive one. The dominant pair engages in playful nips and chases, blending affection with dominance to reinforce their bond, a vital foundation for both successful conception and cooperative parenting. Oestrus in females lasts only a few fleeting days and timing is critical. If the first mating fails, a second oestrus might occur before the winter solstice, but opportunities remain limited. Males' reproductive cycles mirror the females' to a degree, with physiological readiness peaking in synchrony. Testosterone levels rise, testicles enlarge and sperm production reaches its height during the breeding season. Yet unlike females, the male's seasonal pattern is less absolute. Some studies suggest a dramatic drop in sperm production outside the mating window, even temporary infertility, while others indicate low-level reproductive capacity may persist year-round. This variability may serve as a buffer, helping ensure that when the timing is right, the pack is prepared. This intricate blend of social bonding and biological timing is so essential that even captive breeding programs have found that well-established pair bonds must precede successful mating, which may explain why artificial insemination has largely failed. It bypasses the complex social rituals fundamental to dingo reproduction.

As the breeding season intensifies, territory becomes paramount. The dominant male increases his patrolling and scent-marking, while howling reaches its highest frequency of the year. These behaviours serve both to ward off rivals and strengthen pack cohesion, ensuring the pack's breeding success remains uncontested.

WHELPING (MAY–AUGUST)

As the crisp days of winter settle over the Australian landscape, the dingo pack enters a quieter, more focused phase of its life cycle. Whelping season is a time of transformation, as the pack shifts its energy towards nurturing the next generation. For the breeding female, this begins with finding the perfect den to shelter her soon-to-be-born pups, but the effort is far from hers alone. Every member of the family, particularly the breeding male, has a role to play in ensuring the survival of the litter.

Dingoes embody a highly cooperative parenting strategy, deeply investing in a single, well-timed litter each year. This approach is key to the survival of dingo pups, for which survival rates are already low, making the collective effort of the pack essential.

Dens are chosen with care: hollow logs, rocky crevices, abandoned burrows or dense spinifex. Sometimes new ones are dug under trees or shrubs, often into sandy banks or slopes (Fig. 6.2). The dingoes excavate these burrows using their powerful front paws, creating chambers deep enough to protect the vulnerable pups from both weather and predators. Some dens may have multiple entrances, offering quick escape routes if needed, while others are simple but well-concealed single tunnels. Anything that will provide safety and seclusion. Den openings are typically oriented in a south-west direction. This strategic placement minimises exposure to direct sunlight during the hottest parts of the day, helping to keep the den cool and shaded in the harsh desert environment. Such careful orientation reflects the dingo's remarkable adaptability and instinct for survival.

Once the den is prepared, the breeding female minimises risk to the pups by staying close by. The breeding male takes on a protective role, positioning himself nearby to guard the family. His presence deters potential threats, allowing her to devote her energy to her newborn pups.

Nursing occurs in distinct stages. Initially, the pups, born blind and helpless, are completely dependent on their mother, who rarely leaves the den in the first few days, nursing them continuously and keeping them warm. In rare cases, subordinate females may even assist in nursing, a behaviour known as alloparental care, showcasing the communal adaptability of dingo society. By the third week, the mother begins to wean the pups, gradually introducing regurgitated food as a supplement to milk.

Predators remain an ever-present danger. Foxes, goannas and even rival dingoes can easily decimate an unprotected litter. Yet the pack's unity is its strength. While the breeding male acts as a sentinel, other family members patrol the den's surroundings, bring food and keep watch as the pups begin to explore their environment. This cohesive defence system demonstrates the importance of family dynamics in dingo survival.

The mother may rotate her litter between multiple dens during this period. This behaviour minimises predation risk and ensures the pups remain close to reliable food sources. Dens are

Fig. 6.2. An adult dingo maintains vigilance near a den entrance in the Pilbara region of Western Australia. Den sites are carefully selected and often feature multiple entrances for safety. This den's location, like many others, was strategically chosen to provide protection from both predators and environmental conditions. Source: Bradley Smith.

often located near abundant prey populations, such as rabbits and rodents, allowing the pack to hunt efficiently while remaining within proximity to the pups.

Environmental conditions also play a critical role in the success of the litter. Drought periods can reduce litter sizes or delay the dominant female's readiness to reproduce. This flexibility reflects the dingoes' remarkable ability to adapt to their surroundings, aligning their reproductive efforts with the land's capacity to support new life.

The pups' development unfolds quickly. By around day 12–14, their eyes open, and their sensory awareness sharpens. Initially confined to the den, they begin venturing outside during the day by 5 weeks of age. These milestones mark the transition from complete vulnerability to the early stages of independence.

Territoriality intensifies during the whelping season. The dominant male and other pack members increase scent-marking and patrolling to deter rival packs. Disputes, while rare, can erupt over access to food or water sources near the den, further highlighting the heightened vigilance of the pack during this critical period.

PUPPY REARING (AUGUST–OCTOBER)

By late winter, the den hums with new life as the pups venture out to explore their surroundings. Their timing couldn't be better. Spring's arrival brings a seasonal abundance of prey, as many smaller animals are also rearing their young. This synchrony ensures that the pack can secure enough food to support their growing family.

Initially nursed for 7–10 weeks, they are gradually weaned as the adults begin provisioning them with chunks of meat and small prey. The dominant male, along with other pack members, often delivers these meals, ensuring the pups and at times the mother, are well fed. Occasionally, siblings and yearlings contribute by bringing back scraps or engaging in cooperative feeding behaviours, such as sharing portions of prey.

Play dominates the pups' daily lives, but it serves a vital purpose beyond mere recreation. Activities like chasing, mock fighting and pouncing are essential for honing motor skills, practising hunting techniques and understanding social hierarchies. Even their seemingly chaotic antics are deeply instructive, as they learn to navigate the pack's complex dynamics and test the boundaries of acceptable behaviour. Older pack members actively participate, correcting missteps or encouraging cooperation during play.

As the pups grow, they begin to shadow the adults on short excursions beyond the den. These journeys are rich with lessons, offering the pups opportunities to observe hunting techniques, learn to identify prey and become familiar with the pack's territory. Between 9 and 24 weeks, they progressively accompany adults on longer trips, integrating into the pack's daily activities. By 16 weeks, they start to exhibit growing independence, relying less on intensive care and more on the knowledge gained through these formative experiences.

The dominant female's role continues to evolve during this time. Having resumed hunting and patrolling duties, she now leads by example, demonstrating the skills her pups will need to survive. The dominant male and other pack members remain steadfast in their support, maintaining the balance between nurturing the young and sustaining the pack's overall wellbeing.

By October, the pups are ready for greater challenges, exploring farther from the den and engaging more actively with their environment. The transition from dependent, playful juveniles to competent young dingoes is well underway. What begins as exuberant play slowly transforms into the focused behaviours essential for survival.

JUVENILE DEVELOPMENT (NOVEMBER–FEBRUARY)

As summer settles over the Australian landscape, dingo pups enter a pivotal stage of growth and learning. No longer confined to the safety of the den, they now venture into the open world under the watchful guidance of their pack. This period marks their transformation from dependent juveniles to capable young dingoes, poised to take on increasing responsibilities within the pack.

The juveniles' development is a mix of observation, participation and play. By this stage, they are fully weaned and accompany the adults on foraging trips, exploring the pack's territory. These excursions serve as immersive lessons, where juveniles learn to identify prey, interpret the pack's complex vocal and body language, and understand the rhythms of life in the wild. Older siblings often play a pivotal role, acting as patient mentors that demonstrate hunting techniques and model social behaviour.

As juveniles venture farther from the den, they begin to understand the intricacies of their pack's territory. Home-range size is dictated largely by the availability of resources, varying dramatically across Australia's diverse landscapes. In resource-scarce environments, territories can span hundreds of square kilometres as packs cover vast distances in search of prey. However, in areas with concentrated anthropogenic resources, such as livestock or human waste, territories often shrink considerably and pack sizes can swell to unnatural levels. This variability in territory size not only influences how juveniles learn to navigate their surroundings but also impacts the social structure and dynamics of the pack.

Hunting, once a distant curiosity for the pups, now becomes a focal point of their education. Juveniles begin shadowing the adults on hunts, mimicking their movements and strategies. At first, their attempts to catch smaller prey like insects, lizards or young rabbits are clumsy and often unsuccessful. However, these early failures are invaluable; they refine the skills needed for precision, stealth and timing, which are crucial for survival.

Social learning takes centre stage during this period, with play providing a safe and structured outlet for experimentation. Rough-and-tumble play and mock hunting allow juveniles to practice motor skills, test their physical limits and navigate the pack's social dynamics. Even moments of conflict during play are instructive, teaching juveniles how to resolve disputes and establish boundaries. Through these interactions, they begin to understand their place in the pack's hierarchy, preparing them for their future roles.

I've seen this firsthand in captive settings, at the Dingo Sanctuary (mentioned in Chapter 5), where several litters are often born in winter and allowed to mix and play under the watchful eyes of a parent pair. These babysitting adults don't intervene unless they need to, but when a pup oversteps, pestering too much or biting too hard, a sharp growl or quick snap restores order. Watching the pups play is mesmerising. They chase each other endlessly, taking turns as the rabbit, the stand-in prey, while the rest give chase, tackle and wrestle in a flurry of limbs and dust. Once the 'prey' is caught, they pile on, but if it goes too far, you'll hear a sharp yelp and the game ends immediately. There's an unspoken code among them, a boundary not to be crossed. This isn't just chaotic fun. What looks like play is really practice: rehearsals for the real-life skills of hunting, coordination, restraint and social negotiation that will one day define their survival.

The juveniles' activity patterns also begin to align more closely with those of the pack. As the heat of summer peaks, they adopt a crepuscular rhythm: they are most active during the cooler hours of dawn and dusk. This adaptation not only conserves energy but also minimises exposure to predators and increases hunting success in a land where scarcity is the norm.

Despite their growing independence, juveniles remain closely tied to the pack, with adults and older siblings providing protection and guidance during this vulnerable stage. The pack's cohesive structure allows them to gradually take on more responsibility, fostering the confidence and resilience they'll need for their next challenge. By the end of this phase, they have developed the foundational skills required for survival, though they remain reliant on the pack. Their growing competence marks the first steps towards independence.

DISPERSAL AND INDEPENDENCE (MARCH ONWARD)

As juvenile dingoes mature, a critical decision eventually arises: whether to remain with their natal pack or strike out on their own. For some, this transition begins as early as March in their second year, but dispersal doesn't always occur on a fixed timeline. The timing varies widely and may not happen for several years, shaped by environmental conditions, pack dynamics and individual development.

The decision to disperse is influenced by both internal and external pressures. Within the pack, competition for resources and breeding opportunities intensifies as dingoes grow older, while environmental factors such as prey scarcity or drought can prompt individuals to seek opportunities elsewhere. However, not all dingoes disperse. Some remain within the pack they were born into, continuing to live and work alongside their parents and siblings. These stay-at-home individuals (known as philopatric) contribute to the pack's stability and help raise subsequent litters, strengthening the social bonds that hold the group together.

For those who choose to disperse, the journey is fraught with challenges. Success depends on navigating unfamiliar landscapes and securing territory in an already occupied environment. The availability of resources shapes this journey significantly. In resource-poor regions, territories may span hundreds of square kilometres, requiring juveniles to travel vast distances in search of unoccupied land. Conversely, in areas with concentrated anthropogenic resources, such as townships or mine sites, territories shrink and pack densities swell to unnatural levels, intensifying competition.

Despite these challenges, dispersal is essential for population health, preventing overcrowding and maintaining genetic diversity. Environmental conditions heavily influence success rates. In years of prey abundance, juveniles have greater success in establishing territories and forming new packs, while drought periods may delay dispersal or increase mortality. Some secure unoccupied land, while others join lone dingoes or integrate into existing groups.

For dispersing dingoes, independence is a 'trial by fire', testing all the skills honed during juvenile development. For those who succeed, the reward is new territory, a potential mate and the chance to establish their own pack, completing their transformation from playful juveniles into resilient and self-sufficient predators.

POPULATION REGULATION AND THE MYTH OF DOMINANT FEMALE INFANTICIDE

Although dispersal and resource availability have long been recognised as key regulators of dingo populations, a controversial theory from the 1980s has profoundly influenced scientific understanding of dingo behaviour. Known as dominant female infanticide, this theory emerged from experiments by researchers Laurie Corbett and Alan Newsome at a government field station in central Australia. As they documented the social dynamics over several years, they observed what appeared to be a brutal pattern: the dominant female systematically killed the pups of subordinate females, seemingly ensuring her offspring's survival above all others.

The work became widely cited in scientific circles, particularly internationally. It suggested something unique among wild canids. While wolves and African wild dogs maintain pack hierarchy through behavioural suppression of breeding, dingoes appeared to use a more dramatic method. The image of the ruthless dingo matriarch, culling her rivals' pups to secure her genetic legacy, became entrenched in both scientific literature and public imagination. The idea gained traction partly because it was quite novel and seemed plausible in the context of Australia's unforgiving environment, where the constant pressure of drought and resource scarcity might drive extreme behaviours for survival.

However, closer scrutiny reveals a critical flaw in these observations: the entire study was conducted in a confined enclosure measuring just 45 × 25 m – smaller than one-quarter of a football field. In this cramped space, with no opportunity for natural dispersal, social tensions inevitably escalated. Were these behaviours natural expressions of dingo society or artefacts of captivity?

My own fieldwork in Australia's most remote regions tells a different story. One hot and dry afternoon in Western Australia's Pilbara region, I discovered a den site. Inside, 18 pups of varying ages tumbled over each other, some barely old enough to walk, others already showing the confidence of explorers (Fig. 6.3). This wasn't the territory of an authoritarian matriarch, but rather a nursery shared by multiple females. The den entrance was worn smooth by the constant coming and going of adults, all contributing to the pups' care.

Fig. 6.3. A remarkable den site discovered in the Pilbara region of Western Australia containing 18 pups from three different litters, all being raised cooperatively. This observation challenges the long-held theory of dominant female infanticide, demonstrating that when resources are abundant, multiple females can successfully breed and raise pups together. The presence of multiple litters in a single den highlights the cooperative nature of dingo social structures and their adaptive breeding strategies. Source: Bradley Smith.

This observation wasn't an anomaly. Across Australia, researchers are documenting similar patterns of cooperation. Take Rob Appleby, whom you met last chapter; on K'gari he observed a family led by a matriarch known as 'Granny'. Rather than suppressing her daughter 'Aunty's' reproduction, Granny raised her pups alongside her daughter's litter, the two generations cooperating in pup-rearing duties.

The den I observed in the Pilbara was situated near a large mining operation, where year-round access to water and food scraps created abundant resources. This setting revealed a crucial insight: dingo reproductive strategies are shaped primarily by resource availability. In areas of plenty, whether natural or human-modified, multiple females within a pack may breed simultaneously, pooling their efforts in pup-rearing. The density of dens and the proximity of breeding females suggest that cooperation, not competition, emerges as the dominant strategy when resources permit.

During times of scarcity, dingoes demonstrate remarkable population self-regulation through multiple mechanisms. Rather than resorting to infanticide, population regulation occurs through several interconnected processes: subordinate females may suppress their own reproduction through delayed sexual maturity or skipped breeding seasons; young adults are prevented from mating and/or often disperse from their natal territory to establish new ranges; pack size naturally decreases as members split off to form new groups; and breeding pairs may have reduced litter sizes or only breed every second year. Even after birth, natural regulation continues – in drought conditions, fewer pups survive due to reduced milk production and limited prey availability. These flexible responses to environmental conditions showcase the dingoes' sophisticated adaptation to Australia's boom-and-bust cycles, a far cry from the rigid, aggressive behaviour suggested by the dominant female infanticide theory.

A BALANCED EXISTENCE

The seasonal cycles of dingo life reveal a sophisticated interplay between resources, social bonds and territory. Each season brings its own demands and dingoes respond with remarkable flexibility, adjusting their breeding timing to align with prey abundance, coordinating pack efforts to raise pups when resources allow and adapting their territorial boundaries to meet the changing needs of the pack. This adaptability is perhaps their greatest strength. Rather than being locked into rigid behaviours, dingoes read the landscape's rhythms and respond accordingly. In times of plenty, multiple females might breed and support one another in raising pups; during drought, the pack might suppress breeding entirely. Their territories expand and contract with the seasons, tightening around den sites during breeding before expanding to encompass vital hunting grounds as pups grow. Through it all, the pack's social bonds remain crucial, providing the cooperative framework needed to hunt efficiently, defend resources, and ensure the survival of the next generation. This sophisticated self-regulation means that in natural settings, dingo populations remain in balance with their environment, never reaching the problematic densities seen in areas of artificial abundance created by human activity.

7
A wolf in a dog's clothing

AT FIRST GLANCE, A DINGO MIGHT SEEM LIKE JUST ANOTHER DOG. WITH THEIR pointed ears, sleek build and sharp eyes, they evoke comparisons to Australian kelpies or South-East Asian village dogs. But behind this unassuming appearance lies a creature unlike any other. A canid shaped by the wild, not by human hands.

Fully capable of living independently of humans, dingoes embody the essence of wildness. Yet their long history of coexistence with humans, coupled with their dog-like appearance, has fuelled debates about their identity, domestication status and evolutionary story. Are they feral descendants of domestic dogs or were they never domesticated at all? And where do we draw the line between wild and domestic?

These questions lead us to examine what sets dingoes apart: their physical traits, genetic lineage and behaviours. Beneath their outward similarity to dogs lies a deeper story, one that reveals what truly makes a dingo a dingo.

INDEPENDENT BY NATURE

Wild canids, such as wolves, coyotes and dingoes, survive and reproduce by relying on natural hunting, foraging and social strategies, an existence shaped entirely by the pressures of the wild. They hunt, reproduce, communicate and raise their young without any reliance on human intervention. This independence is the cornerstone of what it means to be a wild canid, distinguishing them from domestic dogs, which are behaviourally and biologically adapted to coexist with humans.

As Kathryn Lord, an expert in canid biology, has highlighted, this dependency is a hallmark of domestication. Domestic dogs are inextricably tied to humans, whether directly through care and food provisioning or indirectly by scavenging waste and exploiting human-altered environments. Over thousands of years, dogs have evolved not for survival in the wild but for life alongside humans. Their biology, behaviour and even cognition reflect this profound connection, creating creatures that thrive on human proximity rather than self-sufficiency.

To live independently of humans, wild canids like dingoes require a suite of finely tuned traits. Their reproductive strategies, social structures, communication, problem-solving and physical adaptations all work together to meet the demands of life in the wild. Each trait plays a role in ensuring survival in environments where resources are scarce and competition is fierce. In this chapter, some of these characteristics will be explored, revealing why dingoes are far closer to their wild canid relatives than to domestic dogs and why their independence sets them apart.

THE DINGO FORM

The dingo's anatomy offers a roadmap of its evolutionary journey, showcasing adaptations that allow it to thrive independently of humans. From the skull structure to the sensory systems, each feature reflects the evolutionary pressures of a wild existence. Matthew Crowther, a zoologist and leading expert on dingo morphology and taxonomy, has been instrumental in identifying these defining characteristics.

The skull of a dingo is one of its most striking features. As Crowther has detailed in his 2014 paper, dingoes possess long muzzles and large carnassial teeth, which are specialised sharp shearing teeth essential for processing meat. These traits mark the dingo as a predator reliant on hunting, in contrast to domestic dogs, whose shorter snouts and smaller teeth reflect diets shaped by human provisioning. The dingo's pronounced sagittal and nuchal crests, which are bony ridges on the skull that anchor powerful jaw muscles, enable crushing of bones and accessing the marrow, a critical food source in the wild. Domestic dogs, bred for less demanding diets, lack these robust cranial structures, a stark reminder of how domestication alters not just behaviour but anatomy.

Unlike some domestic dogs, with rounded skulls and brachycephalic features that prioritise aesthetics over function, dingoes possess a flatter skull that enhances their visual field, an essential adaptation for scanning the environment. Adding to this are the dingo's large auditory bullae, the bony chambers housing the middle and inner ear. These structures amplify their hearing, enabling dingoes to detect faint sounds crucial for hunting and survival. By comparison, many domestic dog breeds, particularly those selectively bred for aesthetic traits rather than working abilities, often show reduced sensory capabilities compared to their wild relatives like the dingo (Fig. 7.1).

Interestingly, dingoes' skulls show some 'juvenile features'. That is, they keep some characteristics typically seen in young animals, such as slightly shorter and broader faces compared to wolves (referred to as paedomorphism). Hunting smaller prey, navigating diverse environments and responding to distinct ecological pressures in Australia have likely shaped these cranial characteristics independently of domestication. Dingoes often sit between wolves

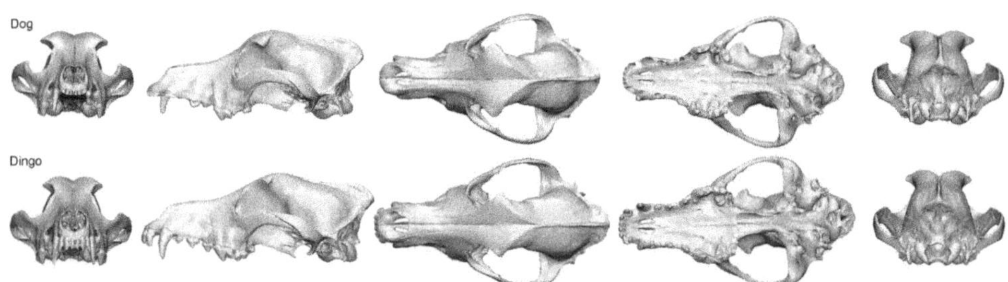

Fig. 7.1. Comparison of a dingo (Bottom) and a free-ranging domestic dog (Top), showing how their skulls differ. The dingo's longer snout, larger teeth and stronger jaw structure are adaptations for hunting and surviving in the wild. In contrast, the domestic dog's skull reflects its reliance on human-provided food. The dog's skull has been scaled to match the dingo's size for easier comparison. Source: created by Will Parr, taken from Smith *et al.* (2019).

and domestic dogs for traits like these, showing subtler paedomorphism compared to the exaggerated forms seen in many domestic dogs. However, their functional cranial features, such as long muzzles and robust jaws, align them more closely with wolves.

A 2025 genetic and anatomical study led by Alice Mouton from the University of California compared the olfactory systems of wolves, coyotes, domestic dogs, and dingoes. They found that dogs, through domestication, have lost a substantial number of functional olfactory receptor genes and exhibit reduced nasal structures associated with smell. Contrary to popular belief, even breeds famed for their legendary noses, such as bloodhounds, show no genetic or anatomical advantage over other dogs – their success is more about training and temperament than biology.

In her study, the dingo was grouped with dogs for the analysis, but stood out as the exception. Among all the dog breeds and lineages tested, dingoes possessed 820 functional olfactory receptor genes, almost identical to wolves at ~826 and coyotes at ~835, and far greater than the domestic dog (average of ~777). In other words, dingoes have held onto the wolf-like genetic blueprint for scent, retaining the capacity to track and hunt. This places them firmly in the company of wild canids, not domestic dogs.

Brain size adds another dimension to the dingo's distinctiveness. After years of studying dingo cognition, I became sceptical of the prevailing assumption that dingoes had smaller brains, like domestic dogs. Determined to uncover the truth, I initiated a collaborative study with an anatomy team (Rachel Norris, Maciej Henneberg and Teghan Lucas from Adelaide University) to measure endocranial volumes in 128 free-ranging dingo skulls, primarily from the CSIRO collection in Canberra, all with documented bodyweights. We used mustard seeds to fill the brain cavity of each skull, a clever method that yields measurements nearly identical to the actual brain tissue weight. These measurements allowed us to calculate the encephalisation quotient (EQ), which is a standardised ratio comparing brain size to body size across species, used to estimate relative intelligence or cognitive capacity. The results were published in 2017 in the *Australian Journal of Zoology*.

We found that the average brain volume in free-ranging dingoes (91.5 mL), combined with their average bodyweight (13.2 kg), produced an EQ of 136.65, which is significantly higher than the average for domestic dog breeds (EQ = 102.02). This elevated EQ suggests that dingoes have retained the brainpower needed for autonomous survival in Australia's demanding environments, where problem-solving and adaptability are vital for securing food and navigating the landscape. By contrast, domestic dogs, whose reduced brain size is a hallmark of domestication across many species, typically 8–33% smaller than their wild ancestors, exhibit neural architecture shaped by dependence on humans and less cognitively demanding lives. Dingoes cluster closer to other wild canids in brain-to-body ratios, including the dhole (EQ = 116.26) and the coyote (EQ = 153.31), hinting at convergent evolution in response to similar ecological pressures.

A few years later, in 2019, Bill Ballard and Laura Wilson used advanced imaging techniques to independently verify our findings. Their research confirmed that dingoes maintain larger brains relative to domestic dogs of similar size and when plotted on brain-to-body graphs, dingoes clustered closer to wolves than to dogs. This independent confirmation strengthens the case that dingoes never followed the typical domestication pathway. Instead, they evolved to meet the mental demands of a life lived wild.

Metabolomic analysis led by Sonu Yadav and colleagues (2021), provides yet another layer of biological distinction. In a comparison of blood plasma metabolites, dingoes were found to have markedly different profiles from domestic dogs, particularly in protein, lipid and carbohydrate metabolism. The dingo's metabolite signature showed lower levels of several amino acid and lipid derivatives, alongside elevated markers related to galactose metabolism. Some differences pointed to variations in immune function and neurological development, hinting that the divergence between dingoes and dogs is not just genetic, but deeply physiological. Importantly, these differences remained even when diet and environment were accounted for, suggesting an innate, evolved distinction.

Even at the microbial level, dingoes are different. A study by Sarah Wigmore (2024) from Federation University found that the dominant skin bacteria on dingoes is *Staphylococcus equorum*, a species rarely seen in domestic dogs. In contrast, *Staphylococcus pseudintermedius*, common on domestic dogs, was found at much lower levels on dingoes. This suggests dingoes may have a distinct skin ecosystem shaped by a different evolutionary history and less exposure to human-associated pathogens.

GENETIC IDENTITY

As introduced in Chapter 2, and explored further in Chapter 13, modern genomic studies, especially those using single nucleotide polymorphism (SNP) data, show that dingoes are genetically distinct from domestic dogs. When their DNA is plotted using principal components or phylogenetic methods, dingoes consistently form a separate cluster from both dogs and wolves.

One striking example comes from a comparison of over 166,000 SNP markers, shown in Fig. 7.2. In this analysis conducted by Kylie Cairns, dingoes form their own cluster, genetically separate from domestic dogs and wolves.

Their genetic signatures also point to a distinct evolutionary path. Dingoes carry a low copy number of the AMY2B gene, which affects starch digestion, which is a hallmark of domestication. Most domestic dogs have multiple copies of this gene, reflecting their adaptation to starchy agricultural diets. The absence of this expansion in dingoes supports the idea that they diverged from domestic dog lineages before the rise of farming. This point was covered in more detail in Chapter 1.

Finally, on the issue of hybridisation, explored fully in Chapter 13; recent genome-wide studies show that we can reliably distinguish dingoes, dogs and hybrids using high-resolution SNP data, which allows researchers to quantify the proportion of dingo versus dog ancestry with much greater accuracy than older methods, helping to separate fact from myth about what lives in the wild.

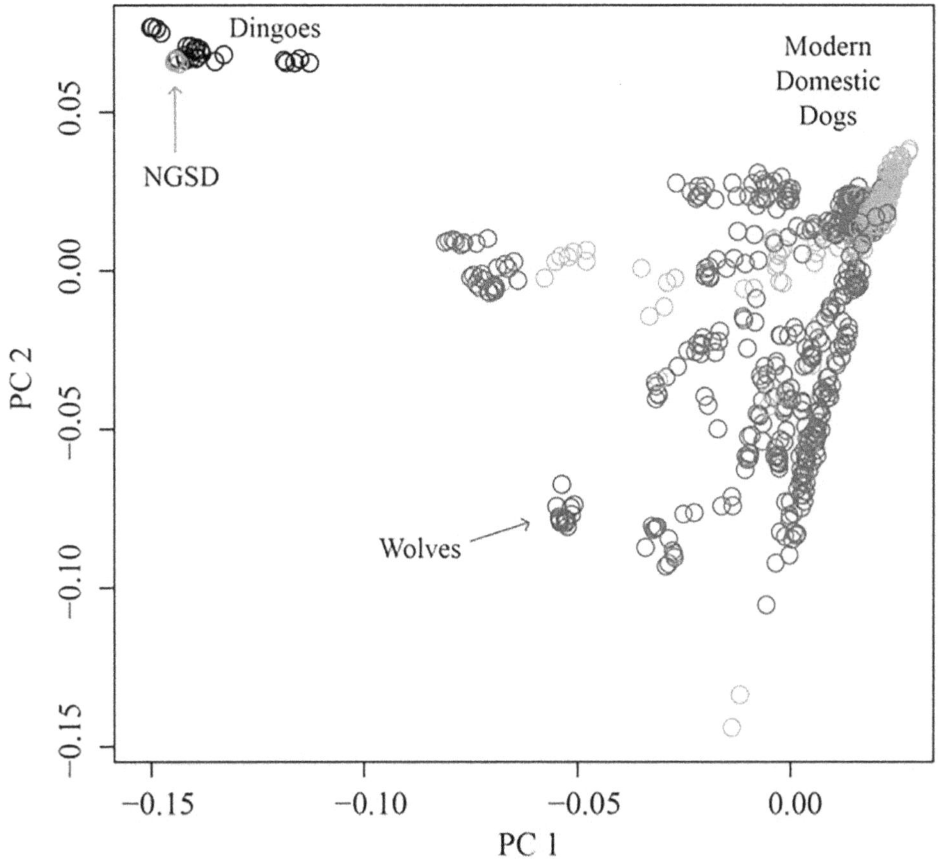

Fig. 7.2. Genetic relationships among dingoes, New Guinea singing dogs, wolves, and domestic dogs based on SNP analysis of over 166,000 markers. Each point represents an individual animal positioned according to genetic similarity. Distinct clusters show dingoes and New Guinea singing dogs grouping closely together, separate from wolves and domestic dogs, confirming that dingoes are not simply feral dogs but a unique lineage with their own genetic identity. Source: Taken from Smith *et al.* (2019).

REPRODUCTION AND PARENTING

Reproduction is where the wildness of dingoes is most apparent, sharply distinguishing them from domestic dogs. As we discovered in the previous chapter, dingoes, like all wild canids, are tightly bound to the seasons, breeding just once a year. Mating occurs in autumn and pups are born in winter, precisely timed with the environment to ensure resources are plentiful during the critical early months of life. Domestic dogs, however, have lost this natural rhythm entirely. Shaped by human intervention, they can have multiple litters per year and their cycles are not tied to seasonality. Thus, it is possible to for dogs to produce litters at any time of the year regardless of season or resource availability.

Dingoes invest in quality over quantity. Each year, they raise a single litter and the whole pack helps raise the pups. It's a family effort and survival rates are high. Domestic dogs take the opposite approach: they produce lots of pups across multiple litters each year, hoping that at least some will make it. It's a numbers game and one that only works because humans provide food and shelter, either directly or through rubbish and scraps.

The role of fathers shows the divide even more clearly (as mentioned in Chapter 6). Dingo dads pull their weight, feeding, guarding and helping to raise the pups alongside the breeding female. It's a pattern also seen in wolves, but not in domestic dogs. Most dog dads are classic deadbeats: uninvolved, uninterested and nowhere to be seen. Back in 2013, when Kathryn Lord, Mark Feinstein, Ray Coppinger and I reviewed parenting across wild and domestic canids, one thing stood out: domestic dogs depend on humans to raise their young, while wild canids rely on each other.

Dingoes also retain a strong social structure, forming stable packs led by a bonded dominant pair. Subordinate members suppress their own reproductive instincts to help raise the pups, ensuring the survival of the pack's next generation. Domestic dogs, in contrast, display chaotic social behaviours. In free-ranging groups, males and females mate indiscriminately, and there's little evidence of pack cohesion or shared parental care. Their survival strategy is shaped not by natural selection but by the human environments they inhabit.

As Lord's work highlights, the differences between dingoes and domestic dogs reflect the divergent pressures of evolution and domestication. Dingoes exhibit all the reproductive and parental behaviours common to wild canids and share little similarity with those of domestic dogs.

TERRITORIAL BEHAVIOUR AND SOCIAL ORGANISATION

Dingoes exhibit territorial and social behaviours that closely align with those of other wild canids, such as wolves. Unlike free-ranging domestic dogs, which form loose and transient associations around human settlements, dingoes maintain distinct territories and cohesive family units, independent of human influence. These tightly knit packs rarely welcome outsiders, maintaining boundaries that help ensure stability and survival. Their strong social bonds support effective territory defence and cooperative hunting.

Dingoes defend specific territories, often centred around core areas of activity, with little overlap between neighbouring packs. Research by Peter Thomson in the Fortescue region of Western Australia showed that dingoes are remarkably loyal to these areas, often returning to the same sites year after year. The size of a dingo's home range, the broader area it uses for hunting, foraging and travel, varies greatly depending on prey availability and landscape productivity. In resource-rich regions, home ranges may span around 100 km^2 or even less, while in the arid interior, they can exceed 2,000 km^2. To sustain their packs, dingoes may travel 10–15 km each day in search of food, a reflection of their endurance and adaptability.

Their hunting strategies are equally flexible. In areas with abundant large prey such as kangaroos, dingoes often hunt cooperatively, using teamwork to outmanoeuvre their quarry. These coordinated hunts reveal their capacity for planning and social cohesion. In contrast, where smaller prey is more common, dingoes tend to hunt alone, showing how they can adapt their behaviour to suit the conditions.

Domestic dogs, including village and feral dogs, do not display the same territorial precision or cohesive pack structure. Their associations are typically loose and short-lived, shaped by proximity to human settlements and access to waste or food scraps. Unlike dingoes, whose survival depends on stable territories and well-coordinated social groups, domestic dogs occupy a human-altered niche where such strategies are unnecessary. This contrast reveals the extent to which domestication has changed the behaviour and social organisation of dogs.

VOCAL COMMUNICATION

The belief that dingoes don't bark is one of the most persistent myths surrounding these animals. Although their vocalisations are less frequent and varied than those of domestic dogs, dingoes do bark and they do so in distinct, purposeful ways. Unlike domestic dogs, which bark prolifically across a wide range of contexts, dingoes reserve their barks for specific functions, such as alarm calls or territorial defence. These barks are often combined with howls to create a bark-howl, a sharp and penetrating sound designed to warn intruders or signal other pack members. This restrained and functional use of barking contrasts sharply with domestic dogs, which often bark to engage humans or seek attention.

Howling, however, is the dingo's most iconic vocalisation. These long, melodic calls carry across vast distances, playing a central role in territorial defence, social cohesion and locating pack members. Research by Laurie Corbett reveals that dingoes can modulate their chorus howls to create the illusion of a larger pack, blending their voices and altering pitch and timing to deter rival groups. This deceptive use of howling highlights the sophistication of dingo communication, aligning them closely with other wild canids like wolves.

In addition to barks and howls, dingoes possess a broad vocal repertoire that includes growls, woofs, yelps and that distinctive 'snuff bark'. Unlike a traditional bark, the snuff bark is a sharp, nasal exhalation used in close-range communication. Often subtle and understated, this unique vocalisation offers a silent warning as described in Chapter 5. Growls and woofs are used in defensive or confrontational scenarios, while yelps often signify submission or distress. These vocalisations, combined with howls and barks, form a sophisticated communication system that helps dingoes maintain social cohesion, defend their territories and navigate the challenges of the wild.

The evolution of these vocal behaviours highlights the difference between dingoes and domestic dogs. Domestic dogs, shaped by millennia of human companionship, have developed a reliance on barking to interact with people and navigate human-dominated environments. Dingoes, unbound by the constraints of domestication, use their voices sparingly and with precision.

WILD INTELLIGENCE

The dingo's mental world is a fascinating blend of instinct, intelligence and independence. Observing a dingo tackle a problem, one is struck by their resourcefulness. Unlike domestic dogs, which often lean on human collaboration, dingoes thrive on self-reliance, a trait shared with their wild counterparts, wolves.

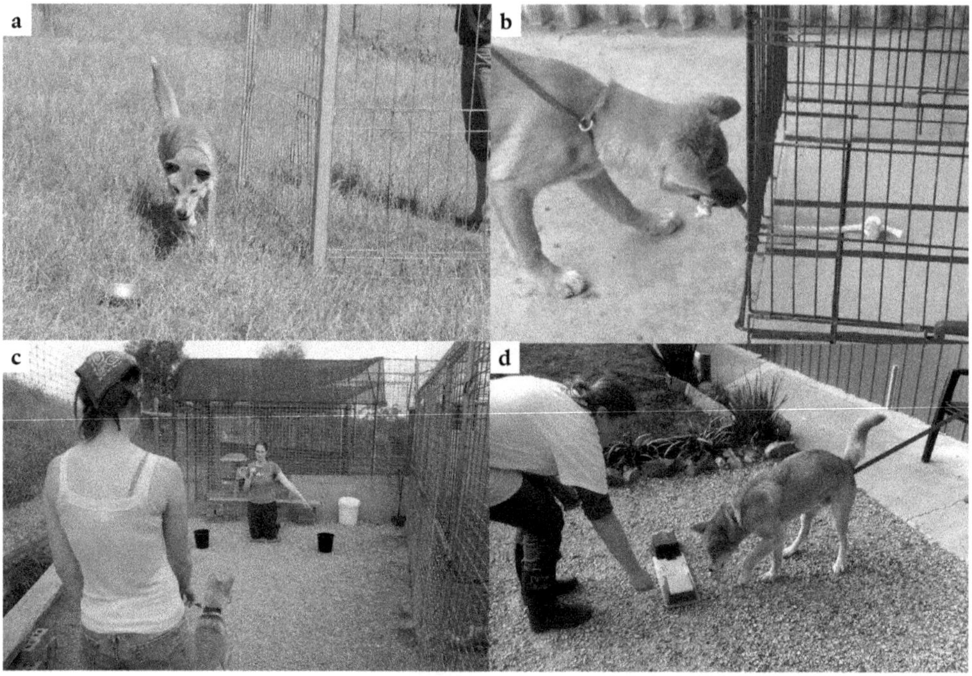

Fig. 7.3. A series of behavioural experiments demonstrates that dingoes behave more like wild canids than domestic dogs. (a) Dingoes navigate natural barriers more quickly and with fewer errors than dogs. (b) Dingoes are less likely to look back to hmans for assistance during challenging tasks. (c) Dogs are more responsive to human social cues, such as pointing. (d) Researcher Angie Johnston (Boston College, USA) testing a dingo with a problem-solving task. Source: Bradley Smith.

Take the detour task as an example (Fig. 7.3a). This spatial problem-solving challenge places a transparent barrier between an animal and a reward. In studies I conducted, dingoes completed the task faster and with fewer mistakes than domestic dogs, even when the setup was reversed to test cognitive flexibility. Wolves performed similarly to dingoes, relying on trial, observation and persistence. Dogs, however, frequently paused to look towards their human companions for guidance, a telling behaviour that reflects the deep reliance forged through domestication. For dingoes, no such dependence exists. Their problem-solving is entirely self-directed.

When it comes to interpreting human gestures, which I first mentioned in Chapter 1, dingoes show a similar mix of capability and independence. A pointed finger, so effortlessly understood by domestic dogs, is met with caution by dingoes. In my research that I alluded to in the preface of this book (Fig. 7.3c), dingoes were able to follow human pointing gestures to locate hidden food, but their responses were slower and more deliberate than those of dogs. It's as though they weighed the value of the signal before acting, reflecting a pragmatic and independent nature. This stands in contrast to the eager compliance often seen in domestic dogs, whose behaviour is shaped by millennia of co-evolution with humans.

The 'looking back' task (Fig. 7.3b) offers another perspective on this independence. In this experiment, an animal is presented with an unsolvable problem, typically a familiar task, such as pulling a rope to access food, but modified so that the solution is no longer possible. Domestic dogs, when faced with this unexpected challenge, quickly seek human assistance, alternating their gaze between the problem and their handler. This behaviour is interpreted as a form of social referencing, shaped by their long history of cooperation with humans. Dingoes, however, tend to persist on their own, rarely engaging in human-directed gazes. As my research demonstrated, their glances are fleeting and task-focused, aligning them more closely with wolves.

Eye contact provides another compelling contrast. Angie Johnston, a researcher at Yale University who specialises in canine cognition (Fig. 7.3d), explored how dingoes compare to dogs and wolves in their interactions with humans. Her findings from 2020 reveal that dingoes initiate eye contact more often than wolves, but for shorter durations than domestic dogs. This intermediate behaviour suggests that the ability to make eye contact with humans emerged early in domestication, but the motivation to sustain prolonged gazes, a hallmark of the human–dog bond, evolved later. Ádám Miklósi, a leading expert in canine behaviour, has further emphasised the importance of sustained mutual gaze in deepening the bond between humans and domestic dogs. Dingoes rarely engage in extended eye contact. Their glances are brief and functional, more about assessing a situation than seeking connection.

Training a dingo is another exercise in understanding their distinctiveness. Domestic dogs thrive on routine and repetition, often eager to please. Dingoes resist traditional training methods. Their cooperation must be earned, not demanded. I've seen firsthand how their intelligence shines in naturalistic settings, such as foraging or hunting simulations, where they adapt and learn with ease. Yet, ask them to follow a command for the sake of compliance, and you'll likely be met with disinterest. I don't believe it is stubbornness; but rather a reflection of their wild lineage and self-reliance.

In experiments designed to test their cognitive limits, dingoes consistently surprise me. They rarely seek help when confronted with a challenge, relying instead on their instincts and problem-solving skills. This independence exemplifies their identity as a species that chooses when and how to engage with the human world.

AS HOUSEHOLD COMPANIONS

There's a reason humans love living with dogs, they've been shaped over thousands of years to meet our needs and expectations. Dingoes are different. My research has consistently shown that dingoes are independent, opportunistic and self-assured. Traits that make them fascinating, but also challenging companions. Unlike dogs, they are less obedient, less trainable and far more prone to escaping. These behaviours reflect their wild ancestry, untouched by the selective breeding that has moulded domestic dogs into cooperative companions.

Through behavioural assessments and owner-reported surveys, it's clear that dingoes fall well outside the typical range of traits we associate with successful household pets. Their autonomy, wariness of strangers and low compliance mean they require a level of care and understanding that few owners are prepared for. Some people do form rewarding bonds with their dingoes, but many struggle. Even among experienced owners, few would recommend dingoes as household companions. They simply lack the traits we've selectively bred into domestic dogs, traits that make dogs easy and enjoyable to live with.

Over the years, I've met many people with companion dingoes and heard countless stories of the challenges they face, dingoes that can open doors, chew through car seat belts, eat through thick-gauge wire or dig massive dens throughout backyards. One memory in particular captures the essence of what it means to live with a dingo.

I was visiting a suburban home in the outer suburbs of Melbourne for dinner when my host opened the front door to greet me. In that brief moment, I saw a tawny blur shoot between my legs and past its owner, one of their dingoes had spotted the split-second opportunity and bolted for freedom.

What followed was illuminating. Despite the owners' experience and immediate pursuit, the dingo remained elusive for 3 days. It stayed within the neighbourhood, occasionally glimpsed but always out of reach. It seemed to operate on its own terms, watching the family's efforts from a distance, ultimately choosing to return home when hunger became more persuasive than freedom.

That experience crystallised what my research has long shown: the fundamental difference between domestic dogs, selectively bred for millennia to cooperate with humans, and dingoes, which retain their independent decision-making no matter how well socialised. That's not to say meaningful bonds can't form between people and dingoes, but these relationships demand an exceptional level of understanding, patience and accommodation. Far more than most pet owners anticipate or can realistically provide.

FULFILLING A SPECIALISED ECOLOGICAL NICHE

Dingoes thrive across most of mainland Australia's habitats, performing a crucial role as apex predators. Through their predation and interactions with other species, they shape ecosystems, affecting everything from prey populations to vegetation dynamics. This pivotal ecological role, which I explore in detail in Chapters 8 and 9, helps demonstrate the dingo's significance as a truly wild canid.

By contrast, domestic dogs are simply not equipped to fulfil such a role. Free-roaming dogs in Australia and elsewhere remain tethered to humans, either as companions or scavengers relying on food scraps around settlements. Despite centuries of opportunity, a few individuals may have been successful for a time, but there is no evidence that feral domestic dogs have established viable, self-sustaining populations in Australia, including in Tasmania, where dingoes have never existed. That absence speaks volumes. It reveals the deep dependency of domestic dogs on human presence and support.

The ecological role of dingoes cannot be replicated by domestic dogs, or other wild canids from elsewhere. It is fair to say that their position in Australia's natural world is not only unique, it's almost irreplaceable.

AUSTRALIA'S WILD CANID

Behind the familiar canine form lies a creature far removed from the world of domestic dogs. The dingo, shaped by millennia of independence, occupies a unique place in the natural world. From its physical adaptations and genetic identity to its behavioural autonomy and ecological role, every feature of the dingo speaks to its wildness.

It's not just that dingoes are different from dogs, they tick every box we use to define a wild canid. They live without human support, breed cooperatively, raise their pups in stable packs, hunt and scavenge with skill, and possess larger brains and sharper instincts than their domestic cousins. They meet the criteria, biological, behavioural and ecological, that distinguish wolves and other wild canids from their domesticated kin.

Far from being just another dog, the dingo is Australia's answer to the wolf: a true wild canid. Its keen mind, finely tuned body and cohesive social structures reveal an animal that has never needed human hands to shape its destiny. In the dingo, we catch a glimpse of what dogs might have been, had they never walked the path of domestication, a thread we return to in a later chapter.

Part 3

Dingoes in the landscape

8
The dingo effect

IN SOUTH AUSTRALIA'S STRZELECKI DESERT, ECOLOGIST KATHERINE MOSEBY manoeuvres her 4WD across the rolling sands. The vast desert stretches endlessly, a landscape shaped by cycles of drought and renewal. But today, something feels off. Scattered across the plains are the carcasses of red kangaroos. Emaciated reminders of a population boom that has pushed this fragile ecosystem to its limits.

This stark scene has become all too common in regions where dingoes have been removed. Without their primary predator, kangaroo numbers swell in favourable conditions, only to crash in times of scarcity. Vegetation, already sparse in the desert, struggles to recover under relentless grazing pressure.

'Dingoes are crucial for keeping kangaroo populations in check', explains Katherine, reflecting on decades of research into Australia's arid ecosystems. 'Without them, the system becomes unbalanced, and we see these boom-and-bust cycles becoming more extreme, leaving the landscape degraded.' The consequences ripple outward: bare soils lose their stability, water retention declines and the biodiversity of plants and smaller animals suffers. The loss of vegetation doesn't just affect herbivores; it reverberates through the entire system, breaking the intricate connections that hold it together.

The same story is playing out in the Mallee woodlands of north-western Victoria, where ecologist Euan Ritchie surveys a landscape under strain. Kangaroos dominate here too, and in many areas their high numbers are suppressing the regeneration of shrubs and grasses, particularly after rain. What was once a rich mosaic of shrubs, grasses and open spaces, a haven for small mammals, reptiles and birds, is increasingly being pushed toward a less diverse and more homogenised state.

'A healthy ecosystem is about balance', Euan observes. 'Dingoes don't just control kangaroo populations; they maintain the conditions that support the survival of other species. When you lose an apex predator like the dingo, you lose more than just a species, you lose the process of predation that helps keep the entire system functioning.'

An apex predator sits at the very top of the food chain, facing no natural enemies of its own. They exert influence throughout entire ecosystems, a process known as top-down control. By regulating populations of large herbivores like kangaroos and feral goats, dingoes prevent overgrazing and enable vegetation to recover. This, in turn, supports the health of the soil and the survival of other species.

Ecologists have observed similar patterns around the world when top predators are removed. Without wolves, elk overgraze riverbanks in Yellowstone. Without sharks, reef fish populations spiral out of balance. In Australia, the dingo fills this critical ecological role, preventing the landscape from being overrun by herbivores and maintaining a delicate equilibrium.

The dingo is what scientists call a keystone species. This term refers to organisms whose influence on their ecosystem is disproportionately large compared to their abundance. Like the keystone of an stone arch, their presence holds the system together. Remove the keystone, and the arch collapses. Similarly, where dingoes persist, ecosystems are healthier, more biodiverse and more resilient.

DINGO DIET AND PREY PREFERENCES

Far from being a picky eater, the dingo is a master of dietary flexibility. Scientists have documented an astounding menu of over 200 vertebrate species in the dingo's diet, including mammals, reptiles and birds (Fig. 8.1), but also insects, fruits and other plant materials. By analysing scats and stomach contents, researchers have pieced together the remarkable adaptability of dingoes to the continent's diverse and often harsh environments. Every sample tells a story. Bone fragments, fur, feathers, scales and seeds all reveal what dingoes eat and how they survive.

A large-scale study led by Tim Doherty (2019) analysed over 32,000 scats and stomach samples from across the continent, providing the most comprehensive insight yet into dingo dietary patterns. Kangaroos and wallabies often dominate, making up the largest portion of

Fig. 8.1. A dingo on 75 Mile Beach, K'gari, with a Bulwer's petrel (*Bulweria bulwerii*) in its jaws. The image illustrates the dingo's opportunistic hunting abilities and dietary flexibility. Source: photo by Uwe-Bergwitz/iStock.

their diet where they are abundant, but this pattern shifts depending on the environment. These dietary preferences reflect ecological principles that govern predator behaviour.

Balancing energy and effort

Ecologists have long been intrigued by how predators make decisions about what to eat and when. To make sense of the dingo's remarkable flexibility, one useful framework is Optimal Foraging Theory. This principle suggests that predators evolve to maximise energy intake while minimising hunting effort; essentially, getting the most nutritional 'bang for their buck'. For dingoes, this means targeting prey that provides a high caloric reward relative to the energy required to catch it.

Kangaroos and wallabies are prime targets when available, as their size and abundance make them efficient energy sources. However, dingoes are opportunists and adapt their strategies when conditions shift. In central Australia, for example, dingoes often exploit the abundance of rabbits, particularly targeting young rabbits as they emerge from burrows. By focusing on this abundant and easily captured prey, dingoes conserve energy while still meeting their nutritional needs.

This balancing act becomes especially apparent during times of scarcity. During droughts, when larger prey is less available, dingoes shift their focus to smaller animals like rodents and even insects. On coastal beaches, they scavenge marine animals such as sea turtles and dolphins, highlighting their ability to adapt to changing conditions. Optimal Foraging Theory explains these shifts, showing how dingoes adjust their behaviour to the availability and energy payoff of different prey.

Linking diet to the environment

But prey availability isn't just about what's easiest to catch; it also depends on the productivity of the surrounding landscape. Another ecological theory, the Exploitation Ecosystems Hypothesis, helps explain how predator diets shift across different environments. It suggests that the productivity of an ecosystem (i.e. how much energy and biomass it can support), influences which species are targeted by predators. In resource-rich areas dingoes focus more on medium and large mammals, where prey populations tend to be stable and predictable. By contrast, where resources are patchy and unpredictable, dingoes broaden their diet to include arthropods, reptiles and carrion.

As shown in Fig. 8.2 and Table 8.1, dingo diets vary significantly across Australia's five bioclimatic zones. The map highlights the distribution of dietary studies and illustrates how dingoes adapt to local environmental productivity. This geographical variation emphasises their role as ecological generalists, adjusting their diet to match the challenges and opportunities of each habitat.

One striking pattern that emerges is the impact of rabbits on dingo diets. In regions where rabbits are abundant, dingoes rely heavily on this invasive prey, often reducing predation pressure on native medium-sized mammals. This efficient use of resources reflects both Optimal Foraging Theory and the Exploitation Ecosystems Hypothesis, demonstrating how dingoes play a stabilising role in ecosystems, even when feeding on introduced species.

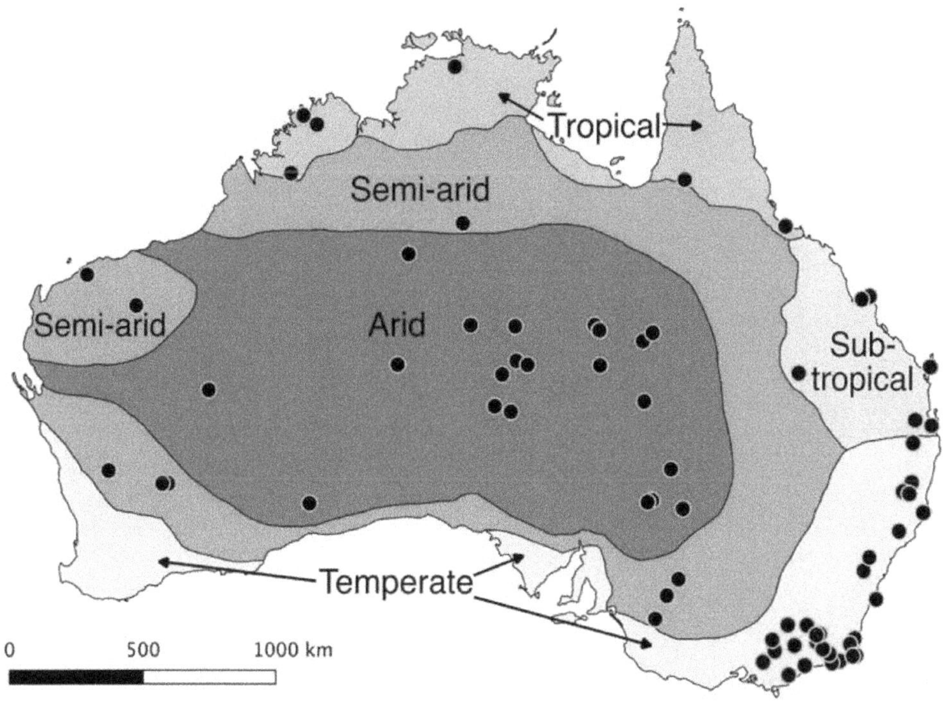

Fig. 8.2. Map of Australia showing the locations of dingo dietary studies (black dots) and the five bioclimatic zones listed in Table 8.1. Source: Doherty *et al.* (2019).

Together, these theories reveal how dingoes fine-tune their behaviour to both the energetic costs of hunting and the broader productivity of the landscapes they roam.

ADAPTABILITY

The dingo's adaptability reveals itself when we look at how these theories play out across real landscapes. Their ability to fine-tune foraging strategies allows them to thrive in environments that challenge even the hardiest species.

The following examples from different regions of Australia show how dingoes put these flexible strategies into action, tailoring their behaviour to the conditions at hand.

In the vast arid regions of central Australia, where rainfall is scarce and prey availability rises and falls with resource pulses (brief periods of abundance following rare rain events), dingoes have evolved to exploit whatever food is available. When kangaroo populations decline during droughts, dingoes switch their focus to smaller prey. In these environments, their scavenging behaviour also plays a critical role. Dingoes are often seen consuming carrion, including livestock carcasses, and even relying on smaller, less energy-intensive prey to bridge periods of scarcity. Their ability to forage efficiently under such conditions highlights their capacity to survive in even the harshest landscapes.

The tropical north presents a different challenge altogether. Here, seasonal rains transform the landscape into a lush paradise, temporarily increasing prey abundance. During this time,

Table 8.1. Food types found in dingo diet across Australia, showing the percentage of scats and stomachs containing each food type

Data from 32,225 samples collected across Australia between 1980 and 2019. Prey types are categorised by size and origin (native *v.* exotic), with regional trends highlighting differences in dietary preferences across bioclimatic zones shown in Fig. 8.2. Source: Doherty *et al.* (2019).

Prey type	How often found	Regional variation
Large native mammals (≥ 7 kg, e.g. kangaroos, wallaroos)	Very common (35%)	Highest in semi-arid, temperate and tropical zones; lowest in arid and subtropical zones
Native medium-sized mammals (0.5–6.9 kg, e.g. possums, bandicoots)	Common (18%)	Most common in temperate and subtropical zones; declines in tropical and arid zones
Small mammals (< 0.5 kg, e.g. native rats, mice, small marsupials, kittens)	Common (15%)	Highest variability in the arid zone; no significant differences noted across other regions
Rabbits	Common (14%)	Highest in arid and semi-arid zones; absent from tropical zone
Reptiles and arthropods (e.g. lizards, snakes, insects)	Less common (12%)	Most common in arid and semi-arid zones
Birds (e.g. emus, waterbirds)	Less Common (9%)	Highest in arid and semi-arid zones
Large exotic ungulates (e.g. cattle)	Occasional (7%)	Highest in arid and tropical zones
Medium-sized exotic ungulates (e.g. sheep)	Occasional (4%)	No significant regional variation
Exotic medium-sized mammals (0.5–6.9 kg, e.g. foxes, feral cats)	Occasional (3%)	No significant regional variation
Marine animals (scavenged remains of sea turtles, dolphins)	Occasional	Coastal regions only

dingoes capitalise on species such as agile wallabies, whose numbers swell after the wet season. This seasonal availability allows dingoes to focus their efforts on larger prey, conserving energy and maximising caloric intake. Notably, in these regions, rabbits, a staple in arid zones, are largely absent, forcing dingoes to adopt different hunting strategies tailored to their surroundings.

In the temperate zones of south-eastern Australia, dingoes turn their attention to medium-sized mammals such as wombats and possums, adapting to prey that thrives in these resource-rich environments. Along coastal regions, dingoes have even been observed scavenging washed-up marine animals, such as sea turtles, dolphins and even whales. This behaviour is particularly common on islands like K'gari, where ranger Linda Behrendorff (2016) has regularly documented dingoes making use of beached carcasses (more about Linda and her work in Chapter 15). Building on these observations, ecologists Emma Spencer and Tom Newsome (2021) recorded dingoes scavenging at least nine different marine species, from turtles to whales, on beaches. Their work highlights how this behaviour not only sustains dingoes but also supports broader nutrient cycling within ecosystems. By breaking down these carcasses, dingoes contribute to nutrient enrichment of coastal soils, which in turn benefits plant communities.

Beyond simply sustaining dingoes, scavenging contributes to ecosystem health in surprising ways. By consuming carrion, dingoes reduce waste as well as play an essential role in nutrient cycling within their ecosystems. This role extends to agricultural regions, where scavenging livestock carcasses has been observed. While this behaviour can bring them into conflict with farmers, it demonstrates their ability to adapt to human-altered environments.

However, this flexibility can bring dingoes into conflict with conservation efforts, particularly in areas where small populations of threatened species persist. Studies of dingo scats have revealed remains of some endangered animals, such as bilbies and bettongs, raising concerns about dingo impact on these vulnerable species. Although this has sparked debates about the risks posed by dingoes, it's important to consider the broader ecological context. Dingoes and these species coexisted for thousands of years before European settlement and it is often the destruction of habitat and the introduction of invasive species like foxes and cats, not dingo predation, that have driven many native species to the brink of extinction.

Research suggests that dingoes may even provide indirect protection to some threatened species. By suppressing populations of invasive predators, dingoes create conditions in which smaller, vulnerable animals can thrive. For example, areas with stable dingo populations often exhibit higher biodiversity and greater abundance of small native mammals compared to regions where dingoes have been removed. These findings challenge the narrative that dingoes are primarily a threat to conservation and instead position them as key as guardians of balance in Australia's wild landscape.

TROPHIC CASCADES: SHAPING LANDSCAPES AND BIODIVERSITY

Imagine the ripple effect of a stone dropped into a still pond. The waves radiate outward, each influencing the next, until the entire surface is transformed. Trophic cascades work in much the same way, with changes at the top of the food chain rippling through ecosystems, influencing herbivores, plants and even soil (Fig. 8.3). As Australia's apex predator, the dingo is the catalyst for these ripples, profoundly shaping the landscapes it inhabits.

This cascading effect begins with changes in herbivore behaviour. In areas where dingoes are present, kangaroos and other herbivores alter their grazing patterns to avoid open spaces where they are more vulnerable to predation. Instead, they move cautiously between patches of vegetation, giving plants a chance to recover and grow. The broader ecological consequences of dingo loss have been documented by prominent Australian ecologists Mike Letnic, Euan Ritchie and Chris Dickman. Their ongoing studies have revealed that areas with dingoes exhibit not only healthier vegetation but also increased biodiversity. Native grasses flourish. Wildflowers and shrubs reestablish themselves, attracting insects and pollinators. Species that rely on dense undergrowth for shelter, such as small marsupials and ground-dwelling birds, fare better in dingo-occupied landscapes.

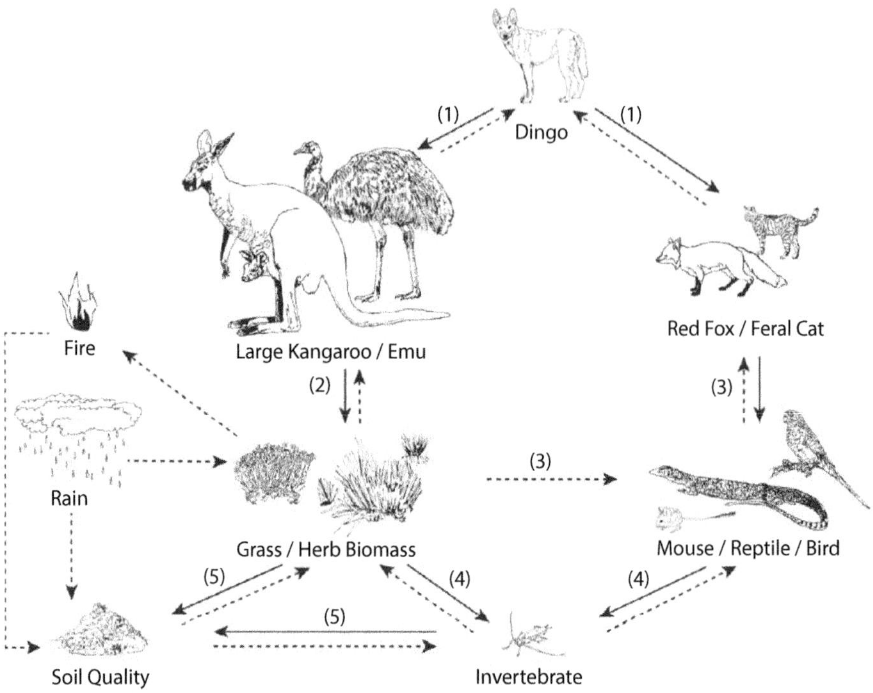

Fig. 8.3. A dingo-induced trophic cascade. Numbers indicate the predicted sequence of events: (1) dingoes suppress large herbivores (e.g. kangaroos and emus), leading to increased grass and herb biomass; (2) dingoes suppress lower-order predators (e.g. foxes and cats), resulting in increased numbers of small mammals (e.g. rodents), reptiles (e.g. goannas) and birds (e.g. parrots); and (3) improved vegetation conditions benefit invertebrates and enhance soil quality. Dashed arrows represent interactions influenced by rainfall and fire. Source: Newsome *et al.* (2015).

The influence of dingoes even extends beneath the surface. In regions where herbivores are kept in check, soils retain more organic matter and nutrients, supporting healthier plant communities. Dingoes, through their regulation of herbivore populations, contribute to a feedback loop of ecological health, where plants, animals and soil all benefit.

Consider the red kangaroo (*Osphranter rufus*), like I described at the beginning of this chapter. In areas free from dingoes, kangaroo populations boom in favourable conditions, consuming vegetation at unsustainable levels. During dry periods, these populations crash, leaving the land stripped and degraded. With dingoes present, however, kangaroo numbers are naturally regulated, preventing overgrazing and reducing the severity of boom-and-bust cycles. Dingoes not only cap kangaroo populations through predation, but also buffer the effects of population booms driven by rainfall, reducing grazing pressure and helping grasslands recover. This balance between predator and prey helps maintain the resilience of Australia's arid landscapes, where resources are often scarce and ecosystems must work hard to recover from disturbance.

As Letnic and his colleague Matthew Crowther (2013) from the University of New South Wales demonstrate, this dynamic is a clear example of top-down-regulation, which is a process whereby apex predators like dingoes help control herbivore populations and prevent overgrazing. In contrast, in the absence of dingoes, kangaroo numbers are driven primarily by bottom-up forces such as rainfall and pasture growth, often resulting in extreme population booms followed by dramatic crashes.

This insight is reinforced by modelling work from David Choquenot and David Forsyth (2013) at the Arthur Rylah Institute, who examined the interaction between pasture productivity, kangaroo numbers and dingo presence. Their simulations showed that dingoes suppress kangaroo population peaks even in resource-rich areas, limiting the kind of boom-and-bust dynamics that can strip vegetation and destabilise soils.

The story of trophic cascades is one of interconnectedness. Dingoes, as apex predators, sit at the centre of this web, their influence extending far beyond the immediate effects of predation. By shaping the behaviour of herbivores, they create conditions that allow biodiversity to flourish, stabilise soils and sustain the delicate balance of life in Australia's unique ecosystems. These ripples may start with a single predator, but their effects are felt across the entire landscape.

THE DINGO FENCE: A CONTINENTAL-SCALE EXPERIMENT

The removal of large predators has transformed ecosystems across the globe, but often, the damage was done long before scientists could measure it. In Australia, however, a unique opportunity exists to observe these effects in real time. Stretching 5,600 km from Queensland to the Great Australian Bight, the Dingo Fence offers one of the world's most remarkable natural experiments in predator ecology.

Fig. 8.4. Ecologist Mike Letnic pictured beside the Dingo Fence. His research has been pivotal in shaping modern understanding of dingo ecology. Source: photo by Wild Pacific Media.

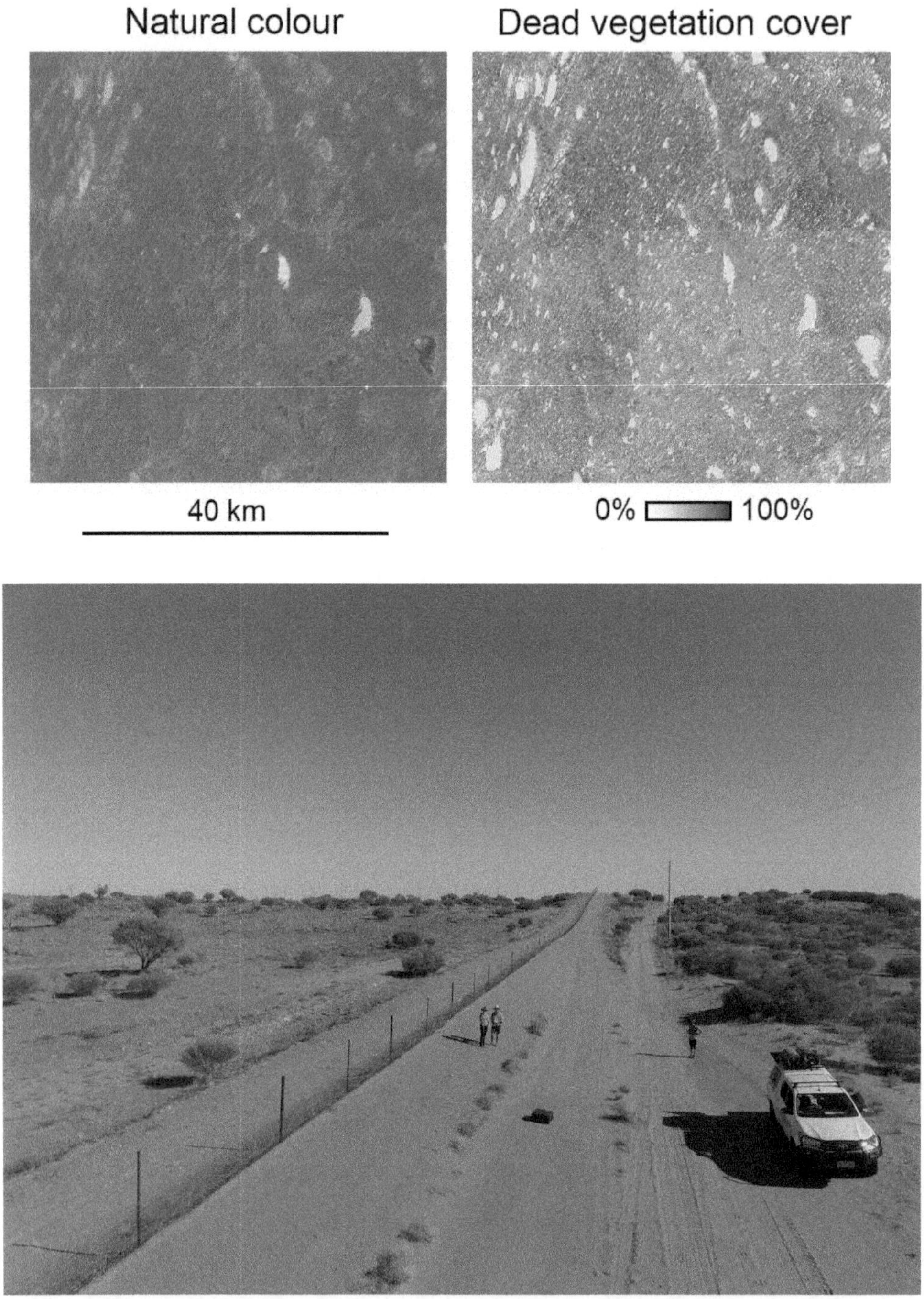

Fig. 8.5. (Top) A satellite image from 2011 shows no visible Dingo Fence in natural colour (left), but the fence becomes clear when viewed as dead vegetation cover (right). (Bottom) A drone image showing the fence in Sturt National Park, NSW, highlighting the difference in vegetation cover caused by grazing pressure in areas with few dingoes (left of fence) from where there are dingoes (right). Source: (Top) Adrian Fisher (data provided by the Joint Remote Sensing Research Program and the United States Geological Survey); (Bottom) Wild Pacific Media.

Fig. 8.6. A fenced plot in Sturt National Park, NSW (March 2018) as part of Letnic's research, highlights the difference in vegetation cover caused by grazing pressure in areas with few dingoes. Source: Wild Pacific Media.

For sheep farmers, the fence works as intended: a formidable barrier protecting millions of dollars in livestock from dingoes. While pastoralists see it as essential infrastructure, ecologists see something else entirely, an unintentional line of demarcation between two vastly different ecosystems. On one side, where dingoes persist, ecosystems remain more intact. On the other side, where dingoes have been eliminated, prey populations like kangaroos have exploded and the landscape has been stripped of its resilience.

Because the fence cuts through the same arid landscapes and soil types, often just kilometres apart, researchers can compare what happens with and without top predators. As Letnic (Fig. 8.4) describes in the scientific journal *Austral Ecology* (2010), the Dingo Fence provides 'a natural experiment' that enables direct investigation of the ecological role of apex predators.

Letnic recalls an early field moment that brought the contrast into sharp focus: driving along the fence line, the difference in vegetation was stark. On the dingo side, the land supported multilayered plant life. On the other side, it was chewed back to bare soil. That's when he realised just how profound the dingo's influence could be.

In later work published in *Landscape Ecology* (2021), Letnic and remote sensing expert Adrian Fisher used 32 years of satellite imagery to quantify the impacts. This wasn't just about green vegetation. Measuring dry leaf litter, twigs and senescent grasses required novel techniques adapted to desert conditions. Fisher and his colleagues devised methods to detect the total biomass response to rainfall events, even when plants weren't visibly green.

The results told a clear story. After rain, both sides of the fence greened up, but the effect was greater and more sustained where dingoes were present. Where dingoes were absent, the explosion in kangaroo numbers brought intense grazing pressure, suppressing regrowth and reducing landscape resilience. In some areas, kangaroos made up more than 95% of the total herbivore biomass, placing immense stress on vegetation.

Over time, the composition of plant communities also began to shift. Palatable shrubs and native grasses that stabilise the soil and retain moisture were grazed out, replaced by tougher, less nutritious species that support fewer native animals.

As Letnic noted in his 2021 study: 'The differences in grazing pressure on each side of the fence were so pronounced, they could be seen from space.'

Satellite data even revealed changes in dune morphology: without dingoes, sand dunes became less vegetated and more mobile, reshaped by wind erosion (Figs 8.5 and 8.6). With less vegetation to anchor them, wind erosion took hold, altering dune shape and movement patterns. These are not just isolated effects. They are cascading, system-wide transformations, set in motion by the absence of one predator.

These changes don't happen overnight. The effects of predator removal accumulate slowly, cascading through the ecosystem over decades, often unnoticed until the damage is done.

SOIL HEALTH AND FUNGAL NETWORKS

As already mentioned, in areas where dingoes are absent, unchecked herbivores consume vegetation at unsustainable levels, depleting the organic matter that would otherwise nourish the soil. Bare soil is more prone to losing essential nutrients, such as nitrogen and phosphorus, and struggles to retain moisture. Over time, this degradation creates a feedback loop: without vegetation to stabilise the soil and recycle nutrients, ecosystems become less productive and more fragile.

As Letnic explains, 'We show that removal of dingoes leads to a trophic cascade that extends right down to the depletion of soil nutrients. Allowing dingo populations to increase could enhance the productivity of ecosystems across vast areas of the country by reducing herbivore numbers.' This finding challenges conventional wisdom, which has often underestimated the indirect impacts of apex predators like dingoes on soil health.

Research in the Strzelecki Desert compared areas on either side of the Dingo Fence and found that where dingoes were rare, herbivores such as kangaroos flourished, leading to overgrazing that reduced vegetation cover and depleted levels of phosphorus, nitrogen and carbon in the soil. By contrast, areas with dingoes exhibited healthier soils enriched by greater vegetation cover. As Letnic observed, 'We need to rethink the idea that kangaroos have benign impacts on ecosystems. Kangaroo numbers are very high across the approximately one third of the continent where dingoes are rare, and are having damaging impacts on soils and vegetation.'

In dingo-present landscapes, healthy vegetation also supports robust fungal networks, which play a crucial role in nutrient cycling and plant health. Research such as that from 2024 led by Todd Elliot from the University of New England demonstrates the unexpected way in which dingoes contribute to these processes: by acting as secondary dispersers of mycorrhizal fungi. Mycorrhizal fungi form symbiotic relationships with plant roots, helping plants absorb nutrients more efficiently while benefiting from the sugars plants produce. As dingoes consume fruits, seeds and plant matter, they unknowingly spread these fungi through their scats, introducing them to new areas and supporting vegetation growth.

ADDRESSING MYTHS: DINGOES AND SMALL NATIVE PREY

A persistent misconception about dingoes is that they pose a significant threat to small native prey, including vulnerable species like bilbies and bettongs. Although dingo scats occasionally reveal remains of these animals, the evidence overwhelmingly suggests that dingoes are not a primary driver of their decline. Instead, habitat destruction, competition with invasive species like rabbits and predation by foxes and feral cats have played far larger roles in pushing many small native mammals to the brink of extinction.

The Top Predator Hypothesis offers a compelling explanation for this dynamic. This ecological theory posits that apex predators, by suppressing smaller predators, create safer conditions for prey species lower on the food chain. In essence, the presence of a top predator like the dingo helps regulate the entire ecosystem, including reducing the impact of mesopredators such as foxes and feral cats (something that will take centre stage in the next chapter).

Research led by ecologist Arian Wallach, published in 2009, provides strong support for this hypothesis. The study investigated two threatened species, the dusky hopping mouse (*Notomys fuscus*) and the malleefowl (*Leipoa ocellata*), finding that both survived and persisted in areas where dingoes were present. These findings demonstrate that species at risk from predation and competition are more likely to thrive when a top-order predator is part of the ecosystem.

Indeed, several studies in dingo-occupied areas consistently show stable or even improved biodiversity compared to regions where dingoes have been eradicated. For example, populations of native rodents and small marsupials are often more abundant in landscapes where dingoes regulate invasive predators and maintain vegetative cover.

Debunking the myth of dingoes as threats to small native prey is crucial for shifting public perception and conservation policy. Rather than viewing dingoes as predators to be eradicated, it is time to recognise their role as protectors of biodiversity and stewards of ecological balance.

9
The feral invaders

IN 1788, THE FIRST FLEET DROPPED ANCHOR IN THE SHELTERED WATERS OF Port Jackson, a deep natural harbour fringed by rocky headlands and eucalypt scrub. It was here, in what would become Sydney Cove in New South Wales, that a new chapter in Australia's story began. Unfortunately, the human arrivals weren't alone. Below decks, crammed into wooden crates and cages, a host of animals, some brought by design, others by accident, waited to disembark. Sheep, cattle, pigs and goats, essential lifelines for the fledgling colony. Rats and mice, stowaways that slipped aboard unnoticed. And among them was one of the deadliest invaders of all: cats.

The history of introduced species in Australia is inseparable from colonisation. Settlers, unfamiliar with the strangeness of the land, sought to remake it in Europe's image. They brought livestock for food and transport, but also animals tied to nostalgia: the ones that made home feel like home. Rabbits were introduced as a food source in 1788, but it was Thomas Austin's release of 24 rabbits in 1859, purely for sport, that triggered one of the greatest ecological disasters in Australian history. Within decades, rabbits were spreading at 100 km per year, stripping vegetation, collapsing ecosystems and fuelling erosion.

Cats were originally brought to control shipboard rats. Many were deliberately released into settlements to keep granaries free of rodents, but they didn't stay where they were put. By the 1850s, cats had spread deep into the interior, their numbers exploding as they preyed on naïve, defenceless native species that had never encountered a predator so efficient.

Foxes came later, but their impact was just as severe. In the 1850s, wealthy landowners imported foxes for hunting, hoping to recreate English sporting traditions. Within a few decades, foxes were thriving across forests, grasslands and even deserts, preying on marsupials and ground-nesting birds with ruthless efficiency. Around the same time, acclimatisation societies introduced the common blackbird to Victoria, one of many European species deliberately released into Australian landscapes.

Settlers saw the rabbits, the sheep-killing dingoes, the encroaching deserts. But they didn't see the real catastrophe: the thousands of tiny extinctions happening out of sight, in the burrows and nests of animals that had once thrived.

The invasion had begun. And it would soon make Australia the extinction capital of the world.

THE SILENT KILLERS

Australia's extinction crisis didn't arrive with a bang. It crept in, quiet and unnoticed, as cats slinked through the undergrowth and foxes prowled the open plains. These predators were small, silent and unrelenting. And they were perfectly suited to a land unprepared for them. One by one, the victims began to vanish, species that had thrived for millions of years, gone within decades of the newcomers' arrival.

Cats (*Felis catus*)

Feral cats are the ultimate stealth predators. They don't chase their prey in open pursuit like a dingo or a wolf. They wait. A twitch of the tail, a single paw step forward, then strike. For Australian wildlife, this hunting style has been catastrophic. Native animals simply never evolved to deal with it. Birds, reptiles and mammals that had spent millions of years in a land dominated by marsupial predators were suddenly outmatched.

Today, Australia's 2–6 million feral cats kill an estimated 1.5 billion native animals each year, including 466 million reptiles and 272 million birds. That's around 4 million native animals killed every single day. A single feral cat can consume up to 1,100 native animals annually and they don't stop. They kill whether they're hungry or not, often playing with their prey in a grim display of surplus killing.

These figures come from comprehensive studies by leading ecologists, including three of Australia's most prominent: Professors John Woinarski, Sarah Legge and Chris Dickman, who co-authored the 2019 book *Cats in Australia: Companion and Killer*. The scale of destruction has led to intense concern among conservationists and policymakers alike.

As those authors put it: 'Cats have been remarkably successful in Australia, spreading pervasively across the continent and many islands, occurring in all environments, and proving to be adept and adaptable hunters.' (Woinarski *et al.* 2019).

And Legge highlights their ecological impact, and places it in a global perspective, telling the *ANU Reporter* in 2019:

> *Australia's mammal extinction rate is by far the highest in the world, and cats have been a leading cause of at least 20, or two-thirds, of our mammal extinctions over the last 200 years. On average, each feral cat in the bush kills a whopping 740 animals per year. In a year with average conditions there are about 2.8 million feral cats, but that figure can double when good rain leads to an abundance of prey animals.*

Some of the casualties are already lost to history. Others still cling on, but only just, with numbats, night parrots and bilbies, all teetering on the edge. Such is their impact that the Australian government has made feral cat control a national priority. In 2015, it set a target to kill 2 million cats by 2020 as part of a broader strategy to reduce predation on native wildlife. Since then, over AU$60 million has been committed to feral cat management, funding more than 50 projects across the country. Some states have taken additional steps. In 2024, New South Wales established a dedicated ground-shooting team to remove cats from national parks. Despite these efforts, feral cats remain widespread and destructive, and the debate continues over whether lethal control alone can deliver meaningful conservation gains.

Foxes (*Vulpes vulpes*)

Whereas cats operate in the shadows, foxes bring a different kind of pressure – speed, cunning and opportunism. After their introduction, they quickly shifted their focus from rabbits to easier prey: ground-nesting birds, small marsupials and reptiles. Unlike dingoes, which tend to hunt larger prey, foxes specialise in the 1–3-kg range, exactly where most of Australia's mammals sit.

Foxes are now everywhere. They occupy 80% of the continent, thriving in deserts, forests, farmland and cities. Their impact is devastating. Research shows they are a primary driver of the extinction of at least 14 native mammal species and a leading threat to many more. In areas where fox control programs have been trialled, native mammal populations often rebound quickly.

AUSTRALIA'S VANISHING MAMMALS

Since European colonisation, more than 30 native Australian mammal species have been wiped from existence, a higher number than on any other continent. Dozens more now teeter on the brink.

Among the forever lost are the lesser bilby (*Macrotis leucura*), the desert rat-kangaroo (*Caloprymnus campestris*), the white-footed rabbit-rat (*Conilurus albipes*), Gould's mouse (*Pseudomys gouldii*) and the crescent nailtail wallaby (*Onychogalea lunata*). Once, they were common across deserts, woodlands and scrub. Today, they exist only in the records of museum drawers and old journals. Some vanished within a single human lifetime.

Others are clinging to survival: numbats, bilbies (*Macrotis lagotis*), central rock-rats (*Zyzomys pedunculatus*), brush-tailed rabbit-rats (*Conilurus penicillatus*) and the northern hopping-mouse (*Notomys aquilo*), hanging on in scattered pockets, behind predator-proof fences or on offshore islands. Many have disappeared from over 80% of their former range.

The biggest driver of these losses has been the arrival of invasive predators. There is no question feral cats and red foxes have decimated small-to-medium-sized mammals, ground-nesting birds and reptiles, but their impact hasn't happened in isolation.

Widespread habitat destruction has stripped away shelter and food, exposing native species to new dangers. More than half of Australia's forests and woodlands have been cleared or severely altered since 1788 and land clearing continues in some states at an alarming pace.

Altered fire regimes have added to the pressure. Fires that were once cool, patchy and predictable under Indigenous management now burn hotter and more intensely, destroying habitat and fragmenting ecosystems. In a climate that's growing hotter and drier, species that were already pushed to their limits are now tipping over the edge.

As Woinarski and colleagues noted in a landmark 2015 review, predation by cats and foxes is the leading cause of mammal declines. However, it is the combination of threats – predators, habitat loss, fire and climate – that so often seals a species' fate.

Case study: ghost kills and conservation sabotage

Even when we try to fix what's been broken, the damage keeps coming.

Across Australia, conservationists are working hard to bring back native animals that once roamed freely. Bettongs, bilbies, quolls; all are being reintroduced into the wild as part of ambitious recovery programs. These are careful, science-led efforts: fenced reserves, predator-proof enclosures, radio-collars, tracking devices. Every animal released is a step towards restoration.

But in many places, they don't survive.

In 2025, a team of scientists led by Ned Ryan-Schofield from the University of Adelaide, along with co-authors including ecologist Katherine Moseby, set out to find out why. They focused on wildlife translocations, those hopeful moments when species are given a second chance. What they found was deeply troubling.

Using DNA testing and forensic necropsies, they confirmed that feral cats were responsible for most of the deaths: 74 confirmed kills, with another 18 strongly suspected. In many cases, the cats didn't even eat what they killed. They cached the bodies, abandoned them or left only fragments behind. Sometimes, all that remained was a radio-collar lying under an acacia tree.

The cats left no tracks. No scats. No cameras caught them. And traditional field methods had missed most of their kills. Without DNA evidence, their role would have remained hidden.

'Feral cat predation has long been suspected as a key cause of translocation failure,' the authors wrote. Thanks to this study, it's now beyond doubt. As Moseby explained:

> *In past releases, we'd be radio tracking animals after release, and if we found dead animals, it would be difficult to determine what caused their death. We would use field evidence like animal spoor, carcass remains, or bite marks on collars to guess whether it was feral cats, birds of prey, fox, or if they just died a natural death. But ultimately, DNA testing highlighted that there are a lot more cat killings than previously thought.*

The study was a wake-up call. Even behind fences. Even in managed landscapes. Even when we do everything right. One feral cat is enough to ruin our best efforts to help recover threatened species.

THE DINGO AS A POTENTIAL SOLUTION

For over a century, conservationists have tried to contain Australia's extinction crisis with poison baits, culling programs and fenced sanctuaries. Millions of dollars have been spent controlling foxes and feral cats, often with limited, short-term success. Eradications are rarely permanent. New individuals move in. Populations bounce back. And the cycle begins again.

But there may be another path. One that doesn't rely on perpetual intervention. Dingoes, Australia's apex predator, have emerged as a potential ecological ally.

To understand why, we need to turn to a key ecological concept: the Mesopredator Release Hypothesis.

The Mesopredator Release Hypothesis

At the heart of most ecosystems lies a delicate balance between predators, prey, and vegetation. Apex predators sit at the top, influencing not just the species they hunt, but also those they compete with and displace. This top-down control is vital to ecological stability.

Without a dominant predator to keep smaller predators in check, mid-sized predators, known as mesopredators (from the Greek *mesos*, meaning middle, or intermediate), increase in number, expand their range and intensify their impact on prey species. The impact of removing an apex predator on these smaller predators is referred to as the Mesopredator Release Hypothesis.

In the Australian context, foxes and feral cats are classic mesopredators. They are larger and more aggressive than most native wildlife, but smaller than the dingo. When dingoes are excluded, through fencing, baiting or ongoing persecution, these smaller predators flourish. They hunt more freely, extend their range into new habitats and exert constant pressure on species already pushed to the edge. The result is an ecological squeeze: small mammals, reptiles and birds collapse under the weight of predation.

This process forms part of a broader pattern that readers will recognise from the phenomenon of trophic cascades described in Chapter 8. When apex predators are removed, the knock-on effects travel down through the food web. Mesopredators increase, prey species decline and ecological processes begin to break down. Seed dispersers disappear. Digging mammals that once turned over soil are lost. Insect populations shift. The structure of the landscape changes, often subtly, but sometimes irreversibly.

The Mesopredator Release Hypothesis also helps explain why conventional predator control often fails in the long-term. Without a functioning apex predator, ecosystems remain out of balance. Foxes and cats may be temporarily reduced through baiting or culling, but in the absence of a sustained top-down pressure, they quickly return. Native species are left in a permanent state of vulnerability.

Where dingoes persist in healthy numbers, this dynamic can shift. Mesopredators are suppressed. Native prey species have more opportunities to survive and reproduce. And the ecosystem as a whole becomes more resilient. The presence of a top predator restores a form of regulation that no amount of human management can easily replace.

In theory, it's a compelling idea. But what happens when dingoes, foxes and cats actually share the same landscape?

HOW DINGOES KEEP FOXES AND CATS IN CHECK

The dingo's influence operates through multiple mechanisms, each reinforcing the others.

Direct killing

Dingoes do not simply compete with foxes and feral cats. They actively hunt and kill them (Fig. 9.1). Foxes are particularly vulnerable. They are smaller, less social and more exposed in open country. A single dingo can overpower a fox with ease. Fox remains have been found in up to 10% of dingo scats, a strong indication that these encounters are frequent and ecologically significant.

One of the clearest demonstrations of this behaviour comes from research conducted by Moseby and her team at Arid Recovery in South Australia. In a carefully designed field experiment (published in 2012), seven foxes were introduced into a 37-km^2 fenced reserve already inhabited by two resident dingoes and six feral cats.

Fig. 9.1. A dingo carrying a dead cat on Cordillo Downs Station in northern South Australia. Source: Steve Eldridge, Desert Wildlife Services.

The outcome was swift and decisive. All seven foxes were dead within 17 days. The cats lasted longer, but within 6 months, more than half had been killed and most importantly, no successful breeding was recorded. Most of the carcasses were left uneaten. The motive was not only hunger, it appeared to be about territoriality, competition and control.

The findings confirmed what many ecologists had long suspected. When allowed to function naturally, dingoes will seek out and eliminate mesopredators.

Feral cats are a more elusive target. They are solitary, secretive and capable of taking refuge in rocky crevices, dense cover or even up trees where dingoes struggle to follow. When encounters do occur, however, the result is often fatal.

I witnessed this directly while working with the Nyangumarta Rangers in the Great Sandy Desert. One of our remote sensor cameras recorded a dingo trotting past with a dead cat dangling from its jaws. That image was only part of the story. When we later examined 36 dingo scats from the same area, cat remains were found in 19 of them, just over half. In that region, cats were the most frequently consumed prey, outnumbering wallabies, reptiles, emus and other birds.

Competition for food

Dingoes and foxes both target small mammals, reptiles, birds and carrion. In landscapes where dingoes are present and active, foxes are often pushed to the margins, where they are forced into suboptimal habitats with fewer resources. They struggle to find food, are excluded from key carcasses and become less efficient hunters. This competitive pressure weakens foxes and limits their impact on native wildlife.

Ecologists describe two main types of competition at play:

- interference competition, where dingoes directly exclude or displace foxes from carcasses and key hunting grounds through dominance or aggression
- exploitative competition, where dingoes reduce the availability of food resources that foxes rely on, leaving them with less to scavenge or hunt.

A 2009 study by Letnic, Koch, Gordon, Crowther and Dickman found that in arid regions of Australia where dingoes were common, red foxes and kangaroos were significantly less abundant, while native rodents and vegetation cover were more prevalent. The authors concluded that dingoes suppress fox populations both by directly excluding them and by dominating the landscape's food supply. Where dingoes were absent, fox numbers were high and rodent populations had collapsed under intense predation. But in areas where dingoes were present, foxes were scarce – and rodents were up to ten times more abundant.

The landscape of fear

Dingoes do not always need to attack to exert control. Their presence alone is often enough to change the behaviour of mesopredators. This is known in ecology as the 'landscape of fear', describing how animals alter their movement, foraging and risk-taking behaviour in response to the threat of predation.

In areas where dingoes are common, foxes and cats become more cautious. They avoid open spaces. They limit their hunting activity. They shift into terrain that is harder to navigate and less productive. This has measurable consequences.

Studies have shown that when dingoes dominate a landscape, mesopredators reduce their activity, spend more time hiding and shift into rocky, scrubby or exposed areas where hunting is less efficient. These changes disrupt the constant foraging pressure that cats and foxes would otherwise place on native prey species, allowing some breathing room for recovery.

Research by Letnic and colleagues Crowther and Koch in the sand dunes of the Strzlecki Desert published in 2009, found that in areas where dingoes were abundant, mesopredators like foxes and cats were significantly less active and endangered native rodents such as the dusky hopping mouse were more likely to persist. The study suggested that dingoes can create a kind of refuge for vulnerable species, both by reducing mesopredator numbers and by altering their behaviour across the landscape.

Further evidence comes from studies such as a 2015 study by Schroeder, Lewis, Kilpatrick and Moseby, published in *Wildlife Research*. They utilised GPS telemetry to analyse interactions between dingoes and introduced mesopredators in arid Australia. They found that dingoes create mesopredator-free spaces on a micro-spatial level, effectively providing refuge areas for threatened mammal and reptile species.

The significance of these findings has not gone unnoticed. In 2013, a group of researchers studying dingo–mesopredator interactions received one of Australia's highest scientific honours – the Eureka Prize for Environmental Research. Often described as the Oscars of Australian science, the prize is awarded for discoveries that shift public understanding of the natural world.

The winning team, led by Chris Johnson (University of Tasmania), Mike Letnic (University of New South Wales), Euan Ritchie (Deakin University), Arian Wallach (James Cook University) and pastoralist Adam O'Neill (Evelyn Downs Station), presented compelling evidence that dingoes act as ecological regulators. Their research showed that dingoes do not merely reduce mesopredator numbers, they change behaviour, restructure predator hierarchies and create space for vulnerable species to survive.

Their work helped ignite a national conversation about the ecological 'value' of dingoes. Should dingoes continue to be controlled as pests or should they be recognised and protected as essential parts of Australia's ecological systems?

That debate continues, but the evidence is becoming harder to ignore.

CASE STUDY: HOW THE DINGO IS HELPING TO SAVE THE NIGHT PARROT

For over a century, the night parrot (*Pezoporus occidentalis*) was one of the great mysteries of Australian ornithology. A nocturnal, ground-dwelling parrot, it was so elusive that some experts feared it had vanished altogether. After decades of unconfirmed sightings, it was declared possibly extinct – a ghost bird lost to history.

That changed in 2013, when wildlife photographer John Young announced he had rediscovered a small population in Queensland. The claim sparked both excitement and scepticism. Was this truly the bird that had been missing for so long? Although doubts later emerged over Young's evidence, the announcement triggered a new wave of research effort. Multiple confirmed sightings soon followed in Western Australia, South Australia and the Northern Territory.

And at the heart of one of the most significant discoveries was the dingo.

In 2024, an extraordinary find was made in the Ngururrpa Indigenous Protected Area (IPA) in the Great Sandy Desert of Western Australia. Ngururrpa Aboriginal Rangers, working alongside scientists, documented the largest known population of night parrots, up to 50 individuals, thriving in a remote, spinifex-covered country.

Why had these parrots survived here, when so many others had vanished? The answer lay in the region's predator dynamics. Surveys using acoustic recorders, camera traps and expert tracking revealed that dingoes dominated the landscape, while feral cats and foxes were rarely detected.

According to the research team, dingo scats frequently contained cat remains, suggesting that dingoes were not just displacing feral cats through competition and fear, but actively preying on them. Ecologist and co-author Rachel Paltridge noted that this behaviour may be creating safer ground for species like the night parrot to persist.

It was, in effect, a kind of natural refuge. Dingoes were creating safe zones where vulnerable species like the night parrot could hold on. In a landscape increasingly reshaped by invasive predators, these dingo-dominated areas may be among the last places where ecological balance still quietly endures.

The discovery at Ngururrpa IPA challenges the assumptions underpinning many of Australia's conservation strategies. Government agencies have spent millions on feral cat control while continuing to cull dingoes, often in the same landscapes. Yet the evidence from places like Ngururrpa suggests that when left alone, dingoes can succeed where poison and bullets have failed.

Could other endangered species be quietly surviving in dingo-protected refuges? The question is now being asked more seriously. Australia's apex predator may turn out to be one of its most powerful conservation tools. One that has been hiding in plain sight all along.

THE DEBATE: DO DINGOES REALLY SUPPRESS MESOPREDATORS?

The idea that dingoes suppress mesopredators, particularly foxes and feral cats, has become one of the most contested questions in Australian ecology. On one side are those who argue that dingoes act as natural regulators, keeping invasive predators in check and indirectly benefiting native species. On the other are those who contend that the evidence is patchy, the mechanisms unclear and the conservation implications oversold.

There is, unquestionably, support for the suppression hypothesis. Across arid and semi-arid landscapes, dingoes have been associated with lower mesopredator abundance, reduced predation pressure and greater persistence of small mammals and birds. Foxes, in particular, appear sensitive to dingo presence. Where dingoes are common and human control is minimal, fox numbers tend to be lower, especially in open terrain, where the dingo's larger size and stamina give it a competitive edge. In these landscapes, prey species like rodents and bandicoots are often more abundant, suggesting that fox suppression creates space for native fauna to persist.

Feral cats are more elusive and behaviourally flexible, but they too appear to respond to dingo presence. In several desert ecosystems, cats shift their behaviour and habitat use when dingoes are active, avoiding open ground, moving into denser vegetation or retreating into rocky areas. They may also alter the timing of their activity, becoming more nocturnal to avoid direct encounters. This kind of suppression is subtle. Dingoes may not always reduce cat numbers, but they change how cats behave, where they hunt and when they move. In ecological terms, that still counts.

Evidence of direct killing supports this picture. Cat and fox remains have been found in dingo scats and stomach contents across multiple regions. In controlled settings, such as enclosed reserve experiments, dingoes have been observed killing both species. These are not isolated incidents. They point to a broader pattern of interference competition and one that reorders predator hierarchies and reshapes the risks faced by smaller carnivores.

But not all studies tell the same story. In 2013, Ben Allen and colleagues reviewed 40 studies on dingo–mesopredator interactions. They found consistent evidence that dingoes suppress foxes, but the effect on feral cats was far less clear. In many cases, cat numbers remained high even in areas where dingoes were abundant. More recently, a large-scale camera-trapping study by Alyson Stobo-Wilson and colleagues (2020), covering vast tracts of northern Australia, found no strong relationship between dingo and cat abundance. Their results suggested that other variables, such as rainfall, vegetation cover and prey availability, played a more significant role. Similarly, Bronwyn Fancourt's work in central Queensland found that dingoes and cats often coexisted, with little evidence of suppression. In some cases, cat activity actually increased where dingoes were present.

These findings challenge the notion that dingoes are a universal solution. If dingo presence automatically reduced mesopredator numbers, the evidence should be more consistent. But suppression, it turns out, is not a binary outcome. It is shaped by a host of interacting factors, such as habitat structure, prey dynamics, predator density and the degree of human interference.

In open deserts and plains, dingoes have the upper hand. But in dense forests, rugged ranges or post-fire landscapes, mesopredators can find refuge. Cats in particular exploit terrain features that dingoes avoid and their behavioural plasticity allows them to persist under pressure. They may not win a head-on confrontation, but they're good at staying out of one.

Temporal avoidance adds another layer. Several studies have documented shifts in cat and fox activity patterns in response to dingoes, especially in places where dingo presence is stable and predictable. Cats may become more nocturnal, hunting during hours when dingoes are inactive. Foxes may shift their foraging to the margins of dingo territories or reduce their movement in high-risk areas. This is the 'landscape of fear' expressed not just in space, but in time.

Human interference complicates things further. Dingoes are persecuted across much of Australia, their numbers reduced by poison baits, traps and exclusion fencing. In heavily managed landscapes, dingoes may be too sparse or too fragmented to exert consistent pressure on mesopredators. Some populations, especially those near human settlements or livestock carcass dumps, have shifted towards scavenging. When their natural behaviours are disrupted, so too is their ecological role.

Finally, the debate over dingo suppression is not just scientific, it's also political. Researchers working in conservation science often highlight the benefits of allowing dingoes to function naturally, arguing that suppression is a useful, low-cost ecological service. Others, especially those affiliated with agriculture or pest control agencies, are more sceptical, concerned about livestock predation, legal responsibilities and the practicalities of coexistence. These opposing perspectives are often rooted not just in different data, but in different worldviews: one sees the dingo as part of the solution, the other as part of the problem.

THE LIMITS AND POSSIBILITIES OF SUPPRESSION

So, do dingoes suppress mesopredators? The answer is yes, though the reality is more complex than simple cause and effect. Foxes appear more vulnerable to dingo pressure than cats and suppression tends to be most effective in sparsely populated, open landscapes where dingoes can roam freely and behave as true apex predators. Feral cats, although harder to displace, still adjust their behaviour in response to dingo presence. They may alter where they move, when they hunt or how they use the landscape. These shifts are not always visible in abundance data, but they reflect an ecological influence that matters.

Even so, dingoes are not a universal fix. Suppression alone will not rescue every at-risk species. In some areas where dingoes are present and active, animals like the bilby, central rock-rat and kowari continue to decline. Many persist only in predator-free reserves or on islands where foxes and cats have been entirely excluded. These cases show that top-down control, while helpful, is not always sufficient. Some species are already so diminished that they need direct, targeted intervention to survive.

Feral cats, in particular, remain an especially difficult challenge. Their adaptability allows them to exploit cover, change their hunting strategies and thrive in a wide range of environments. Even in landscapes where dingoes are relatively abundant, cats often find ways to persist. Where habitat has been cleared or fire regimens disrupted, the vulnerability of prey species can become so great that even modest predation has devastating effects. In these settings, the benefits of dingo suppression may come too late, or not at all.

This is why the debate about suppression can sometimes miss the point. The real question is not whether dingoes always control mesopredators. It is whether they can, under the right conditions, and whether we are prepared to let them. When dingoes are allowed to maintain stable populations and natural behaviours, they can help restore balance. But when they are relentlessly baited, shot or excluded, their ecological function is diminished. Suppression depends not just on their presence, but on how they are managed.

BEYOND SUPPRESSION

Dingoes are not a silver bullet. In some habitats, cats continue to hunt and persist even where dingoes are present. And some native species, already hanging by a thread, may need more direct protection than a predator–prey balance can offer.

But to suggest that dingoes don't help is to ignore a growing body of evidence and stories like the night parrot, which show just how powerful their presence can be. In the right conditions, dingoes suppress invasive predators, create space for vulnerable species and help ecosystems hold together. They are not the only solution, but they are a key part of it.

Still, the value of dingoes should not rest on a single ecological function. Their worth is not conditional. Even if future research found that their effect on cats was limited, it would not change the fact that dingoes belong. They are Australia's native apex predator. They shape the land, connect ancient stories to present-day science and remind us that not every species must earn its place by being useful to us.

10
Rewilding

DAVID POLLOCK INHERITED A PROBLEM.

Wooleen Station, a sprawling 152,000-ha property in Western Australia's southern rangelands, had been pushed to its limits. When his father bought the station in 1989, government assessments painted a bleak picture – 38% of the land was classified as poor, another 36% as fair and more than 25% was suffering from erosion. The worst damage was in the floodplains, areas once lush and productive, now scoured raw by overgrazing and the slow, grinding loss of topsoil. Some believed the damage was permanent.

For over a century, pastoralism in the region had followed a familiar pattern. The land had once supported thick stands of perennial grasses, but relentless grazing pressure had stripped them away. The sheep and cattle that had been introduced since European settlement played their part, but the real problem was something else entirely – unmanaged grazers. By the early 1990s, the majority of grazing pressure came primarily from feral goats and kangaroos, devouring any vegetation before it had the chance to recover.

It wasn't as if pastoralists weren't trying to stop it. They shot kangaroos by the thousands, mustered goats for sale and kept stocking whatever livestock their land could support. But there was a flaw in the system. Pastoralists could remove their sheep and cattle if conditions got bad, but the unmanaged animals stayed. The land never got a break, and so it never recovered.

In these parts, dingoes had long been eliminated. They were treated as the enemy of pastoralism and the government poured money into their destruction. Entire landscapes were reshaped around their absence. The result was what Pollock saw at Wooleen: a land stuck in an irreversible decline, stripped of resilience and on the verge of collapse.

When David Pollock and his partner, Frances Jones (Fig. 10.1), took over the running of Wooleen in the early 2000s, they knew something had to change. The station was locked in a slow death spiral, one that no amount of conventional pastoral wisdom seemed able to stop. If they were to have any hope of saving the land, they had to rethink everything.

Then, in 2007, the dingoes returned. At first, they were just shadows on the horizon, flickers of movement at the edges of the station. For over 100 years, dingoes had been absent from Wooleen, but now, slowly, their numbers grew. Where most pastoralists would have reached for the rifle, Pollock and Jones let them stay. While they didn't actively bring dingoes back, their decision to reject lethal control was what ultimately allowed dingoes to re-establish their ecological role at Wooleen.

For years, Wooleen had been overrun with feral goats. No matter how many were mustered and sold, up to 2,000 a year, a core breeding population always remained. Once dingoes were re-established, the goats began to disappear. By 2010, commercial goat mustering stopped entirely and by 2014, not a single goat remained on Wooleen.

Kangaroos were even harder to control. At their peak, over 20,000 red kangaroos crowded a single floodplain on the station, competing with livestock for every blade of grass. In 2010, Pollock himself shot 3,434 kangaroos in a desperate attempt to reduce numbers, while a professional kangaroo shooter culled another 1,500. It made no difference. Over the next 7 years, the dingoes did what bullets could not, kangaroo numbers dropped by 90%.

With the biggest drivers of overgrazing removed, something extraordinary happened. For the first time in a century, new seedlings began to appear along the Murchison and Roderick rivers. River red gums, swamp she-oaks and bottlebrush, species that had been unable to regenerate for over 100 years, were now pushing their way through the soil. The damage done by grazing pressure had been so absolute that many believed these species had disappeared from the landscape altogether. Now, they were coming back.

Perennial grasses, saltbush and bluebush, plants critical for stabilising the fragile soil, returned to areas that had once been stripped bare. Erosion slowed. The floodplains, which had once bled away their soil at every rainfall, began to retain water again.

One of the clearest signs of change was House Creek, a waterway that had for years run a murky brown, carrying away Wooleen's last scraps of topsoil. Now, with vegetation returning, the water ran clear.

For the first time in 130 years, the land could rest.

To most pastoralists, the return of dingoes is a threat. If left unchecked, surely they would turn on the cattle next? But Pollock watched closely. He waited for evidence of predation, but it never came.

Over 12 years, he found no evidence of dingo attacks on calves.

What was happening at Wooleen defied conventional wisdom. On degraded stations, cows were forced to walk up to 12 km each day between water and food, leaving their calves behind, vulnerable and exposed. But with healthy pastures returning, food and water were always close. Cows stayed with their calves and the dingoes, with an abundance of easier prey, left them alone.

With the unchecked grazing pressure finally under control, Wooleen could also implement rotational grazing, a strategy that had been impossible before. Instead of keeping cattle in the same areas indefinitely, they were moved systematically, giving pastures time to recover. The land was healthier, and so were the livestock.

Of course, not everyone agreed. Pastoralists across the region still viewed dingoes as a scourge. Government funding still prioritises dingo culling. But Pollock had seen what happened when the fight against dingoes continued unchecked. Stations collapsing under the weight of relentless grazing, landscapes stripped to the bone, pastoralists caught in a cycle of never-ending control programs that never quite worked.

Dingoes were not a threat to his operation, but an asset.

His experiment at Wooleen documented in his 2019 book *The Wooleen Way* and published papers in 2021 and 2022, is convincing evidence that keeping apex predators around can work, not just as an ecological strategy, but as a viable pastoral practice. The lesson from Wooleen is that sometimes, the best thing we can do for the land is to stop fighting it.

Fig. 10.1. David and Frances Pollock at Wooleen Station. Source: Wooleen Station.

THE CONCEPT OF REWILDING

Rewilding is more than simply bringing back lost species. It is about restoring the fundamental processes that shape ecosystems. At its core, rewilding seeks to re-establish self-regulating natural systems, allowing landscapes to heal by reducing human intervention and letting nature take the lead. It is an idea that challenges conventional conservation, shifting the focus away from simply protecting what remains to actively restoring what has been lost.

The concept has gained traction globally, with projects aiming to reintroduce large herbivores, restore degraded landscapes and, perhaps most controversially, bring back apex predators. These top-level carnivores sit at the peak of the food chain, influencing everything beneath them.

One of the most famous examples of rewilding comes from Yellowstone National Park in the United States. For much of the 20th century, Yellowstone's ecosystem had been unravelling in ways that weren't immediately obvious. Without wolves, which had been exterminated by the 1930s, elk populations grew unchecked, over-browsing streamside vegetation and preventing the regeneration of key species such as willows and aspens. The landscape became overgrazed, waterways eroded and biodiversity suffered. It wasn't until wolves were reintroduced in 1995 and 1996 that ecologists began to see a reversal of this decline.

This concept, known as a trophic cascade (discussed in Chapter 8), suggests that predators shape entire ecosystems. With wolves back in Yellowstone, elk changed their foraging patterns, no longer lingering in open areas where they were vulnerable to predation. This allowed vegetation to recover, beavers to return and entire food webs to realign.

This dramatic transformation has inspired rewilding projects worldwide. In Europe, Eurasian lynx, brown bears and wolves are being reintroduced to forests where they were once wiped out. In Africa, conservationists are exploring the reintroduction of cheetahs into regions where they historically roamed.

But what about Australia?

CAN DINGOES HEAL LANDSCAPES?

If letting dingoes return to Wooleen Station could transform it from dust-raked paddocks into a recovering ecosystem, might the same principle apply more broadly? It's a question increasingly asked not just by ecologists, but by land managers, Traditional Owners and conservationists looking for scalable, cost-effective ways to restore degraded country. The idea of using dingoes to 'heal the land' has moved from fringe speculation to serious scientific consideration, supported in many cases by mounting ecological evidence.

A major review by Al Glen and colleagues in 2007 made the case for the dingo as a trophic regulator. They argued that its decline from much of southern and eastern Australia has coincided with the rise of invasive predators and widespread biodiversity loss. Glen's team pointed to the need for controlled reintroductions to properly test the dingo's ecological function, an idea also supported by Chris Johnson, who described the disappearance of top predators like the dingo as a tipping point for entire ecosystems. Andrew Claridge and Rob Huntin a 2008 article in *Ecological Management and Restoration* suggested that dingoes should be considered for reintroduction into conservation areas where threatened species are at risk, not just for their effects on foxes and cats, but because of their broader ability to reestablish natural ecological processes. In 2015, Thomas Newsome and his team, drawing on the Wooleen example and other case studies, presented dingoes as ecosystem engineers, capable of shaping not only animal communities but also the physical structure of the landscape. Their review outlined how the presence of dingoes could lead to improved vegetation growth, greater soil stability and more predictable ecosystem dynamics.

They did not stop at theory. Newsome and his colleagues proposed a real-world experimental reintroduction in Sturt National Park in far western New South Wales, a region of roughly 3,000 km^2, large enough to sustain multiple dingo packs. Importantly, the park already supports the full cast of characters needed to observe potential ecological shifts: feral goats, kangaroos, rabbits, foxes, cats, emus and threatened native rodents. By modifying a section of the Dingo Fence and implementing a Before–After–Control–Impact study design, they argued it would be possible to test, for the first time under rigorous scientific conditions, how the reintroduction of an apex predator like the dingo might reshape the broader ecosystem.

The idea was bold but grounded. Rather than speculate about the dingo's role, they wanted to test it openly, systematically and at scale. 'The best way forward,' they wrote, 'is that a controlled reintroduction experiment be undertaken', which they described as offering a 'unique opportunity' to assess interactions between dingoes, invasive mesopredators, overabundant herbivores, native prey and vegetation simultaneously. In short, it was a chance to evaluate whether dingoes could, in practice, deliver the kind of top-down ecological regulation often attributed to apex predators elsewhere.

For Newsome, this was not about trying to restore the missing link in a system now defined by imbalance. Apex predators structure ecosystems around the world. The question, as they posed it, was simple: why would dingoes be the one exception?

Underlying all of this is a shift in ecological thinking. Much of Australia's conservation policy has been shaped by assumptions based on stable, equilibrium-style ecosystems, ideas that rarely apply to arid and semi-arid zones. As David Choquenot and David Forsyth (2013) argue in *The Rangeland Journal*, non-equilibrium models, those that incorporate disturbance, variability and climatic unpredictability, better reflect how Australian ecosystems actually function. Within this framework, dingoes are key regulators that help smooth out boom-and-bust cycles, particularly in systems driven more by rainfall than by simple predator–prey dynamics.

And it is not just about plants and animals. As described in Chapter 8, dingoes influence the very ground beneath our feet. Studies comparing areas with and without dingoes have found higher levels of organic matter, phosphorus and nitrogen in soils where dingoes still roam. Dingoes may help also disperse beneficial soil fungi through their scats, supporting healthier plant communities in ways we are only beginning to understand. This is in part due to reduced grazing pressure, which allows vegetation to cycle nutrients more efficiently.

In this view, rewilding with dingoes is not a short-term fix for environmental problems, but catalysts for re-establishing natural feedback loops, many of which have been disrupted or broken altogether by centuries of predator removal.

Of course, healing landscapes with predators comes with challenges. But the Wooleen example, now backed by ecological studies, and in bold proposals like that for Sturt National Park, suggests that giving dingoes space to do what they do best might be one of the most promising and underused tools in our restoration toolkit.

DEBATING THE EVIDENCE

It's a compelling idea: that dingoes could restore balance to degraded Australian landscapes. Of course, not everyone's convinced. Although a growing number of ecologists argue that dingoes play a vital role in suppressing invasive predators and regulating herbivores, others remain sceptical, arguing the science isn't strong enough, the ecosystems too complex and the risks too great. And unlike most conservation debates, this one has become unusually heated. This is all perhaps less about data, and more about the very lens through which we view Australian ecology.

One of the most common criticisms comes down to evidence, or, depending on who you ask, the lack of it. Ecologists such as Peter Fleming, Ben Allen, and Guy Ballard (2013) have argued that the idea of dingoes as ecological saviours is based more on correlation than causation. They say many studies claiming benefits from dingo presence are flawed, failing to control for confounding factors like rainfall, habitat differences or land management. And it's true that Australia's ecology is hard to pin down. In arid regions especially, things don't follow neat patterns. Populations of kangaroos, rabbits and small mammals can skyrocket after rain and then collapse just as quickly. Teasing apart what's caused by dingoes, and what's driven by weather, fire or grazing pressure, isn't easy.

A 2023 review led by Castle, Kennedy and Allen took aim at dozens of dingo studies, arguing that only a handful used rigorous designs capable of testing cause and effect. They concluded that the science of dingo-driven restoration was still 'stuck in the mud' – highly suggestive, but far from settled.

Yet critics don't always practice what they preach. They may call for better experiments, but many of the studies used to challenge dingo effects suffer from the same weaknesses: short timeframes, poor controls and overreliance on anecdote or correlation. In some cases, researchers failed to significantly reduce dingo numbers during control trials, making it impossible to test their ecological impact in the first place. As Chris Johnson and Euan Ritchie (2013) pointed out in their replies to papers by Fleming and Allen, if the studies used to argue against dingo rewilding don't actually remove dingoes, or control for confounding variables, they're not in a position to dismiss the positive effects observed elsewhere.

And then there's the matter of bias. Most scientists declare conflicts of interest when publishing research, but it's less common for them to declare professional affiliations or sources of funding. That matters. Many of the strongest critics of dingo ecology are tied to government departments or agriculture-backed programs tasked with reducing dingo numbers. Their research is often shaped, directly or indirectly, by the same institutions that have historically supported lethal control. That doesn't make their science invalid, but it does mean we need to read it with care.

This gets to the heart of the debate: two very different worldviews colliding. On one side are ecologists, conservationists and Traditional Owners who see dingoes as essential players in a functioning ecosystem. On the other are scientists and land managers focused on protecting livestock, who view dingoes, regardless of their ecological function, as a threat to agricultural productivity. Both sides make scientific arguments, but grounded in different values, priorities and experiences. That's one reason why the debate so often gets personal.

Even the idea that dingoes suppress cats and foxes, a central claim in many dingo-positive studies, remains controversial (as discussed in Chapter 9). Some research shows strong suppression effects. In parts of arid Australia, for instance, foxes and cats are found in lower numbers where dingoes are active. Other studies suggest more complex dynamics. In Queensland, for example, feral cats have been observed increasing their activity during times when dingoes are less active, suggesting behavioural avoidance rather than outright exclusion. Similarly, foxes may coexist with dingoes in some environments, but be outcompeted in others. The relationship clearly shifts depending on conditions, prey availability and landscape features.

There's also the idea of a 'landscape of fear', borrowed from Yellowstone and the reintroduction of wolves as mentioned earlier. Some scientists have proposed that dingoes might create similar fear effects in Australian herbivores, like kangaroos and goats, prompting them to avoid certain areas and reduce grazing pressure. While some studies support the possibility of these behavioural shifts, the evidence remains limited and context-dependent. In arid regions, for instance, red kangaroos may continue to frequent open floodplains despite the presence of dingoes, particularly when resources are scarce. Critics argue that the strength and consistency of fear-driven effects in Australian ecosystems have not been well demonstrated and that predator–prey dynamics here may differ substantially from those in more temperate, forested environments like Yellowstone.

Indeed the critics often seize on this to argue that Yellowstone comparisons don't hold up. And they have a point. Australia isn't Yellowstone. Australian landscapes are older, drier and more variable. It doesn't have tight predator–prey systems like wolf and elk or forested valleys with clear cascade effects. However, the theory might still apply. In fact, as Thomas Newsome and others have argued, Australia's boom-and-bust ecology may make apex predators even more important. During droughts or bust periods, when native prey populations crash, cats and foxes often intensify their hunting around waterholes, exactly the time when dingoes can act as a regulating force. If anything, Australia's ecological volatility makes the case for predator regulation more urgent, not less.

This is where some global context helps. Even the original Yellowstone study, led by ecologist William Ripple, has faced challenges and pushback. Other researchers have questioned the strength of the trophic cascade, the role of beavers and climate, even the actual mechanics of fear-based behavioural change. These critiques haven't invalidated the broader idea, however. They've highlighted how hard it is to study complex systems and how cautious we need to be in making sweeping claims. Australia's critics sometimes forget this. They demand perfect evidence from dingo studies while leaning on incomplete evidence themselves. It's a double standard that doesn't help the science, or the land.

At its core, the debate over dingoes is a debate about uncertainty. How much do we need to know before we act? Should we wait for decades of controlled experiments or make use of the evidence we already have? Should we let landscapes continue to unravel while we argue over probabilities and sample sizes?

As ecologist Euan Ritchie put it to me while we were both attending a forum discussing dingo reintroductions:

> *We might never be able to conduct the perfect experimental study, but the weight of ecological evidence already suggests dingoes can help biodiversity in some contexts. If we delay management actions until such a perfect study exists, we risk missing real opportunities to learn and protect what's left.*

If anything, the fierce back-and-forth in the literature – the rebuttals, counter-analyses and letters to the editor – show just how important this question has become. Dingoes are at the centre of a growing scientific conversation about how we restore ecosystems, who gets to define 'healthy country' and what role predators should play in a changing Australia.

THE 'BOSS OF COUNTRY'

While the scientific debates over the role of dingoes in Australian ecosystems continue, an alternative perspective has existed for thousands of years. For Aboriginal Australians, the dingo is a fundamental part of Country, a guardian of balance, a regulator of life and a cultural totem. Across many language groups, the dingo is recognised as the 'Boss of Country'.

This knowledge is based on thousands of years of continuous observation, experience and interaction with the land. As one Nyangumarta elder Augustine Badal (in a paper we published together in 2024), explains:

> *Dingo be here all the time in Australia. Had a relationship with Aboriginal people in Dreaming and is a spirit animal – a companion and a totem for people...They protect the land and watch it. Dingo is a sign of healthy Country.*

In this worldview, dingoes are not just another species within the ecosystem, but an integral part of a network of relationships, between animals, plants, land and people. Just as rivers shape the land and fire regenerates forests, dingoes help structure the landscape by controlling herbivore populations, influencing where grazing occurs, and regulating other predators. Their role in maintaining this balance is embedded in Aboriginal management practices, where Country is understood as a living system that requires active care.

As representatives from the Bundjalung, Yuin Walbunja and Southern Yuin Tribes explain in their submission to the Royal Zoological Society of NSW 2019 dingo forum:

> *All plants and animals have those roles and responsibilities to each other. That's kinship and that's a big part of the story for us – understanding that there is a cultural landscape.*

This deep-time ecological understanding aligns with the growing body of Western research suggesting that dingoes can regulate herbivores and suppress invasive predators. But while scientists continue to debate the details, arguing over sample sizes, study design and statistical models, many Aboriginal communities see the answer as obvious: when dingoes are present, Country is healthier.

The principle of *healthy dingo = healthy Country = healthy people*, is deeply embedded in First Nations ecological knowledge. Many Traditional Owners have observed firsthand how removing dingoes leads to declining biodiversity and a loss of balance in the landscape.

At the 2023 First Nations Dingo Forum, Aboriginal representatives from across the country came together to issue a powerful statement on the role of dingoes in maintaining Country. The First Nations Dingo Declaration reaffirmed that dingoes are not pests, but essential protectors of the land. The suppression of dingoes, through baiting, shooting and exclusion fences, was seen as a direct attack on both ecological health and cultural heritage. Representatives from the Bundjalung, Yuin Walbunja, and Southern Yuin Tribes again put it best:

> *Aerial baiting has the potential to cause catastrophic impacts to native dingo populations and may potentially lead to localised extinction. This will impact on our Aboriginal cultural values. Dingoes maintain their ecological function in our Country. They suppress the abundance or activity of other pest animals such as cats and foxes and may help protect smaller mammals.*

Aboriginal knowledge offers a different way of thinking about rewilding. It does not view dingoes as a species to be reintroduced, but as a part of a system that has always been there, one that should be allowed to function as intended. For many Aboriginal people, landscapes should be managed with dingoes, not against them. Dingoes have always been part of the landscape and their disappearance leads to an imbalance not just in ecosystems, but in the cultural and spiritual health of Country itself.

DINGO REINTRODUCTIONS: PROGRESS AND PUSHBACK

Only in the past decade have suggestions of dingo reintroductions started, often to much outrage. For example, efforts to reintroduce dingoes into parts of Victoria have sparked heated debate, pitting conservation goals and Aboriginal cultural values against deep-seated fears from the agricultural sector. While many see dingoes as essential for restoring ecological balance, farmers and livestock groups have launched fierce campaigns to prevent any formal reintroductions, often invoking exaggerated or misleading claims to maintain the status quo.

The Grampians (Gariwerd) proposal: a cultural and ecological vision

One of the most significant reintroduction proposals came in 2021, when a 15-year draft management plan for the Grampians (Gariwerd) National Park recommended investigating the return of culturally and ecologically significant species, including the dingo. The plan was developed by Parks Victoria in consultation with the Gunditj Mirring Traditional Owners Aboriginal Corporation, the Eastern Maar Aboriginal Corporation and the Barengi Gadjin Land Council Aboriginal Corporation.

For these Traditional Owner groups, restoring dingoes to Gariwerd was not just an ecological goal, but a cultural necessity. The proposal recognised dingoes as a totemic species that had coexisted with Aboriginal people for thousands of years. Their absence was seen as a disruption to both ecological balance and cultural traditions. As the management plan noted: 'Dingoes have for thousands of years had a symbiotic relationship as semi-wild companions to Aboriginal people.'

The plan also made a scientific case for dingo reintroduction, suggesting that returning a native predator could help control pest species like foxes and feral cats, as well as reduce overabundant kangaroo and wallaby populations that were damaging vegetation. However, it acknowledged the need for further research on how dingoes might interact with the existing landscape.

The moment the plan was released, it provoked an intense reaction from local farmers, many of whom feared that reintroduced dingoes would pose an existential threat to their livestock (Fig. 10.2). John Crawford, a sheep farmer in Victoria Valley, spearheaded a petition against the proposal, gathering over 800 signatures within weeks. He dismissed the idea outright, stating:

> *No matter where I walk, everyone is horrified. Most people thought it was a hoax, that it wasn't fair dinkum.*

Fig. 10.2. (Top) A public notice in the Hamilton Spectator newspaper (April 9, 2022). (Bottom) A meeting of farmers in western Victoria in April 2022 over plans to introduce dingoes to the Grampians National Park. Source: Stock Journal/ACM.

Farmers amplified fears about tourism impacts, arguing that hikers would refuse to walk in the park if dingoes were present. Others claimed that fencing dingoes within the park was unrealistic, despite the fact that no official plan had suggested containment fencing as a solution.

Livestock industry representatives, including the Victorian Farmers Federation and wool brokers, ramped up the rhetoric, warning that dingoes would destroy regional economies and using Queensland's declining wool industry as a cautionary tale.

In response to the community backlash, the final management plan released in December 2021 excluded the dingo reintroduction proposal, abandoning the idea indefinitely. Instead, it focused on raising awareness about the cultural significance and ecological role of dingoes, stating that any future reintroduction would require broader community support.

North-west Victoria: the first real shift in dingo protection

While reintroductions in places like the Grampians have failed under political pressure, 2024 saw a major policy shift in north-west Victoria. Not a reintroduction per se, but something equally significant.

Dingoes in north-west Victoria have almost forever been classified as unprotected wildlife, allowing landholders to trap, poison and shoot them without restriction. But in March 2024, the Victorian government placed a stay on the unprotection order, effectively banning lethal control in parts of the state where dingo numbers had declined drastically. Some estimates suggested as few as 40 dingoes remained in the area.

If the Grampians proposal was controversial, the 2024 dingo protection decision sent farmers into a full-blown panic. The Victorian Farmers Federation, rural media outlets and conservative politicians immediately framed the decision as a direct attack on agriculture. Bill Tilley, MP for Benambra in the affected region, issued a furious statement titled: 'Barking mad activists leading to wild dog carnage.'

Tilley's rhetoric escalated the fear campaign with a mix of exaggerations, emotional appeals and outright misinformation. His statement included nearly every argument used against dingoes that have been used for decades (and all of which are systematically debunked by science, and covered in this book):

- that dingoes are introduced and therefore have no conservation value
- that all wild canids in Victoria are hybrids
- that dingoes wiped out the thylacine and Tasmanian devil
- that dingoes kill sheep 'for fun'
- that dingoes kill hundreds of sheep in a single night
- that dingo control programs should be judged on the number of dingoes killed, not livestock saved.

Farmers and industry groups circulated anecdotal horror stories, including one claiming a single farm lost 200 sheep in 6 weeks due to dingo attacks. These claims, while dramatic, often lacked verification and were strategically deployed to pressure the government into reversing its decision.

The Victorian government's decision to halt lethal dingo control marks the first real challenge to the long-standing war on dingoes in Victoria. Although it is not a reintroduction, it is as good as one, allowing dingoes to naturally restore their presence. It also represents a shift towards a different model of management, where dingoes are no longer treated as indiscriminate pests.

For Aboriginal groups, conservationists and ecologists, this represents a crucial test case. If the stay on lethal control remains in place and leads to no catastrophic increase in livestock losses, it will undermine the long-held belief that dingoes must be eradicated to protect farming.

However, the government has made this decision with little consultation or practical support for farmers, especially in terms of viable alternatives. That's a missed opportunity. For strategies like this to succeed, they must be developed with farmers, not against them.

Farmers and their political allies are determined to prevent this experiment from succeeding. If they can manufacture enough fear, they may force the government to backtrack, just as they did with the Grampians reintroduction proposal.

MOVING BEYOND THE DEBATE

Whether dingoes are good or bad for Australian ecosystems cannot be answered with a simple yes or no. Their effects vary depending on environmental conditions, prey availability and land use practices. In some places, they suppress invasive predators and reduce overgrazing; in others, their impact is less predictable.

What is clear is that dingoes are highly adaptable. Unlike wolves in Yellowstone, which exert consistent predation pressure, dingoes respond flexibly to boom-and-bust cycles, shifting their diet and hunting strategies as conditions change. This ability to adjust their ecological role could become even more valuable as climate change drives greater environmental unpredictability.

When all the available evidence is considered, scientific studies, small-scale trials, observational data and even anecdotal accounts, a pattern begins to emerge. Taken together, the weight of evidence points to dingoes playing a largely positive ecological role, contributing to biodiversity and ecosystem function across much of the continent.

If dingoes are to play a role in ecosystem restoration, conservation efforts must move beyond simplistic narratives and instead focus on practical, evidence-based approaches. Carefully managed reintroductions could provide insights into their long-term ecological role, but scientific monitoring is needed to track both their benefits and potential risks. Aboriginal ecological knowledge must be central to this process, recognising that dingoes have long been part of the cultural landscape. At the same time, coexistence strategies, especially non-lethal livestock management, will be essential in ensuring that conservation efforts work alongside, rather than against, agricultural interests.

We cannot continue to avoid restoring ecosystems purely for the benefit of one industry. Healthy landscapes support everyone. Working collaboratively with landholders is not only possible, it is necessary. As shown in Chapters 19 and 20, coexistence models offer promising ways to ensure dingoes can benefit both nature and people.

Part 4

An identity in crisis

11
Taxonomic tangles

WHAT IS A DINGO? IT SOUNDS LIKE A SIMPLE QUESTION, UNTIL YOU TRY TO answer it. But it is one of the most persistent, consequential, and controversial questions in Australian science. As the zoologist Frederic Wood Jones observed nearly a century ago:

> *That the dingo is some sort of a dog is practically the only point upon which all writers have agreed. Whether he is a domestic dog run wild, a feral dog of ancient introduction, or a true native species of Australia, are matters upon which wide divergences of opinion have been expressed.*

And even that now seems up for debate.

Look at wildlife field guides, government documents, museum labels or scientific papers and you'll find dingoes variously described as *Canis familiaris*, *Canis lupus dingo*, *Canis dingo* or even *Canis antarcticus*. One animal. Multiple names.

Some argue the dingo is a feral dog, a descendant of domesticated animals that reverted to the wild. Others say it's a unique species of wild canid, evolved in isolation and adapted to the Australian continent. New genetic and ecological studies have deepened the debate, but haven't resolved it.

For most species, classification is fairly straightforward. A lion is a lion. A kangaroo is a kangaroo. The dingo resists this simplicity. It occupies a strange space in the animal kingdom because it doesn't fit neatly into any of the categories we've created (such as wild and domestic; native and foreign; natural and unnatural). Because of that, the debate over what a dingo *is* remains alive and well.

You might be wondering why does this even matter. It can seem like the sort of debate taxonomists quietly have in the back rooms of museums while everyone else gets on with their lives. At its core, scientific classification presents a way for us to make sense of the natural world, but practically it can influence legislation, conservation policy and public perception.

For example, some Australian governments treat dingoes as pests, basing policy on the assumption that they're simply feral dogs. Others acknowledge them as native wildlife. As such, scientific uncertainty gives politicians and managers room to choose the definition that suits their agenda.

THE EARLY EUROPEAN CLASSIFICATIONS

From the moment Europeans encountered dingoes, there was confusion about what to call them. The first formal scientific names were not the result of field studies or systematic comparison, but drawn from engravings (drawings) and second-hand reports of captive animals in the colonies.

Fig. 11.1. 'Portrait of a Large Dog from New Holland (Dingo)' by George Stubbs, 1772. Painted from preserved skins and written descriptions provided by Cook's expedition. This is the earliest representation of the dingo in Western art. Source: National Maritime Museum, Greenwich, London.

One of the most iconic, and probably the most accurate early portrayal of the dingo appeared in 1772, by painter George Stubbs. He was commissioned by Joseph Banks to illustrate a 'large dog' from New Holland, using preserved skins and written descriptions from Captain Cook's expedition. His image (Fig. 11.1), along with a kangaroo, became the first portrayal of Australian mammals shown to a British and European audience. Both were later included in a 1777 global map of mammal distribution, securing the dingo a place in scientific history.

In 1789, Governor Arthur Phillip published *The Voyage of Governor Phillip to Botany Bay*, which included an engraving titled 'Dog of New South Wales', drawn by artist Peter Mazell. The image was the first published image of the dingo, which was based on a dingo pup captured near Sydney Cove. Surgeon John White followed in 1790 with *A Journal of a Voyage to New South Wales*, which included a similar illustration (Fig. 11.2).

Fig. 11.2. The dingo, from John White's *A Journal of a Voyage to New South Wales* (1790), is one of the first published depictions of the dingo and was used to classify them. Source: digitised by the Biodiversity Heritage Library.

One of the more striking visual records from this era came in 1797, depicting a snarling dingo sitting by a dog kennel and chained to a post. This drawing likely depicted the same animal featured in White's 1790 illustration. It was reportedly owned by a servant of John Shea. Descriptions surrounding such accounts tell how dingoes could not be tamed or cured of their savageness. For me, this reinforces how Europeans viewed dingoes at the time – unruly, half-tame and only barely under control.

These drawings, sketched from dingoes captured as pups, formed the foundation for how the dingo would be described and named in Europe. In 1792, Scottish biologist Robert Kerr provided the first formal scientific name: *Canis antarcticus*, or the 'New Holland dog'. His description, which appeared in his English translation of Carl Linnaeus's *Systema Naturae*, was based entirely on Phillip's brief notes and the engraving shown in Fig. 11.1. In it, Kerr described the dingo simply:

> *The tail is bushy, and hangs downwards; the ears are short and erect; and the muzzle is pointed.*

A year later, German naturalist Friedrich Meyer entered the discussion. In 1793, he published a rival name, *Canis dingo.* Meyer offered a slightly more detailed description, similarly noting it as 'having short, erect ears, a straight, hairy tail', but adding claims about its savageness, diet and supposed inability to bark. These additions were based largely on anecdotal accounts from the growing collections from colonial diaries, including Phillip and White's, and reports from the several dingoes that were sent back to European private menageries.

These were hardly scientific accounts. Kerr had no preserved specimen, no measurements and no context for how the animal lived or behaved in the wild. Meyer's account went further than Kerr's, but it still lacked firsthand observation. They were simply piecing together an identity from notes, rumours and sketches.

Other names were proposed in the decades that followed. For example, in 1826, British zoologist John Edward Gray suggested *Canis australiae*, a name that emphasised the dingo's geographical isolation. None of these names stuck. The dingo's classification remained unsettled. The early confusion persisted throughout the 19th century and in many ways, it still lingers, leaving dingoes in a kind of taxonomic limbo that modern science is still trying to resolve.

In 1957, the International Commission on Zoological Nomenclature (ICZN) stepped in to resolve a naming conflict, issuing an opinion that conserved Meyer's *Canis dingo* over Kerr's earlier *Canis antarcticus*. Although Kerr's name had chronological priority, the Commission suppressed it to maintain stability, noting that 'dingo', the Aboriginal word recorded in the Sydney region, had long been entrenched in both common and scientific use.

Importantly, this ruling did not declare *Canis dingo* to be the biologically 'correct' or 'valid' species name. The ICZN does not determine taxonomy or decide what organisms should be called in a scientific sense; it simply regulates the rules of zoological nomenclature to preserve consistency and avoid confusion.

THE MISSING HOLOTYPE

In biological classification, every species is supposed to have a holotype. A holotype is a single preserved specimen stored in a museum, used as the definitive reference for identifying that species. When disputes arise, scientists return to that example, compare new findings and confirm whether an animal belongs to the species in question. A holotype provides clarity. It anchors a name to something tangible and objective.

But the dingo has no such anchor.

As is now clear, unfortunately its classification was built on impressions rather than physical evidence. There was no preserved animal, no scientific measurements taken from a wild individual and no biological material kept for future study. Neither Kerr nor Meyer designated a holotype and no reference specimen has ever been created.

In 2014, zoologist Matthew Crowther and colleagues published a landmark paper that brought the issue back into the spotlight. Until then, the missing holotype had been regarded as a historical footnote – an unresolved technicality known mainly to taxonomists and wildlife managers. But Crowther's team made it clear that this was the root of a centuries-long identity crisis with real-world consequences for how dingoes are treated and managed in Australia.

The absence of a holotype, while unusual, does not invalidate a species or name, but unfortunately it has created three persistent problems. First, it allows for inconsistent definitions. Scientists often disagree on what distinguishes a dingo from a domestic dog because they are working from different baselines. Second, it makes it harder to define and conserve 'pure' dingoes, especially in regions where hybridisation with domestic dogs is suspected. Third, it opens the door to political manoeuvring. Without a fixed biological standard, definitions can be adjusted to suit particular policies or agendas.

Some researchers, like Crowther and the geneticists we met in Chapter 2, have tried to resolve the issue by studying 19th-century museum specimens collected before widespread hybridisation. These skins and skulls offer important clues, but they are not holotypes. Their origins and quality vary and without official designation, they cannot settle the debate.

BIOLOGICAL CLASSIFICATION

Perhaps the best way to understand the naming debate is to rewind for a moment and start at the top. All living things can be placed somewhere on a taxonomic tree, a kind of ancestral map that traces how species are related to one another through evolution (Fig. 11.3). The higher levels hold the broader groups, like mammals, reptiles, and birds. The lower branches become more specific, narrowing into family, genus and species.

For dingoes, most of this is straightforward. They are mammals, like us: warm-blooded, covered in hair and giving birth to live young, which they nourish with milk. They're placental mammals, members of the eutherian lineage with long pregnancies and sustained development in the womb. This is unlike the marsupials (short gestation, pouch development) and monotremes (egg laying) that dominate much of Australia's native fauna. That alone sets dingoes apart.

Eukaryotes
└── Animalia (Kingdom)
└── Chordata (Phylum)
└── Vertebrata (Subphylum)
└── Mammalia (Class)
└── Eutheria (Placental mammals)
└── Carnivora (Order)
└── Canidae (Family)
└── Canis (Genus)
├── *Canis latrans* (Coyote)
├── *Canis lupus* (Wolf)
└── *Canis familiaris* (Domestic dog)
├── *Canis dingo* (Dingo) **Classification debated*

Fig. 11.3. The taxonomic tree of canids. The question remains: where exactly does the dingo belong? Source: Bradley Smith.

Fig. 11.4. A visual comparison of dingoes and other wild canids from around the world. (Top left to right) grey wolf (*Canis lupus*), coyote (*Canis latrans*) and Ethiopian wolf (*Canis simensis*). (Bottom left to right) black-backed jackal (*Canis mesomelas*), dingo (*Canis dingo*) and domestic dog (*Canis familiaris*). Despite their similarities, only the domestic dog has undergone extensive artificial selection. Sources: wolf - Marcel Langthim (Pixabay); coyote - Veronika Andrews (Pixabay); Ethiopian wolf - Charles J Sharp (Wikimedia); black-backed jackal - Margo Tanenbaum (Pixabay); dingo - Bradley Smith; blue heeler (dog) – Zingpix (Wikimedia).

Dingoes belong to the order Carnivora and more specifically, to the family Canidae, the group of dog-like carnivores. Within the Canidae, they're part of the genus *Canis*, a familiar club that includes wolves, coyotes, jackals and domestic dogs (Figs 11.3 and 11.4).

So far, so good.

It's only when we reach the species level that things fall apart.

Kerr and Meyer both got it right at the genus level. The dingo clearly belongs in *Canis*. But the species-level classification has never been fully resolved. Whether the dingo should be recognised as its own species, a subspecies of dog or simply a feral offshoot remains a point of contention. The uncertainty that began in those first accounts in the 1790s, and never disappeared.

THE MANY WAYS TO DEFINE A SPECIES

Classifying living things might seem like a neat and tidy business, slotting plants and animals into named boxes, each with its place on the great tree of life. In reality, taxonomy is messy, and far more subjective than I ever imagined. It is shaped by human interpretation, judgement calls, historical conventions, and philosophical preferences.

There is no universal agreement on what makes something a distinct species. In fact, there are more than 30 different species concepts used by scientists today, each with its own criteria. Some focus on physical features, others on genetics, behaviour, ecology or whether animals can successfully interbreed. These concepts are not interchangeable and different researchers favour different ones depending on their field, philosophy or purpose.

This diversity of approaches has given rise to two broad camps in taxonomy: lumpers and splitters. Lumpers prefer broader definitions, grouping similar populations together as one species unless the differences are extreme. Splitters, on the other hand, emphasise distinctions, especially when those differences are consistent or meaningful, and are more inclined to classify populations as separate species. Neither approach is inherently right or wrong. They reflect different values about what matters most when defining biodiversity.

The dingo sits squarely in the middle of this divide.

Some scientists, particularly those working in wildlife management or policy, treat the dingo as a feral dog (*Canis familiaris*) because it can interbreed with domestic dogs. This view aligns with the Biological Species Concept, which defines species by their ability to produce viable and fertile offspring. But this concept quickly falls apart when applied to canids. Wolves and dogs can interbreed, yet are treated as separate species. So can wolves and coyotes, jackals and dogs, wolves and jackals, coyotes and dogs – you get the idea. Even grizzly bears and polar bears can produce fertile hybrids in the wild (known as pizzly or grolar bears!), but no one suggests they are the same species. In fact, if the Biological Species Concept were applied consistently across the genus *Canis*, it would collapse most canid species (wolves, coyotes, jackals and dogs) into a single species due to their capacity to interbreed. Yet these animals are universally recognised as distinct. The logic used to deny species status to dingoes on this basis is not applied elsewhere, exposing a clear inconsistency.

Critics of the domestic (or feral) dog label instead apply the Phylogenetic Species Concept, which defines species by unique genetic or morphological traits. By this definition, dingoes clearly qualify. They differ from domestic dogs in key ways, as we examine across several chapters (particularly in Chapters 2, 7 and 12). They reflect a unique evolutionary path.

The Ecological Species Concept is yet another lens. It defines species based on the role they play in an ecosystem (covered in Part 3). Dingoes are Australia's top predator. They behave nothing like domestic dogs. They hunt cooperatively, regulate prey numbers, suppress invasive predators like cats and foxes and live in stable, cooperative packs. They fill a niche no other canid does in Australia.

By both the phylogenetic and ecological species concepts, which are the two most relevant frameworks for understanding canids, the dingo qualifies as a distinct species. It is genetically distinct, ecologically unique and has a well-defined evolutionary lineage.

Some opponents of *Canis dingo* argue that dingoes have not been isolated long enough to count as their own species. They often cite a conservative estimate of 4,000–5,000 years since arrival. However, more recent genomic studies suggest a deeper divergence, up to 10,000 years ago (see Chapter 2), placing the split well before the rise of agriculture and the modern domestication of dogs.

There are clear precedents for recognising distinct species after relatively short periods of isolation. The Channel Island fox (*Urocyon littoralis*), for example, diverged from its mainland ancestor only around 9,000 years ago. Not only that, but archaeological evidence shows these foxes were transported by Indigenous peoples to the islands, meaning they were deliberately introduced. Yet this human involvement did not disqualify it from species status. Despite the short timeframe and anthropogenic movement, the island fox is recognised as a separate species, protected under conservation law and managed independently.

In any case, there is no minimum time threshold for speciation. Some species diverge over hundreds or thousands of years, others over much longer periods. Divergence is driven by the accumulation of meaningful differences, not the age of the split.

Even if dingoes arrived in Australia through human agency, it should not undermine their status as a distinct species. What matters is what they have become, ecologically, genetically and behaviourally, not how they got here, which might be something we never completely figure out.

The dingo has been geographically and reproductively isolated on the Australian continent for millennia, living apart from other canids until the arrival of domestic dogs with European settlers. This extended isolation, combined with genetic, ecological and behavioural distinctiveness, meets or exceeds the thresholds used in other cases.

In my opinion, had the dingo been discovered today, assessed with modern tools and free of historical baggage, the dingo would almost certainly be recognised as a distinct wild canid.

But the dingo was not named under modern scientific standards. It was classified in an era before Charles Darwin, before the theory of evolution and long before genetics, under a system that relied on physical appearance and colonial assumptions. That confusion has lingered, with real consequences. The Biological Species Concept, the oldest and most traditional definition mentioned earlier, was developed during that time. It defines species solely by their ability to produce fertile offspring and was the dominant framework for centuries. As science has progressed, its limitations too have become clear, especially for canids, which routinely interbreed across species lines.

Taxonomy itself was never culturally neutral. It was developed in the 1700s by European scientists such as Carl Linnaeus, who not only categorised plants and animals, but also ranked human beings into racial groups. He placed white Europeans at the top and assigned stereotyped traits to others. His system was ultimately about hierarchy and control. The same mindset shaped how European naturalists interpreted animals in colonised lands. The dingo was caught in that framework.

Many modern taxonomists continue to cling to these outdated frameworks, relying on rigid definitions and century-old concepts that no longer reflect the complexity of nature. Despite major advances in genetics, ecology and evolutionary theory, taxonomy has been slow to evolve. Instead of adapting, it often defaults to tradition, treating classification as a fixed science rather than the dynamic, interpretive process it really is.

If taxonomy was truly objective, it would recognise that species definitions need to be flexible, and evolve with the species it attempts to define. It would acknowledge that hybridisation does not erase identity, and sometimes can act as a benefit to that species' survival. And it would accept that the dingo meets multiple valid definitions of a species, even if it does not fit neatly into one.

THE MODERN BATTLE OVER THE DINGO'S NAME

A name is never just a name. In taxonomy, the name assigned to a species shapes how governments define it in law, how managers treat it on the ground and whether it is protected or killed. For the dingo, the naming issue has long simmered in the background, but in recent years it has erupted into a public scientific debate.

That 2014 paper by Crowther and colleagues mentioned earlier reignited the issue by calling for the reinstatement of *Canis dingo* as the proper species name, citing the need to recognise its historical and scientific standing. Their position gained traction, particularly among ecologists and conservation biologists.

But not everyone agreed.

In 2017, Stephen Jackson, a taxonomist working with the NSW Government, and Peter Fleming, a researcher with the NSW Department of Primary Industries in the vertebrate/invasive pest research unit, led a team that challenged the view proposed in the Crowther paper. Drawing on the Biological Species Concept and broader assumptions about the dingo's history, they argued that dingoes should be classified as *Canis familiaris*, the same as domestic dogs. Their claim rested on several points: that dingoes were introduced to Australia by humans and had not been isolated long enough to evolve separately; that they were likely domesticated before arriving, perhaps even by Aboriginal Australians; that their genome places them within the domestic dog lineage; and that their capacity to hybridise with dogs means they cannot be reproductively isolated.

Two years later, in response, I gathered together a diverse group of dingo researchers, many of whom appear throughout this book, in a formal response published in the international taxonomy journal *Zootaxa*. We came from a range of disciplines: ecology, genetics, anatomy, palaeontology, psychology and anthropology. Our argument was straightforward: the dingo is genetically, ecologically, morphologically and behaviourally distinct from domestic dogs. That it makes more sense to recognise it as a unique, wild canid. We supported *Canis dingo* not as a sentimental gesture, but as the most scientifically coherent classification.

This disagreement reflects the deeper philosophical divide introduced earlier in this chapter: lumpers versus splitters. Jackson and his colleagues are lumpers. They prefer broader classifications, grouping closely related animals together. Our team are splitters, arguing that dingoes have diverged far enough to warrant their own designation.

In a 2021 follow-up, Jackson and colleagues outlined four criteria that they argued must be met for the dingo to qualify as a species separate from domestic dogs. So in 2025, a team led by geneticist Kylie Cairns, including me, published a detailed rebuttal, addressing each point directly.

Criteria 1. Reproductive isolation and genomic separation over millennia

Jackson argued that because dingoes can interbreed with domestic dogs, they cannot be a separate species. But this logic has long been discredited in canid taxonomy and it ignores the fact that dingoes and domestic dogs have remained genomically isolated for thousands of years.

Importantly, dingoes differ from domestic dogs in reproductive timing, biology and behaviour. As already noted, dingoes have an annual breeding season and form monogamous pairs that raise pups cooperatively. In contrast, domestic dogs do not follow a seasonal breeding pattern and have no such cooperative rearing system, opting for a quantity over quality approach to reproduction. These biological and behavioural differences reduce the likelihood of hybridisation in natural environments (covered in more depth in Chapter 13).

Moreover, reproductive isolation does not have to be absolute to justify species status. The mere possibility of hybridisation does not override consistent genetic, ecological and behavioural divergence, especially when these differences persist despite human-induced contact.

Criteria 2. Dingoes are genetically distinct from domestic dogs

There is clear evidence that dingoes form a genetically distinct population, separate from both ancient and modern domestic dog lineages (see Chapter 2). For example, dingoes lack the high-copy-number variant of the AMY2B gene, which allows domestic dogs to digest starch, an adaptation to agricultural environments. This alone signals an early divergence from the domestic dog lineage, likely before the rise of agriculture.

Whole-genome studies also show that dingoes cluster separately from domestic dogs, even ancient ones. They represent a unique branch of the canid family tree and have retained ancestral traits that are no longer present in most modern dog breeds.

Criteria 3. Dingoes are more genetically distinct from dogs than many dog breeds are from each other

Jackson's third criterion argued that dingoes must be more genetically divergent from domestic dogs than dog breeds are from each other. This sets an arbitrary and unnecessary bar. Dog breeds are the result of intense artificial selection over short periods, often just 200–400 years, and primarily on cosmetic traits, not long-term evolutionary divergence. Dingoes, by contrast, diverged from the domestic dog lineage thousands of years ago and have remained isolated ever since.

Even so, Cairns has shown that dingoes are in fact more genetically distinct from domestic dogs than many recognised dog breeds are from one another. This difference is consistent across the entire genome and is supported by both mitochondrial and nuclear DNA (Fig. 7.2).

Furthermore, genetic structure within the dingo population itself, such as the differentiation between south-eastern and north-western forms, suggests that dingoes are not only distinct from dogs, but also evolving independently within Australia.

Criteria 4. The evolutionary history of dingoes is well understood

Jackson called for a clear understanding of how dingoes relate to South-East Asian dogs and New Guinea singing dogs. That understanding now exists.

Genomic evidence confirms that dingoes and New Guinea singing dogs share a common ancestor that diverged from East Asian village dogs approximately 8,000–10,000 years ago.

After arriving in Australia, dingoes became reproductively and geographically isolated for millennia, evolving along a separate trajectory.

Modern phylogenetic analyses consistently place dingoes in their own distinct cluster, separate from both modern domestic dogs and wolves. Their evolutionary story is not murky, but becoming well established.

The problem with comparing canids

One of the reasons this debate persists is that canids are a notoriously flexible and variable group. There is more phenotypic variation within domestic dogs than across many entire mammalian families.

That's why species classification cannot hinge on one trait. For example, you cannot say that howling, or having an annual breeding seasons, or brain size to body ratio alone are traits that split domestic dogs to dingoes, because there are always exceptions. Dogs are so artificially varied, that any one dog breed may share such traits. These are almost always the exception rather than the rule. These exceptions don't invalidate the broader patterns.

You must consider the whole puzzle with all the pieces: genetics, morphology, behaviour, ecology. Like a legal case built on circumstantial evidence, it is the accumulation of consistent differences that builds a compelling argument. A single thread might be weak. Together, they make a rope.

THE CASE FOR *CANIS DINGO*

The case for recognising the dingo as *Canis dingo* rests on an integrative assessment across genetics, morphology, behaviour, ecology and taxonomy. This approach aligns with modern species frameworks and reflects how taxonomic decisions are increasingly made: not by a single criterion, but by a convergence of evidence from multiple fields. This case was first laid out in our 2019 *Zootaxa* article and is now reaffirmed with even stronger evidence in a comprehensive 2025 review. Below is a synthesis of the strongest reasons to recognise the dingo as a distinct species. Some of these points build on ideas raised earlier, but here they are brought together to show how strongly the evidence converges.

1. The name is already in use and conserved. The International Commission on Zoological Nomenclature (ICZN) has formally conserved *Canis dingo* (Meyer 1793) to maintain stability of usage. Although the ICZN does not recommend separate names for wild and domestic forms, the Code permits authors to name them distinctly, and zoologists frequently do so. In this context, *Canis dingo* clearly functions as the name for the wild form, distinct from the domestic dog.
2. The dingo has a unique evolutionary lineage. Genetic evidence places the dingo on a branch of the canid family tree that diverged from ancient dogs around 10,000–12,000 years ago – well before the rise of agriculture and modern domestication. There is little to no evidence of artificial selection during its time in Australia. Dingoes retain ancestral traits and lack domestication-associated genomic features such as the AMY2B gene

duplication for starch digestion. Recent studies also report structural chromosomal differences between dingoes and domestic dogs, further supporting their distinct lineage (e.g. Field *et al.* 2022).

3. It is geographically and reproductively isolated. For thousands of years, dingoes lived in isolation from other *Canis* species. Hybridisation with domestic dogs has occurred only recently due to European colonisation ~230 years ago. However, population-genetic studies from 2023 to 2024 (see Cairns *et al.* 2023, Kreplins *et al.* 2024, Weeks *et al.* 2024) demonstrate limited gene flow and effective reproductive isolation in contemporary populations, even where dingoes and domestic dogs occur in close proximity. In remote parts of Australia, particularly arid and alpine zones, dingoes maintain self-sustaining populations with all the hallmarks of wild canids. Notably, Tasmania, which has had no dingoes, also has no viable population of wild-living domestic dogs, suggesting that dingoes are uniquely capable of surviving and reproducing in wild conditions without human support.
4. The Biological Species Concept is inconsistently applied. Many *Canis* species can interbreed, yet are still recognised as distinct. Arguing that dingoes are not a species because they can hybridise with dogs ignores this broader precedent and reflects a double standard. As previously mentioned, applying this concept consistently across the genus *Canis* would collapse most members into a single species, a conclusion few taxonomists accept.
5. Dingoes meet multiple species definitions. Under both the phylogenetic and ecological species concepts, dingoes qualify as a distinct species. They are genetically and morphologically diagnosable, fill a unique ecological niche as Australia's apex predator and show consistent behavioural and reproductive traits that separate them from domestic dogs. Dingoes breed seasonally, form stable monogamous pairs and live in cooperative social groups, patterns far more typical of wild canids than domestic dogs. Recent genetic studies have also confirmed that dingoes are very different from domestic dogs at the DNA level. Scientists use a kind of scorecard to measure how different two animal populations are from each other. Think of it like a scale from 0 to 1 – where 0 means they're genetically identical, and 1 means they're completely unrelated. A score above 0.25 is considered a strong sign of separation. Dingoes and domestic dogs score above that line, showing they've been on separate evolutionary paths for a long time – more different than many species that are officially treated as separate in science (see synthesis in Cairns *et al.* 2025).

   Moreover, where hybridisation does occur, the dingo phenotype often re-emerges over generations, suggesting strong natural selection for wild traits.
6. They show regionally adapted variation. Distinct south-eastern and north-western dingo populations have been identified, with subtle differences in morphology and genetics. This suggests that dingoes are not only distinct from domestic dogs, but are also diversifying within Australia, as would be expected of a wild species adapting to varied environments. Additionally, recent phylogenetic studies have shown that Asian village

dogs are heavily admixed and are not part of the same evolutionary lineage as dingoes and New Guinea singing dogs, further clarifying the dingo's unique status.

THE POLITICS OF NAMING

The scientific name given to the dingo carries weight.

Although Jackson and collaborators frame their position as purely taxonomic, the implications stretch far beyond scientific classification. Their use of *Canis familiaris*, whether intentional or not, supports agricultural and bureaucratic interests that benefit from the freedom to bait, trap and shoot dingoes. In contrast, conservationists, including myself (an ethologist), Crowther (an ecologist), Cairns (a geneticist) and many others, argue that dingoes meet multiple legitimate species concepts and deserve to be recognised as *Canis dingo*. Not only because it suits conservation policy, but because the evidence points clearly to their biological and ecological distinctiveness.

This is not a new pattern. Taxonomy has long been wielded as a political tool. Throughout history, the way animals are classified has been shaped by human agendas, from colonial expansion and resource control to conservation priorities. In North America, for example, the American bison was long treated as a type of cattle (*Bos bison*), rather than a distinct genus. That classification suited colonial interests: reducing the animal to livestock status helped justify its mass slaughter during westward expansion, clearing the plains for agriculture and undermining Indigenous cultures that depended on it. The dingo is no exception.

Tracey Kreplins in 2018 led an analysis of how the name given to dingoes directly influences their treatment under law and management policy. When referred to as *Canis familiaris* or simply 'wild dogs', they fall under pest control frameworks – subject to legal baiting, shooting and trapping. When identified as *Canis dingo*, however, they are seen as part of Australia's native fauna and may receive protection.

Governments have often chosen the classification that best suits their agenda. Jackson's work is regularly cited to justify control policies, while alternative peer-reviewed research is overlooked. This cherry-picking of science undermines the principles of evidence-based decision-making. When agencies actively ignore compelling evidence and sideline credible alternative perspectives, taxonomy ceases to be a scientific tool and becomes a political weapon.

A comparable situation occurred with the red wolf (*Canis rufus*) in the United States. This canid, believed to have arisen through natural hybridisation between wolves and coyotes, faced shifting classifications that directly impacted its conservation. Reintroduction programs were launched and later abandoned. The uncertainty around its taxonomy stalled protection.

Recognising the limitations of rigid taxonomic criteria, conservation geneticist Bridget vonHoldt and others (2017) have called for more flexible policies. They argued that hybridisation should not automatically disqualify a species from protection. Instead, each case should be evaluated in context, especially given growing recognition that hybridisation is a natural part of evolution.

In the case of dingoes, it is often difficult to distinguish hybrids in the wild. A more effective approach would be to prioritise safe zones where dingoes can re-establish stable social groups.

This would not only support their ecological function, but also reduce hybridisation rates by reinforcing natural behaviours.

Some Australian jurisdictions protect dingoes, while others classify them as pests (e.g. Table 1.1). This patchwork approach lacks a consistent scientific or legal foundation. In Europe, protections have even been extended to hybrids between wolves and dogs to avoid mistaken killings and to support conservation. These policies reflect ecological reality rather than arbitrary genetic purity.

The dingo deserves the same logic. How we define it, and its relationship to domestic dogs, shapes everything from management strategies to public perception. And nowhere is this more visible than in the legislation that governs biodiversity in Australia.

The EPBC Act and the species problem

The *Environment Protection and Biodiversity Conservation (EPBC) Act* recognises the dingo as Australian wildlife. Yet when it comes to legal protection, the Act's species-based framework creates a deeper problem: dingoes are assessed within the same biological species concept that treats them as part of the domestic dog lineage, *Canis familiaris*. What seems like a technical detail has far-reaching consequences.

This tension arises from the EPBC Act's reliance on the Biological Species Concept, a framework that defines species primarily by their ability to interbreed. Because dingoes and domestic dogs can produce fertile offspring, they are treated as belonging to the same biological species, despite their profound ecological, behavioural and evolutionary differences.

The Act does include a caveat that allows for the recognition of 'distinct biological entities' within a broader gene pool. In principle, this exception should apply to dingoes.

Like wolves and coyotes, dingoes can interbreed with domestic dogs, but that does not erase their identity. And as outlined in Chapter 13, the level of hybridisation across Australia is negligible. And let's be honest, no one confuses a dingo with a pug or expects to see a chihuahua in a dingo enclosure at the zoo.

By forcing dingoes within *Canis familiaris*, Australian law creates a legal grey zone that leads to conflicting interpretations of their status. This produces a series of contradictions:

- either all domestic dogs (*Canis familiaris*) would have to be considered native, which they are not;
- or a legal distinction would need to be made between native *Canis familiaris* (dingoes) and domesticated *Canis familiaris* (dogs);
- or dingoes risk losing native recognition altogether if grouped too closely with their domesticated relatives.

Together, these contradictions place dingoes in a kind of bureaucratic limbo. A classic case of Australia's 'conservation with a catch' approach, where a species is legally acknowledged, but practical protections remain absent or conditional.

The result is inconsistency at every level. The EPBC Act offers only incidental protection, applying mostly to Commonwealth lands, while leaving day-to-day management of dingoes to the states. In practice, this has resulted in wildly divergent policies. In some jurisdictions, dingoes are listed as wildlife and enjoy limited protection. In others, they are classified as pests and subject to lethal control measures like aerial baiting, even within national parks.

As such, the ecological, genetic and behavioural uniqueness of the dingo is obscured by a classification system that fails to reflect its true status as a wild, native species. And as we will see in Chapter 14, this scientific ambiguity translates into political convenience, administrative contradictions and real-world threats to the dingo's survival.

But the legal and regulatory contradictions go even deeper than conservation policy.

If dingoes are legally classified as *Canis familiaris*, then they are, by definition, domestic dogs. This has cascading practical consequences. For example, you don't need a wildlife permit to keep one. Dingoes can be exported overseas as pet dogs rather than native fauna. In zoos, they would be displayed as *Canis familiaris*, the very same species name as the family dog at home. But how can you give two animals the same scientific name and treat them so differently? Where exactly is the line between a dingo and a dog? How would a government agency prove the difference in court if it were ever tested? It is hard to see how such distinction could withstand legal scrutiny.

WHY IT IS MORE THAN JUST A NAME

In 2017, Jackson and his co-authors claimed that '...denial of formal taxonomic status for the dingo and New Guinea singing dog does not in any way diminish their scientific, ecological and cultural significance.' But this sidesteps the issue. It ignores the real-world implications of taxonomic decisions, especially when those decisions influence policy, funding, public perception and whether an animal is legally protected or destroyed.

The dingo is a distinct wild canid. Even the criteria emphasised by Jackson and colleagues – ecological independence, consistent morphology, behavioural distinctiveness and the existence of a long-standing, self-sustaining wild population – collectively support that conclusion. The continued use of *Canis familiaris* is not the result of scientific uncertainty, but of historical inertia and selective alignment with policy interests that favour lethal control. In that context, taxonomy has been weaponised.

The dingo deserves better. It deserves a name that reflects what it is, not a label that can be used to rationalise its destruction.

12
Between worlds

AS DISCUSSED IN THE LAST CHAPTER, THE DINGO HAS LONG DEFIED EASY classification. At the heart of this uncertainty lies a deceptively simple matter: of determining whether the dingo has ever been domesticated.

If dingoes were domesticated (at any point), they can be classified as introduced animals that went feral, a biological and legal framing that often justifies their control or eradication. If they never crossed that threshold, they represent a rare case of a free-living canid with ancient ties to humans, yet shaped almost entirely by natural rather than human hands.

To figure out whether dingoes qualify as domesticated, we first need to understand what domestication actually entails. And as with most things in the dingo's story, the answer is anything but black and white. Domestication, in this sense, is a spectrum. A gradual and complex process with many stages and degrees. Understanding this continuum is essential to recognising where dingoes sit, and how they challenge conventional definitions.

WHAT IS DOMESTICATION?

Domestication is often misunderstood. Many assume that if an animal lives alongside humans, tolerates their presence or even forms a bond, it must be domesticated. But tameness and domestication are not the same. Taming affects individual behaviour. Domestication reshapes a species.

Unlike taming or captivity, domestication is a long, gradual process. Over thousands of years, selective breeding alters an animal's genetics, physiology and behaviour. Their brains shrink, faces shorten and coats change. They become more docile, more trusting and, crucially, more dependent on humans for survival. A pug cannot survive in the wilderness. A farmyard chicken would not last long in the jungle. These are creatures shaped by humans, bound to a world they can no longer leave.

Different disciplines define domestication in slightly different ways. Early researchers like Frederick Zeuner (1963) focused on controlled breeding, whereby humans decide which animals reproduce. More recently, evolutionary geneticists Keith Dobney and Greger Larson (2006) have emphasised the genetic and morphological changes that result from this process, increasingly separating domesticated species from their wild ancestors.

From a cultural perspective, scholars like Juliet Clutton-Brock (1999) describe domestication as a kind of social contract between humans and animals. In this view, domestication is not just about genetic change, but about how these relationships reshape human culture, transforming how we live, work and relate to the natural world.

Regardless of the perspective, one truth remains constant. Domestication is a species-wide transformation that leaves an evolutionary signature. It produces animals genetically adapted to human-altered environments, with traits that persist across generations.

The Self-domestication Hypothesis

Not all domestication needs to begin with human intent. One influential theory, the Self-domestication Hypothesis, challenges the traditional top-down model. Proposed by Raymond and Lorna Coppinger (2002), this theory suggests that some animals, especially canids, initiated the process themselves.

Rather than being captured or bred from the outset, early wolves may have been drawn to human settlements, scavenging food scraps and refuse. The boldest and least aggressive individuals were more likely to thrive in these spaces and their traits, such as shorter flight distance, sociability and reduced aggression, became more common over generations.

In this model, domestication begins as an unplanned partnership. Humans did not set out to tame wolves. They simply tolerated those that were least troublesome. The animals, in turn, adapted, not through human-directed breeding, but through natural selection in a human-shaped environment.

This is how early dogs likely emerged: canids that scavenged on the edges of camps, followed nomadic groups and occasionally formed loose bonds with people. They were not selectively bred nor controlled or shaped for human purposes. They occupied an evolutionary transition zone between wild canids and the domestic dogs that would evolve much later.

Domestication versus related concepts

Domestication is often mistaken for familiarity with humans. As just noted, tolerance of humans or even close bonds do not equal domestication. Similar behaviours can arise from very different evolutionary histories and kinds of human relationship.

Taming is perhaps the most common source of confusion. *Taming* is an individual-level process. A wild animal can be tamed if it learns to tolerate human presence through socialisation or training. A circus tiger, a falcon trained for hunting or a hand-raised magpie may all appear tame, but they remain wild. They retain the instincts, physiology and genetics of their ancestors. Their offspring will not inherit their behaviour.

Domestication, by contrast, is a species-wide transformation. It occurs over many generations, driven by selective breeding. Domesticated animals, like the domestic dog, show reduced aggression, altered reproductive cycles and physical changes such as smaller teeth, shorter snouts, floppy ears or a range of coat colours. These changes are heritable, not learned, and persist even without human training.

Feralisation is different again. This refers to the process by which domesticated animals return to a wild or semi-wild state. Feral animals may recover behaviours that allow them to survive independently, but their genetics still reflect their domesticated origins. Feral horses, for example, differ in key ways from prehistoric wild horses and feral pigs still carry markers of domestication. Even after generations in the wild, they don't simply become wild again. The imprint of domestication remains.

Understanding these distinctions is essential when it comes to dingoes. Many animals live near humans, exploit our resources or form bonds with people, but that does not make them domesticated. These blurred boundaries are what make the dingo's story so compelling – and so easy to misinterpret.

The grey zone

Some human-animal relationships sit between categories. These cases challenge the idea that an animal must be either fully wild or fully domestic – and they help illuminate where the dingo sits.

Take giant pandas (*Ailuropoda melanoleuca*) for example. Their survival today depends heavily on conservation programs, including breeding in captivity. But they haven't been selectively bred for human use. They don't show the typical physical or behavioural traits of domesticated species. Despite their dependence on humans for survival, they remain fundamentally wild.

Urban coyotes are similar. These animals den under buildings, raid bins and grow bolder around people. They are adapting to our landscapes, but no one bred them for these behaviours. They remain wild canids, exploiting human environments without becoming dependent.

Even birds complicate the picture. A wild magpie might learn to accept food from humans, recognise faces and even play with children. But this behaviour is learned, not inherited. The magpie's offspring will not be born friendly to humans. In contrast, domesticated birds like chickens and canaries retain their traits even when raised in isolation from people.

These examples reinforce a central truth: coexistence with humans does not make an animal domesticated. Without heritable genetic and behavioural changes, an animal may be habituated or opportunistic, but it remains wild.

Adaptable, not domesticated

Animals may tolerate people or form temporary bonds, yet this says little about domestication. Such behaviours are better understood as evidence of behavioural flexibility, a trait shared by many wild species. In fact, it is often the key to survival.

Urban foxes, crows and coyotes all exploit human landscapes without being tame or domesticated. They are adaptable, opportunistic and quick to learn.

Dingoes are similar. In some parts of Australia, they live on the fringes of mining camps, townships and rubbish tips. In others, they are raised in homes as companion animals. These behaviours may resemble domestication on the surface, but they are better understood as signs of the dingo's extraordinary adaptability. The same traits that help them survive in deserts or alpine forests also allow them to navigate these human spaces.

This flexibility likely helped their ancestors coexist with people in South-East Asia thousands of years ago, without ever crossing the threshold into domestication. Like the earliest dogs, dingoes can form temporary or partial associations with people, yet retain their independence, complex social structures and strong instincts.

In short, behavioural flexibility enables coexistence, but it does not always imply domestication.

To try and determine whether dingoes ever crossed the domestication threshold, we can look for the signs this process leaves behind.

THE HALLMARKS OF DOMESTICATION

Across species, animals that have lived under sustained human influence show predictable changes. These include behavioural shifts, physical modifications, altered reproduction, dietary adaptations and increasing reliance on humans. Together, they form the hallmarks or signatures.

Human-directed selection

One of the clearest markers of domestication is artificial selection, whereby humans control which animals reproduce in order to promote traits like tameness, obedience or usefulness. All domesticated species show evidence of this process, in their behaviour, appearance or function. Dogs were bred to herd, guard or retrieve. Chickens were selected for egg-laying or meat production. Horses were shaped for strength, endurance or speed.

Dingoes show no such signs. Their populations have always replenished naturally. Even among Aboriginal Australians, who sometimes raised pups in their camps, there is no sign of structured breeding. These pups typically returned to the wild once mature. As discussed later in this chapter, there was no attempt to select for traits or maintain controlled lineages.

This absence of human intervention also applies to the regions where dingo ancestors likely originated. In South-East Asia and New Guinea, there is no indication that early communities bred canids for specific purposes. These animals lived near people, but they reproduced freely and were not shaped to serve particular roles.

Physical changes

Domestication often alters an animal's physical form. Traits like shorter snouts, smaller skulls, reduced teeth and floppy ears are common in domestic species. Many dog breeds also show exaggerated juvenile features, a condition known as paedomorphism. These changes are side effects of selecting for tameness and sociability.

Dingoes do not fit this pattern. Their skulls are long and narrow, built for a carnivorous lifestyle. They have large carnassial teeth for cutting and crunching and their skull crests are well developed, allowing for powerful jaw muscles. Their body proportions are consistent and their coats remain largely within a natural range of colours – ginger, sandy and black-and-tan.

Although dingoes do show some mild paedomorphic features, such as slightly shorter snouts compared to wolves, these are subtle. They likely emerged through natural selection in response to Australian environments, not from human intervention. If domestication had shaped their appearance, we would expect to see more variation, smaller skulls and traits associated with tameness. None of these are found.

Reproductive patterns

A reliable sign of domestication is a shift in reproductive biology. Wild canids, like wolves, typically breed once per year, timed with environmental conditions to maximise survival of pups. Domestic dogs, by contrast, follow no seasonal cycle and can have multiple litters per year. This increased fertility was favoured by humans seeking to expand their animal populations, and it aids in the dogs reproductive strategy of quantity over quality first mentioned in Chapter 6.

Dingoes follow the wild pattern. They breed once per year, usually between March and June. This cycle is strongly seasonal and closely linked to rainfall, food availability and climate. There is no sign of the more flexible reproductive timing seen in domestic dogs. They remain tied to natural cycles like other wild canids.

Dietary adaptations

Domestication often influences what animals eat, normally shifting to a human-centric diet. One of the best-known genetic markers of dog domestication is the duplication of the AMY2B gene, which allows dogs to digest starch. This change emerged as early dogs scavenged on human agricultural waste.

As we already discovered in Chapter 2, dingos lack this adaptation. Like wolves, they have only a single copy of the AMY2B gene. Their digestion is geared towards a high-protein, meat-based diet. They are not well equipped to process starches or grains.

This is strong evidence that dingoes were never part of farming societies and were not shaped by a scavenging lifestyle. They were hunters, not animals being fed or living off human scraps.

Reliance on humans

Perhaps the most important marker of domestication is the dependence on humans.

Domesticated species, even when feral, usually remain tied to people. They may live outside the home, but they continue to rely on human infrastructure, rubbish dumps, villages and farms for food and shelter. Feral dogs, for example, are rarely found in truly remote areas. They cluster around human settlements and depend on access to human waste to survive.

Dingoes live across the Australian continent, including some of the most remote, inhospitable regions in the world. They hunt for themselves, raise pups in dens, defend territories and form strong social bonds, all without any reliance on people.

As first discussed in Chapter 7, this independence is arguably one of the most important defining features of a wild animal. Dingoes are not tethered to human environments in the way feral dogs are. They do not need people and they do not rely on human resources – deliberate or otherwise.

They are self-sufficient. Although we cannot completely rule out the possibility that dingoes were once tamed or partially associated with humans in their deep past, there is no clear evidence that they were ever fully domesticated. The absence of key markers across morphology, genetics, reproduction and behaviour strongly suggests that dingoes likely never crossed that threshold.

BEFORE AUSTRALIA, WERE DINGOES EVER DOMESTICATED?

Whether dingoes were already partially domesticated before arriving in Australia is still debated, mainly because we lack a detailed understanding of this relationship. Some researchers argue that they were part of early human–animal networks, living around camps and travelling with people. Others suggest they never crossed into domestication, but instead belonged to a transitional group often described as proto-dogs.

Proto-dogs, as characterised in Ray Coppinger's self-domestication theory, are considered those early canids, that interacted with humans without becoming dependent on them. They scavenged around settlements, followed nomadic groups and may have helped with hunting, but they were not selectively bred. If you were to place dingoes in any of these categories, this is where they best fit. They differ from both fully domesticated dogs and wild species like wolves who have not been altered by human hands. Their ancestry links them to the semi-feral village dogs still found across Asia, but I do not believe they were ever pushed across the threshold into domestication.

The most consistent picture is that dingoes stem from an ancient dog lineage and humans likely helped move them around parts of Asia, but they largely remained wild. Their early split from this dog lineage pre-agriculture, their lack of human-driven traits and their long isolation all point to wild ancestry. If there was any domestication, it was shallow and short-lived and its effects were erased by natural selection over time.

A DIFFERENT KIND OF BOND: DINGOES AND ABORIGINAL AUSTRALIANS

The relationship between Aboriginal Australians and dingoes is one of the oldest and arguably longest known human–canine partnerships, but it did not follow the same path as canine domestication elsewhere in the world. Although dingoes lived closely with Aboriginal people across the continent, forming strong social and spiritual ties, they were never selectively bred or genetically altered in the way domestic dogs were.

Even Charles Darwin , reflecting on domestication in the decades following his 1836 visit to Australia during the *Beagle* voyage, emphasised that true domestication requires sustained human control over breeding. He wrote that ancient peoples must have paid close attention to pairing animals, because the inheritance of desirable and undesirable traits was so evident. From this perspective, the absence of deliberate breeding by Aboriginal Australians stands out. Anthropologist Mervyn Meggitt, who worked extensively with Aboriginal people in central Australia, found it remarkable that a people could associate so intimately with an animal of such cultural and practical importance for millennia without ever domesticating it.

Modern research supports the idea that this relationship was not one of domestication, but of commensalism. Dingoes lived alongside people, but retained their autonomy. In many Aboriginal groups, dingoes were not treated as property or tools, but were seen as spiritual beings, companions or even family members.

One of the first questions I explored in my own research was whether dingoes had ever been domesticated at all. It mattered for the cognitive and behavioural comparisons I was making between dingoes, dogs and wolves. In my first published research paper in 2009, we found no evidence that Aboriginal Australians domesticated dingoes. Several key factors support this conclusion:

Dingoes reproduced in the wild: unlike domestic dogs, which were selectively bred by humans over generations, dingoes retained their natural breeding patterns. Aboriginal Australians would take pups from dens and raise them in their camps, but these animals were never bred in captivity. Once they reached sexual maturity, they typically left the camps and reintegrated into the wild. This cycle, of raising pups, but allowing them to return to nature, ensured that dingoes remained genetically and behaviourally independent.

Dingoes were resistant to training and control: even when raised by humans, dingoes displayed strong instincts that made them difficult to train, contain or breed selectively. Their independent nature, strong prey drive and lack of dependence on humans set them apart from domesticated animals. Unlike dogs, which evolved to follow human cues and cooperate, dingoes retained a level of autonomy that resisted domestication.

Spiritual and cultural beliefs discouraged domestication: the role of the dingo in Aboriginal Dreaming placed it in a unique position, one that was both physical and spiritual. Many Aboriginal groups distinguished between 'camp dingoes' and 'wild dingoes', suggesting that while some dingoes lived near humans, they were not considered property in the way that domestic dogs were. In many traditions, dingoes were viewed as totemic animals, ancestors or spiritual beings. This reverence may have contributed to the absence of attempts to control or alter them through breeding.

Nomadic lifestyles made domestication impractical: many Aboriginal groups were semi-nomadic, which posed practical challenges to the domestication process. Domesticated animals typically require sustained human intervention, controlled breeding and physical containment – none of which aligned with Aboriginal Australians' way of life. Instead, they developed a relationship with dingoes that provided practical benefits without the need for full domestication.

In many ways, this relationship is really difficult to categorise within existing frameworks. Aboriginal Australians achieved something few other cultures did: a way of living with a wild animal that neither broke its spirit nor reshaped its nature. Doing so would go against natural lore. Instead they built a bond grounded in respect, tolerance and deep ecological understanding.

THE MYTH OF THE REWILDED (FERAL) DOG

Some international researchers, such as Hungarian canine scientist Adam Miklósi and others working in the field of comparative cognition, have suggested that dingoes reverted to the wild as feralised dogs. In this view, dingoes are considered to have been at some point fully domesticated animals that later escaped into the wild and evolved back into something more wild or wolf-like. It's an intriguing idea and on the surface, it seems to fit: dingoes share a common ancestor with domestic dogs, they have a historical association with humans (in Asia and Australia), and they resemble wild canids in both form and function.

I was curious about the theory at first, but the more I examined it, the less convincing it became.

Let's consider what feralisation actually looks like. Some animals, when released or escaped from human care, can survive and reproduce in the wild. Over generations, they may begin to express more wild-type traits. Feral pigs are a common example. Domestic pigs that go feral can grow coarser hair, their snouts elongate and they may even develop tusks. But they still carry the genetic fingerprints of domestication. Their brains are smaller, their skulls differ from those of wild boars and their reproductive biology reflects their domestic ancestry.

The same is true for feral dogs. Even when free-ranging, they tend to remain near human settlements, scavenging from rubbish tips and relying on human waste for survival. Their social behaviour, reproductive patterns and digestive systems still reflect their domesticated past. They do not thrive in remote or resource-scarce areas without human influence.

Dingoes are different. As we have seen, they show none of the expected signs of past domestication. Their skulls are large and robust. Their brain size is closer to that of wolves than dogs. They reproduce seasonally like wild canids. They lack the genetic adaptations that allow domestic dogs to digest starchy foods, such as the duplication of the AMY2B gene. Behaviourally, they are cautious, independent and capable of cooperative hunting. Crucially, they live and thrive in some of the most remote and harsh environments in Australia, far from any human resources.

Dingoes are close relatives of domestic dogs, they have history of living with humans both in Asia and Australia, and they can live as household companions and read human cues. But proximity to humans does not equal domestication. From what I can see, dingoes represent one of two possibilities.

The first is that they were never domesticated at all. They may have been wild-living canids in Asia and New Guinea, similar to New Guinea singing dogs or early proto-dogs, that followed humans across landscapes, but retained their independence.

The second, more nuanced idea is that they may have undergone a shallow form or initial stages of domestication, such as taming or occasional association, before arriving in Australia, but they never crossed the threshold into full domestication. They were not bred for docility or friendliness or for hunting assistance. They retained their social structures, hunting behaviours and reproductive cycles. Whatever minimal influence humans had was either negligible (e.g. isolated to small areas and not a population wide change), or erased by thousands of years of natural selection in the Australian bush.

Either way, from what I can deduce, and what all my research has shown, is that dingoes are not dogs, and neither am I convinced that they were at some point domestic dogs that reverted back to a wild state. I can only see evidence that dingoes are continuing a long evolutionary history of living independently from, but alongside humans. They are a distinct lineage of wild canid. Calling them feral dogs is quite likely inaccurate and obscures their evolutionary history and their ecological role.

LIVING BETWEEN THE LINES

The dingo likely descended from canids that interacted with people in South-East Asia or Melanesia. These animals may have accompanied humans during migrations as opportunists – scavengers, camp-followers and loosely affiliated companions. Much like the semi-feral village dogs found across much of the world today.

When dingoes reached Australia, whether brought by people or arriving independently, they did not become domesticated. Instead, they spread across the continent, living as independent predators. They were not shaped for herding or guarding, nor bred for obedience. Yet they retained traits that enabled coexistence with humans: social intelligence, cognitive flexibility and behavioural plasticity. These were the same traits that once allowed proto-dogs to thrive near humans without becoming domesticated.

In this way, the dingo represents an evolutionary fork in the road. Where other canids were slowly moulded into domestic dogs, the dingo veered off. It remained wild-living, shaped by ecological pressures rather than human preference. If the story of the dog began with self-domestication, then the dingo is the version that never turned the final page.

If dingoes are not feral dogs, and they are not wolves, then what are they?

Several researchers have explored this ambiguous space. Loukas Koungoulos describes the dingo as a proto-dog; that is, an early canid that interacted with humans, but never crossed the threshold into domestication. These animals lived near people, scavenged and followed nomadic groups, but retained their independence, free-breeding behaviour and wild instincts. Dingoes may offer a glimpse into what early dog domestication looked like before artificial selection became dominant.

Adam Brumm suggests a slightly different view. He sees the dingo as having the best of both worlds: neither fully wild nor fully domesticated. Through the Aboriginal practice of raising dingo pups, a unique ecotype emerged – one that lived close to humans, but remained genetically and behaviourally autonomous. Brumm argues that this challenges the traditional view of domestication as a process requiring strict human control over breeding.

My own research supports this interpretation. In studies comparing dingoes, domestic dogs and wolves, I've found that dingoes often fall in between. They are capable of using human social cues, such as gaze direction or pointing, but are less likely than dogs to rely on humans for help during problem-solving tasks. In cognitive tests, they behave more like wolves: curious, persistent and independent, yet they are more tolerant of humans than wolves typically are.

This same balance plays out in real life. Dingoes can live alongside humans, exploiting modified environments like rubbish tips or mining camps, but they do not become dependent on them. They remain capable of surviving and reproducing entirely on their own. I've met people who live with dingoes as companions, and although many fail, many also succeed. Dingoes are not domestic animals. They may form bonds (indeed, sometimes really strong connections with humans, many dingo carers arguing much stronger than with a dog), but they never surrender their autonomy.

To me, dingoes truly do represent the best of both worlds. They are completely wild and free-ranging, yet uniquely able to coexist with humans when it suits them. A wolf cannot do that. Wolves avoid people entirely. Domestic dogs, on the other hand, rely on us. Dingoes sit in between, being independent, adaptable and socially intelligent.

Their existence forces us to rethink what it means to be domesticated. The dingo blurs the boundary between wild and tame, between self-sufficiency and association with people. It reminds us that not all animals fit neatly into our human-made categories.

It is not a dog. Not a wolf. It is something else. And that something else is what makes the dingo so extraordinary.

13
The hybrid myth

FOR MOST OF AUSTRALIA'S HISTORY, THE DINGO WAS SIMPLY CALLED A DINGO. From around the 1970s, a new term began creeping into government documents and pest control strategies: wild dog.

Today, 'wild dog' is a fixture in the language of land management. Farmers report wild dog attacks on livestock. Government programs outline wild dog control strategies. The barrier to keep them out is now called the Wild Dog Fence rather than the Dingo Fence. For most Australians, the phrase conjures an image of feral domestic dogs rather than dingoes and a native wild animal.

The danger in this kind of rebranding is that it erases the dingo's identity altogether. As dingo ecologist Thomas Newsome once warned, if we lose the ability to distinguish dingoes from domestic breeds, 'there is a risk that we shall lose the dingo as an entity in its own right, and thus consider it no different to a pug.'

It's a provocative image, but an accurate one. Pugs represent the extreme outcome of human-driven domestication – animals highly dependent on us for survival. Treating dingoes as equivalent to such breeds obscures their identity as wild predators and undermines recognition of their distinct evolutionary lineage as a wild predator.

This shift in terminology was a direct consequence of scientific debates about hybridisation. Before the 1970s, the question of whether dingoes were pure or hybrids was rarely discussed. But in the late 20th century, CSIRO researchers Alan Newsome and Laurie Corbett began investigating whether Australia's wild canids showed signs of domestic dog ancestry.

At the time, hybridisation was framed as a threat, something that could dilute the 'purity' of the dingo and ultimately erase the species. Yet this perspective ignores a fundamental reality of canid evolution.

Hybridisation is common in the *Canis* family and has played a significant evolutionary role in wolves, coyotes and even domestic dogs, sometimes offering adaptive advantages in changing environments.

Despite this natural tendency for canid lineages to mix, the genetic identity of dingoes has remained remarkably stable.

These early findings, particularly in south-eastern Australia, offered a convenient rationale for lethal control. If hybridisation was erasing pure dingoes, then protection seemed unnecessary.

By the 1980s, the idea had taken firm hold in management circles. The term 'wild dog' gave government agencies a way to sidestep tricky legal and ethical questions about killing a native species. The term functions as a kind of linguistic camouflage, more closely aligned with the image of a domestic dog gone wild and therefore more palatable to the public than admitting it was dingoes being killed. As geneticist Kylie Cairns puts it, "Wild dog' isn't a scientific term, it's a euphemism.' A way to disguise dingoes as less native, less important and more like a pest, to make their killing easier to justify.

The irony is that even as this narrative took hold, the scientific methods for identifying hybrids were deeply flawed. Early morphological classification methods, and later DNA analysis, overestimated hybridisation. By the time better techniques emerged, however, the wild dog framework was so entrenched that it didn't matter.

FROM SCIENTIFIC CURIOSITY TO MANAGEMENT DOGMA

By the late 20th century, determining whether a canid was a dingo, a domestic dog or a hybrid depended almost entirely on physical traits. With no genetic tools available at the time, researchers relied on morphology (referring to form, shape and function) to make distinctions and no one was more influential in shaping this approach than Corbett and Newsome. They developed a formula-based skull classification system, which became the dominant method for distinguishing dingoes from hybrids (Fig. 13.1).

The CSIRO method involved meticulously measuring skull features, such as cranium height, jaw breadth and tooth structure. Based on these characteristics, a scoring system was created that defined a 'dingo range'. If a skull fell within this range, the animal was classified as a dingo. If it deviated too far, it was labelled a hybrid. At the time, this method was considered a scientific breakthrough, offering what seemed like an objective way to identify dingoes. However, as we found out much later, it was fraught with problems.

To begin with, the process required the animal to be dead. Skull measurements were only possible after the specimen had been boiled, stripped of flesh or chemically treated, which aligned conveniently with lethal control practices but not so much for those who had a captive dingo and wanted to know its ancestry. Even beyond this logistical issue, establishing a baseline for a 'pure' dingo skull was inherently problematic. No one knew for certain what an undisturbed, genetically pure dingo population looked like and regional variation in skull shape was never fully accounted for. Genetics have since shown that dingo populations form distinct regional clusters, each with subtle morphological differences that earlier methods might have mistaken for signs of hybridisation.

Morphology itself is a highly unreliable trait, influenced by factors such as age, diet and environment, which could all affect the shape and size of a skull. This meant that hybrids did not always present an obvious mix of dingo and domestic dog traits, making classification difficult and, at times, completely arbitrary. And over all of this hung a simple but unavoidable reality that these measurements were taken by people, leaving plenty of room for human error, inconsistency and subjectivity in how they were recorded and interpreted.

Despite these flaws, Corbett in particular, had a profound influence on dingo ancestry and 'purity'. His research suggested two critical conclusions that shaped future dingo management. First, he observed that dingoes bore a striking resemblance to village dogs from South-East Asia, particularly those from Thailand and Indonesia. This conclusion was based on skull comparisons with dogs sourced from Asian dog meat factories, reinforcing

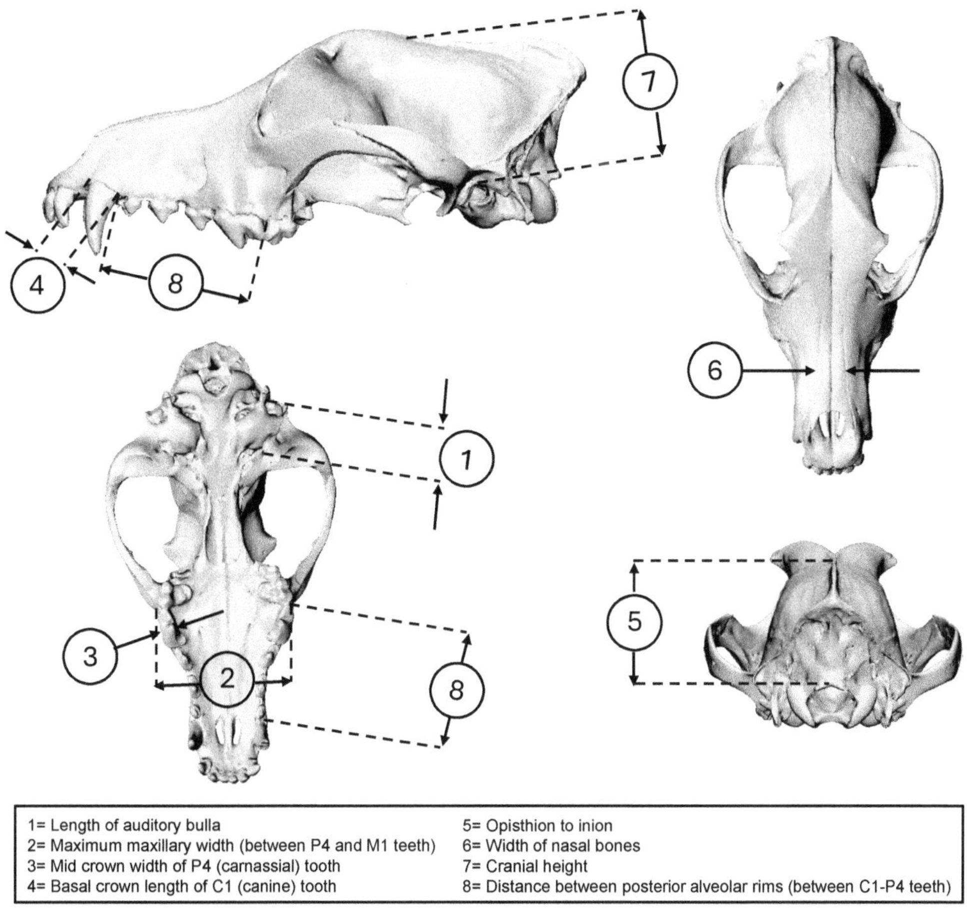

Fig. 13.1. The eight skull measurements once used to distinguish dingoes, domestic dogs and their hybrids. Researchers recorded these straight-line dimensions and entered them into a discriminant formula that produced a single score: values within a certain range were classified as dingoes, while those outside it were labelled dogs or hybrids. This cranial-measurement method was an early attempt to quantify hybridisation in wild canids, but it is no longer considered reliable. Measurements shown here are based on those described in Corbett (2001).

the idea that dingoes shared an ancient lineage with regional pariah dogs (this is reflected in the title of his 1995 book *The Dingo in Australia and Asia*). Second, and more importantly in the Australian context, he found evidence of widespread hybridisation, particularly in south-eastern Australia, below the Dingo Fence, where domestic dogs and dingoes were more likely to interact.

It was this second finding that proved the most influential. If dingoes were disappearing due to hybridisation, their conservation value came into question. This idea was quickly embraced by government agencies as a convenient tool for policy, because it provided justification for large-scale lethal control. If hybrids were taking over, and pure dingoes were all but gone, then efforts to protect dingoes were unnecessary and largely in vain.

THE TROUBLE WITH PHYSICALLY IDENTIFYING DINGOES

Because skull measurements are only possible after an animal has died, those working in the field have always had to rely first on appearance. Identifying a dingo seems straightforward: the classic sandy coat, pointed ears and lean build. If a canid doesn't match that template, for example if it is black, white, brindle or too stocky, or even missing a white tip on its tail, it is often dismissed as a hybrid. Evidence, supposedly, that domestic dog genes have crept in.

This confidence in visual identification has shaped public perception, management strategies and lethal control programs. Farmers, land managers and even scientists have relied on physical traits to distinguish dingoes from hybrids, often believing that hybridisation is something that can be seen.

I've lost count of how many times I've been told by farmers, by pest controllers, even by managers at remote mine sites, that 'those aren't real dingoes around here'. They'll point to a dark-coloured animal or one that looks a bit heavier and say, with absolute certainty, that it's a hybrid. Everyone seems so sure they know what a dingo looks like, but in truth, most people, including many who work with wild canids, don't actually know how to tell them apart with 100% certainty. I always treat anyone reporting they can with a great deal of scepticism.

The problem isn't just misidentification, but the mistaken belief that hybridisation is visually obvious. It isn't. Determining whether a wild canid is a pure dingo or a hybrid based on appearance alone is far more difficult than many assume. Scientific research has repeatedly shown that physical traits, including coat colour and body structure, are not reliable indicators of ancestry.

Dingoes exhibit a wide range of natural variation in coat colour, body shape and size. And there is little correlation between colour and genetics. A 2008 study by Amanda Elledge, and later work by Cairns, shows that coat colour is not a valid indicator of ancestry. Dingoes have carried diverse coat colours (black, white, brindle, tan, patchy) for thousands of years, and these variations do not necessarily indicate hybridisation. The widespread belief that only tan-coloured dingoes are real dingoes has no scientific basis, yet it continues to justify lethal control in areas where non-tan individuals are assumed to be hybrids. Simply put, what an animal looks like does not always reflect its genetic reality.

Beyond coat colour, other physical traits vary significantly between individuals and across regions. Size, skull shape and body proportions can be influenced by environment, diet and genetics, making visual classification unreliable. Hybridisation, when it does occur, does not always produce obvious physical differences. Many supposed hybrids look exactly like dingoes, yet are still labelled, and killed, as wild dogs based purely on the assumption.

Even as the scientific evidence showing that dingoes remained largely distinct from domestic dogs mounts, many in government and industry cling to the outdated belief that hybridisation was and still is a major problem. Such is the lingering influence of those early identification studies.

THE GROWING OBSESSION WITH PURITY

By the early 2000s, DNA testing had begun to replace skull measurements as the primary tool for identifying dingoes. The transition was led by Alan Wilton, introduced in Chapter 2, who dedicated much of his later career to dingo conservation. Wilton developed the first microsatellite markers designed to distinguish dingoes from domestic dogs in the late 1990s, offering a genetic approach to what had previously relied on subjective physical assessments. His research was groundbreaking at the time, marking the first attempt to quantify hybridisation rates using DNA.

Wilton's method used a small set of genetic markers, just 23 microsatellites, to detect traces of domestic dog ancestry in wild canids. For the first time, scientists had a tool to measure how much genetic material had been exchanged between dingoes and domestic dogs. His research suggested that hybridisation was particularly high in south-eastern Australia, which supported CSIRO's skull measurements, a finding that quickly influenced government policies.

It's important to recognise that Wilton's motivation was conservation, not extermination. Deeply concerned about the future of dingoes, he hoped that identifying hybridisation would help protect genetically distinct populations from further dilution. In his own writings, he cautioned against overinterpreting the results and acknowledged the limitations of his marker set. But, as with so many conservation tools, the science was quickly divorced from its context. Once governments had a number – a so-called 3Q score that quantified an animal's 'dingo-ness' – it became all too easy to draw hard lines and justify hard measures. What began as a pioneering tool for understanding dingo identity was soon wielded as a mechanism for exclusion and control.

The obsession with dingo purity, arguing over what percentage of dog DNA is too much, carries uncomfortable reminders of Australia's colonial past. As cultural scholar Fiona Probyn-Rapsey points out, talk about 'hybrids' being a threat to 'pure' dingoes isn't just about science. It taps into old fears about mixing, and ideas once used to justify racist policies and control over Aboriginal people. Even if it's unintentional, the way we talk about dingoes can carry the same kind of thinking. In 2021 she put it bluntly: 'the lament over the extinction of the dingo is a dog-whistle – it feeds off race panic, it situates the hybrid as the already dead, already gone, ungrievable.'

Building on Wilton's work, Danielle Stephens, a geneticist co-supervised by Wilton during her PhD, conducted the first large-scale genetic survey of dingoes across Australia in 2015 (Fig. 13.2 top). Using the same 23 microsatellite markers, Stephens concluded that many dingoes, particularly in south-eastern Australia, carried significant signatures of domestic dog ancestry. Her study reported that dingoes in some regions had less than 65% dingo DNA, leading to the conclusion that interbreeding was readily and frequently occurring, resulting in genetic swamping and the erosion of dingo identity. These results were widely circulated by government agencies and land managers, reinforcing the belief that hybridisation was widespread, growing,

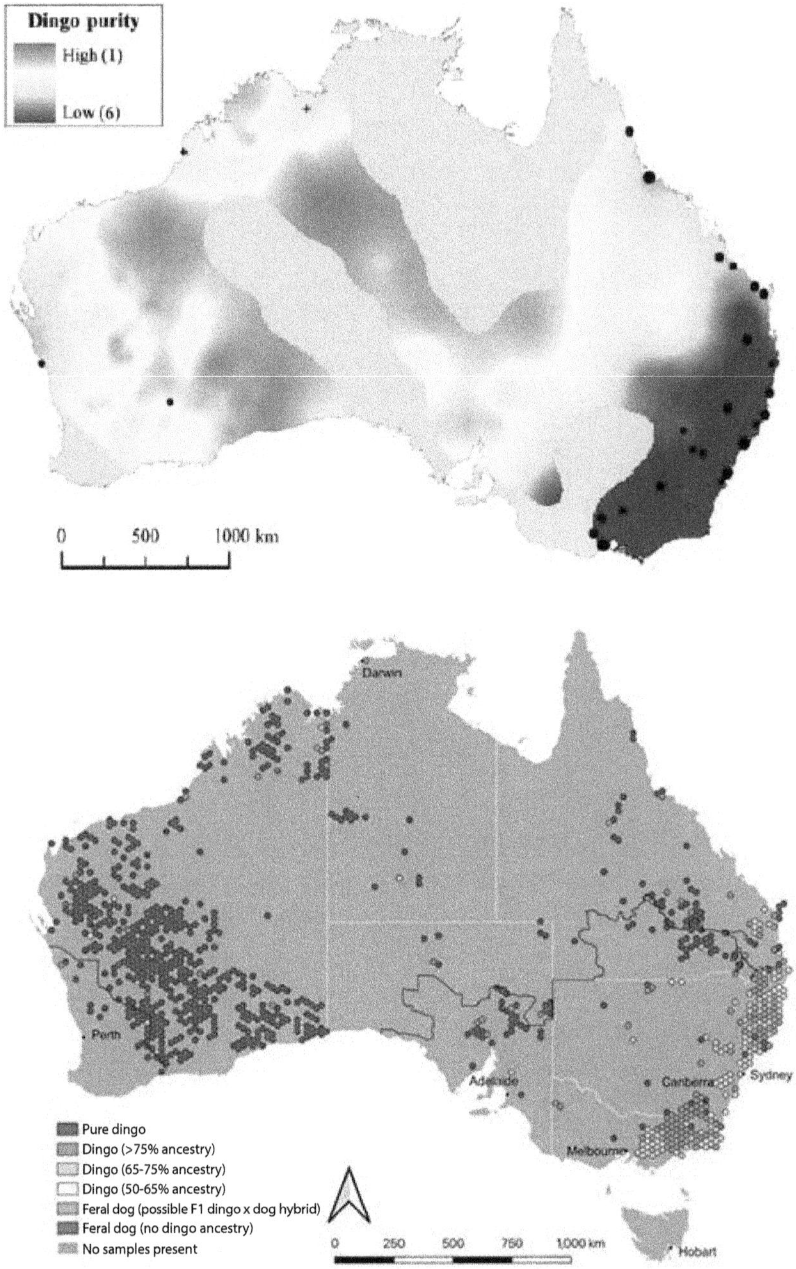

Fig. 13.2. Comparisons of early and updated assessments of dingo–dog hybridisation. (Top) Early estimates of dingo hybridisation based on a limited 23-microsatellite test (Stephens *et al.*, 2015) suggested widespread interbreeding with domestic dogs (darker shading), particularly in south-eastern Australia. These findings fuelled the belief that dingo genes were being swamped by domestic dog genes. (Bottom) Updated genomic analysis using the same microsatellite test but broader sampling of wild dingoes across south-eastern Australia (Cairns *et al.*, 2021) revealed a very different picture: dingoes remain genetically distinct across the continent and earlier claims of mass hybridisation were significantly overstated.

and irreversible – especially below the Dingo Fence, where contact between domestic dogs and wild canids was thought to be more common.

However, both Wilton's and Stephens' studies shared a critical flaw: they relied on a small, outdated set of microsatellite markers. These markers were never designed to detect subtle patterns of hybridisation and were biased towards identifying recent admixture within three generations. More importantly, they failed to account for natural geographical variation within the dingo population. Regional differences, such as those between south-eastern, desert and northern dingoes, were often misinterpreted as signs of hybridisation. As Cairns and colleagues later showed, these early tools consistently overestimated the presence of dog ancestry in dingoes, particularly in the south-east (Fig. 13.2 bottom).

Despite these limitations, the findings were again swiftly adopted by government agencies eager to justify ongoing dingo control programs. If hybridisation was as widespread as the Stephens study suggested, then most wild canids were no longer 'true' dingoes. This bolstered a growing narrative: that protecting dingoes was no longer practical or even possible because, genetically speaking, there was little left to protect.

But here is the contradiction. If dingoes were so thoroughly mixed with domestic dogs that pure ones hardly existed, then the idea of protecting dingoes from hybridisation made little sense. In that scenario, there was nothing left to save. Yet lethal control continued unabated. The solution was not to rethink the policy, but to change the language. Instead of acknowledging the implications of their own claims, authorities leaned more heavily on the convenient catch-all label wild dog, a term broad enough to justify killing any canid, while sidestepping uncomfortable questions about dingo identity or protection. As Cairns has argued, the findings of early genetic studies were not used to protect dingoes, they were instead weaponised to justify their removal.

THE GENETIC REVOLUTION

The turning point came with the application of modern genomic tools – technology capable of finally testing decades of hybridisation claims. Kylie Cairns used these tools to investigate dingo ancestry in unprecedented detail and what she found overturned the dominant narrative.

Wilton and Stephens had relied on just 23 microsatellite markers, but in 2021 Cairns was able to use genome-wide sequencing of more than 195,000 single nucleotide polymorphisms (SNPs), providing a much clearer and more reliable picture of ancestry. Her findings were groundbreaking.

1. Dingoes remain overwhelmingly genetically distinct from domestic dogs.
2. There is no evidence that dingoes are being genetically 'swamped'. Admixture has occurred, but it is rare across Australia and where it does occur, hybrids tend to backcross into the dingo population rather than dilute it.
3. Earlier estimates of hybridisation were massively inflated, the product of outdated methods and a failure to consider natural population structure.

Cairns analysed over 5,000 wild canid samples from across Australia and found that feral domestic dogs were almost completely absent. The genetic 'crisis' was overstated and her finding directly contradicted the hybridisation myth. If wild populations were supposedly full of hybrids, as earlier studies claimed, then where were the dogs responsible for producing them?

As Cairns explained in a media release relating to her findings:

> *We don't have a feral dog problem in Australia. They just aren't established in the wild. There are rare times when a dog might go bush, but it isn't contributing significantly to the dingo population.*

This ecological reality reinforces the genomic data that dingoes persist in the bush, but domestic dogs do not. The dingo remains the dominant and enduring wild canid across the Australian landscape.

In a 2023 follow-up, Cairns directly compared the older microsatellite-based tools with genome-wide sequencing. Again, the contrast was clear: the older methods couldn't distinguish between recent mixing and ancient shared ancestry and they often misread natural regional variation as evidence of hybridisation. Many animals previously assumed to be hybrids were, in fact, pure dingoes.

As Cairns explained in the same media release:

> *For years, we were told that wild dogs were replacing dingoes, but the data just doesn't back that up. When we test these so-called hybrids, we find that they are overwhelmingly dingoes.*

This didn't mean that earlier researchers got it wrong on purpose. They were working with the best tools available at the time. But as the science improved, so did the answers and those answers told a very different story.

A second opinion

Cairns' findings triggered a broader re-evaluation of hybridisation claims and fair to say upset a lot of people. Criticism and dismissal of her research and results became common. However, work by independent geneticists has continually reinforced and extended her work.

At the University of Melbourne, evolutionary geneticist Andrew Weeks led a major genomic study in 2024. Analysing over 400 wild canid samples from around Australia using high-resolution genomic tools, Weeks and his team found no evidence of pervasive hybridisation between dingoes and domestic dogs.

What Weeks' study showed was that earlier genetic methods had confused shared ancestry with recent interbreeding. Because dingoes and dogs both came from the same ancient lineage, their DNA still shares some similarities. Older tools weren't sharp enough to tell the difference, so they mistook this for signs of hybridisation. With modern high-resolution techniques, Weeks and his team were able to clearly identify real hybrids (i.e. known first-generation crosses and low ancestry backcrosses) when they were present.

Crucially, Weeks found that previous claims of widespread hybridisation, like those in Stephens where 99% of samples in Victoria were deemed hybrids, were the result of flawed assumptions and biased reference populations. His reanalysis of that microsatellite dataset found clear genetic structuring and patterns of isolation by distance, not admixture.

Most significant of all were the policy implications. In Victoria, dingoes are listed as a threatened species, yet they continue to be killed under wild dog control programs, based on a myth. As Weeks concluded, 'lethal management is currently leading to the culling of a legally protected threatened species', with no measurable presence of domestic dog DNA to justify it.

Ancient DNA

If contemporary dingoes are this genetically distinct, the next question is whether that distinctiveness is a recent development or a legacy stretching back centuries. When researchers turned to ancient genomes for answers, the pattern became even clearer.

In 2024, Yassine Souilmi and colleagues from the University of Adelaide analysed ancient dingo genomes ranging from 400 to 2,746 years old, revealing that the genetic structure of modern dingoes was already established thousands of years ago, long before European arrival. Their findings show that dingoes have remained genetically distinct despite centuries of coexistence with domestic dogs. Not only did the ancient genomes show genetic continuity, they also revealed the same patterns of population structure seen in modern dingoes, confirming that dingo diversity today is not the result of recent mixing, but a long-standing feature of the species.

'The ancient DNA shows us that dingoes have been genetically stable for thousands of years,' Souilmi explained in a media interview. 'There is no evidence that they are being genetically erased by hybridisation.'

These findings dismantle the hybrid myth. While some regions show limited traces of dog ancestry, the broader picture is one of continuity, not replacement. There is no evidence of widespread post-colonial genetic mixing and no support for the idea that dingoes are being bred out of existence. They were never 'just' wild dogs to begin with. This research reinforces the dingo's status as a unique, free-living canid, not a feral derivative of domestic dogs. As Souilmi's team concluded, management strategies should reflect this reality.

Science versus policy

Despite overwhelming evidence from modern genome-wide analyses, population genetics and ancient DNA, the hybridisation myth remains deeply embedded in government policy.

Lethal control continues to be justified using outdated assumptions about hybridisation, despite growing evidence that such interventions may actually be part of the problem (see later in this chapter). The wild dog label has never been about genetics; it has been a convenient political tool.

BARRIERS AGAINST HYBRIDISATION

The genomic findings align perfectly with what we know about dingo biology, behaviour and ecology. Hybridisation is rare in the DNA, as well as in the real world.

Studies on wolves and domestic dogs in other parts of the world show a similar pattern. That is, hybridisation can occur, but there are natural barriers that make it uncommon. The same barriers apply to dingoes. Their social behaviours, reproductive strategies and ecological pressures all act to limit mixing with domestic dogs. Even when they occupy the same areas, they do not mix freely, and the few hybrids that do appear rarely persist.

A 2024 study by Kreplins and colleagues, for example, genetically tested camp dogs from Aboriginal communities in Western Australia and found that only 3.5% carried any dingo ancestry, while pure dingoes were almost absent from these domestic dog populations. This suggests that hybridisation is not happening at a significant rate, even in places where dingoes and domestic dogs regularly encounter each other.

Dingoes are not socially compatible with domestic dogs

Dingoes live in structured, territorial packs, where social order is essential. Domestic dogs, on the other hand, do not form the same kind of highly coordinated units. This mismatch makes sustained interbreeding difficult.

To a dingo pack, an unfamiliar dog is an intruder. In many cases, when domestic dogs enter dingo territory, they are attacked and driven out rather than incorporated into the group. Long-term population studies have shown that feral domestic dogs rarely establish themselves in wild dingo populations. In several cases, domestic dogs that encroach on dingo territory have been killed outright. This makes it extremely difficult (but of course not impossible) for a domestic dog to infiltrate a dingo pack, let alone breed within it.

The breeding barrier

Even if a domestic dog were to break through social barriers and survive in dingo territory, there is still a major biological hurdle: Dingoes have a strict seasonal cycle, with mating occurring only between March and May. For hybridisation to occur, a domestic dog would need to be in the right place at the right time when the dingo female is reproductively active.

There is also a behavioural asymmetry that makes some pairings far less viable than others. For a litter to survive in the wild, the mother must be capable of raising her young without human support, often relying on help from the rest of the pack. If a female dingo mates with a male domestic dog, she may still raise her pups successfully using her instincts and cooperative care. But the reverse, a female domestic dog giving birth in the bush after mating with a male dingo, is far less likely to succeed. Domestic dogs are not adapted to raise young in harsh, wild conditions and male dingoes, unlike male domestic dogs, play an active role in parenting. A male dingo will not help raise a litter born to a domestic dog and she will not have the benefit of a dingo pack to assist her.

Even when hybrid pups are born, many struggle to survive. Studies have shown that hybrids often lack the full suite of traits needed to thrive in the wild, further reinforcing the idea that dingoes have remained genetically distinct despite thousands of years of potential exposure to domestic dogs.

Physical and ecological barriers

Beyond social and reproductive barriers, geography itself plays a major role in preventing hybridisation. The idea that hybridisation is widespread across Australia assumes that there are large numbers of free-roaming domestic dogs in dingo habitats, but we know from the latest genetic research that this not the case.

As mentioned earlier, both of the Australia-wide genetic studies by Cairns (2021) and Stephens (2015) found that feral domestic dogs (or dogs with no dingo ancestry) were virtually absent in most of the Australian bush. That is, there are no signs that dogs are surviving on their own away from people. In addition, there are massive areas of Australia where no one lives and there are no domestic dogs. Without a source population of dogs, hybridisation cannot occur.

Most claims of high hybridisation rates come from south-eastern Australia, where domestic dogs are more common in rural areas. But even there, genetic studies show that dingoes remain largely distinct, with hybridisation occurring at much lower rates than previously thought.

A human-made pathway for hybridisation

If natural barriers prevent most hybridisation, human intervention can undermine them.

Large-scale baiting and culling programs disrupt dingo pack structures, removing dominant breeding pairs and creating social instability (see Chapter 18 for more). In stable dingo populations, a single dominant pair controls breeding, minimising the chance of interbreeding with outsiders. But when lethal control removes these dominant individuals, the resulting social chaos can increase the chances of crossbreeding with wandering domestic dogs.

Paradoxically, the very practices long justified on the basis of protecting dingoes from hybridisation may be making hybridisation more likely. Instead of preserving dingo genetic integrity, lethal control can dissolve the social structures that normally safeguard it.

Dingoes are still dingoes

Every piece of scientific evidence suggests that dingoes are not being replaced by hybrids. In sum:

- Dingoes' social structures and seasonal breeding naturally limit hybridisation.
- In most of Australia, free-ranging domestic dogs are virtually absent.
- Where overlap does occur, genetic studies show hybridisation remains minimal.
- Lethal control may increase hybridisation by disrupting dingo pack structure.
- As noted earlier, assumptions based on appearance continue to justify lethal control, despite their unreliability.

As Cairns put it:

> *The dingo has survived in Australia for thousands of years, and there is no evidence to suggest that it is disappearing now. The claims of widespread hybridisation simply don't hold up against the data.*

RESILIENCE IN FORM AND FUNCTION

Beyond genetics, the dingo's physical form and biology tell a consistent story. In 2016, anatomist Will Parr and his colleagues used 3D scans to compare the skulls of dingoes, domestic dogs and known hybrids.

They discovered that even where some hybridisation between dingoes and dogs had occurred, the hybrids retained the dingo's cranial shape. Instead of drifting toward the more variable and domesticated dog forms, they clustered close to the dingo's distinct cranial profile, so much so that hybrids and pure dingoes could not be reliably distinguished by skull shape alone. Even the proportions of the skull (the long muzzle, narrow cranium and powerful jaw) remain unmistakably dingo.

Parr's team concluded that the dingo's cranial morphology is dominant, an adaptation so well suited to life in the wild that even when dog genes enter the population, the offspring tend to revert to the dingo's characteristic head shape within a few generations. As the authors put it, hybridisation 'does not spell the end for native morphology.'

This inability to distinguish hybrids from dingoes by skull shape has sometimes been misinterpreted as evidence that dingoes are simply dogs. But that was not Parr's conclusion. His findings showed the opposite: that dog ancestry fails to significantly disrupt the dingo's morphology. In other words, it is the dog that becomes more dingo-like, not the dingo that becomes more dog-like. The authors also noted that this morphological stability likely preserves the dingo's feeding eco-niche – its ability to hunt and process prey in the same way as pure dingoes.

The same resilience shows up in reproduction. Researchers Marina Cursino and colleagues (2017) set out to determine whether, when dingoes and domestic dogs interbreed, their offspring breed like dogs or like dingoes. They studied the reproductive cycles of female hybrids and compared them with both pure dingoes and domestic dogs. Hybrid females were found to retain the dingo's strict seasonal breeding pattern, coming into oestrus only once a year, usually in autumn, just as pure dingoes do. Domestic dogs, by contrast, cycle throughout the year.

Even with a measure of dog ancestry, the rhythm of reproduction stays anchored to the dingo's natural cycle. Hybrids were not adopting the dog's continuous breeding pattern, which would allow for greater mixing and gene flow. It suggests that the environmental and social cues shaping dingo reproduction are deeply ingrained, placing a biological brake on further hybridisation.

Taken together, these findings show that even where genetic exchange has occurred, the essence of the dingo remains intact. Its form, function and life cycle are shaped by the demands of the wild, and those pressures continue to mould anything that survives in that landscape. Hybrids may exist, but natural selection keeps drawing them back toward the dingo form.

HYBRIDISATION AND ECOLOGICAL ROLES

Many people fear hybridisation not just because it mixes DNA, but because they assume it dilutes what makes dingoes unique – that hybrids won't behave like dingoes or functional predators. However, a study led by Mathew Crowther, a dingo ecologist at the University of Sydney, argues that hybridisation does not significantly alter dingo behaviour, hunting patterns or ecological impact.

Crowther's analysis of the literature shows that dingo-hybrids continue to hunt the same prey species, suppress foxes and feral cats and maintain territorial behaviours essential for ecosystem balance. His findings challenge the assumption that hybridisation is an ecological crisis, revealing that even where some genetic mixing occurs, dingoes are still fulfilling their natural role.

The fear that hybrids might behave more like domestic dogs, scavenging instead of hunting or relying on human food sources (namely, sheep), is not supported by the evidence. Hybrids remain self-sufficient hunters, capable of taking down kangaroos, wallabies and other native prey. They also exhibit the same cooperative pack structures, which are crucial for maintaining their role as top predators.

This pattern is not unique to Australia. Wolf–dog hybrids in Europe and North America continue to function as apex predators, maintaining strong hunting behaviours and ecological interactions. Just as wolves remain wolves despite occasional hybridisation, dingoes remain dingoes. Hybridisation does not trigger the ecological collapse that some management policies might suggest.

As Crowther puts it:

> *The dingo is still the apex predator, regardless of hybridisation. Culling programs have treated hybridisation as an ecological disaster, but there's no evidence it alters how dingoes function in the environment. If we're serious about conservation, we need to stop using outdated genetic arguments to justify killing them.*

IF IT LIVES LIKE A DINGO, IT'S A DINGO

The debate over 'purity' has done more to obscure the dingo than to clarify it. For decades, governments classified dingoes using outdated genetic markers and arbitrary percentage cut-offs. These cut-offs were based on methods now shown to have dramatically overstated hybridisation.

The deeper problem is that this approach reduces an ecological role to a number on a lab report. A wild canid living independently on the land, hunting native prey, suppressing foxes and cats, and defending territory behaves as a dingo, regardless of whether a fraction of its genome once came from a domestic dog. Ecology, not ancestry, determines its place in the landscape.

For many Aboriginal people, this is obvious. Dingoes have been part of Country for thousands of years, woven into cultural stories, practices and relationships with the land. As clearly articulated by representatives from the Bundjalung, Yuin Walbunja, and Southern Yuin Tribes (New South Wales) in their article in *Australian Zoologist*:

> *If it's out there in the wild, performing the same role as a dingo, then it is a dingo. This obsession with purity is a colonial construct that ignores the reality of the animal itself.*

This perspective challenges the justification for lethal control programs, especially those that claim to be protecting the dingo by removing hybrids. In practice, culling targets all wild canids, so 'pure' dingoes are killed alongside hybrids, destabilising populations and making hybridisation more likely rather than less. If the goal was really to protect the dingo, then indiscriminate culling should be the last thing we do.

So what's left of the wild dog myth? Very little.

If a wild canid is living, hunting and shaping the land like a dingo, then it is a dingo.

The real threat to dingoes is not hybridisation but the policies built on outdated assumptions. It's time to stop arguing over DNA percentages and start recognising dingoes for what they are: Australia's apex predator, still holding its place in the landscape, despite everything thrown at it.

14
True blue

IN THE AUSTRALIAN VERNACULAR, 'TRUE BLUE' MEANS SOMETHING unmistakably, undeniably Australian, a marker of genuine belonging. It's a phrase steeped in national identity, used to describe people, places, even things, that seem to embody the spirit of the country.

But what does it mean to be considered truly Australian? For the dingo, like many of the topics already discussed, that question is anything but straightforward.

Some argue that because dingoes are believed to have arrived with humans, they can never be native, but rather lumped in with rabbits, foxes, camels and the long list of introduced species. Or that because they descend from an ancient lineage of Asian dog-like canids, they are nothing more than domestic animals gone feral. Both perspectives presume the dingo's status is fixed and final, irreversible and permanently denying it the right to be considered part of Australia's natural heritage.

Others contend that time and adaptation matter profoundly. That dingoes have been here long enough, shaped the land long enough and played such an integral role in Aboriginal cultures that they are native in every meaningful sense. That their ancient genealogy and arrival story are either irrelevant or, as we've previously discussed, fundamentally misunderstood.

This tension is the bleeding heart of the debate around dingo identity, a debate that follows naturally from our examination of taxonomy, domestication and hybridisation.

These are not just public or political questions. Scientists themselves remain divided. As noted in the introduction (Chapter 1), the Delphi study of 28 dingo experts conducted by Donfrancesco and colleagues in 2023 revealed deep disagreement over key issues, including the dingo's nativeness, identity and how it should be managed. These divisions reflect deeper tensions around values, priorities and even the language used to describe the dingo rather than simple disputes over evidence.

Some might dismiss the question as academic, a debate for scientists and policymakers. But the answer carries real consequences. If accepted as native, it deserves the same protections as koalas, echidnas and bilbies. If deemed non-native, it becomes a target. A pest. An invader suffering the same fate as foxes and feral cats, horses, pigs, goats and camels: shot on sight, poisoned en masse and systematically erased from the landscape without legal consequence or moral hesitation.

This rejection of the dingo's right to belong filters into everyday culture. At an agricultural field day, I was handed a stubby holder printed with the phrase: 'I like dogs, just not wild ones.' A harmless joke to some, perhaps. But beneath its surface lies a potent message. We embrace the animals we control, the ones that sit at our feet or work our farms. The others, the truly wild ones, are treated as outsiders. The slogan captures the everyday casualness with which dingoes are dismissed, not for what they are, but for what they are not: tame, owned, obedient.

The question of dingo nativeness, then, is less about biology, and more about law, conservation and belonging. A question that circles back to a simple idea of what it means to be true blue.

LEGAL DEFINITIONS

If belonging was as simple as meeting the legal definition of nativeness, the dingo's place in Australia wouldn't be questioned. By Australian federal law, it is native. Despite this formal recognition in national legislation, dingoes remain among the most legally ambiguous and inconsistently managed species in the country.

The *Environment Protection and Biodiversity Conservation (EPBC) Act*, Australia's primary environmental legislation, defines a native species as one that was established in Australia before AD 1400. I am not quite sure how this number was determined, but seemingly a conservative time that ensures it predates arrival of Europeans (from the 1600s onwards). By this measure, dingoes qualify without dispute.

Oddly, being classified as native does not automatically provide protection. Nowhere in the EPBC Act are dingoes explicitly listed as a species requiring conservation, which is another function of the Act. Unlike iconic native species, dingoes exist in a legal grey zone, recognised as native in theory yet denied protection in practice. There is no conservation advice, listing advice or recovery plan for the dingo under the EPBC Act, meaning there is no federal strategy for their conservation or management. The Act provides no clear national directive on dingo management, leaving their fate in the hands of individual states, where their status varies wildly.

A HISTORY OF UNCERTAINTY

The question of whether the dingo is native has been debated since colonisation. Early European settlers and naturalists frequently questioned the dingo's place in Australia, often comparing it to wild dogs and assuming its arrival was recent. In the late 19th and early 20th centuries, biologist Baldwin Spencer and ethnographer Walter Roth documented the dingo's deep cultural significance to Aboriginal Australians. Spencer, conducting fieldwork in central Australia in the 1890s and early 1900s, and Roth, researching Indigenous societies in Queensland during the same period, recorded the dingo's role in Aboriginal life.

Their work ultimately reinforced the prevailing European belief that dingoes were introduced rather than native. This was shaped by their assumption that any animal living closely with Aboriginal people must have been brought by them, combined with a broader scientific bias that equated nativeness with species existing independently of humans. As a result, dingoes were often (but not always) categorised alongside domestic animals rather than as Australia's native predators.

This uncertainty quickly spilled into law. Unlike marsupials or monotremes, which were readily accepted as native due to their evolutionary distinctiveness, the dingo's resemblance to domestic dogs led to persistent confusion about its status. Because Spencer, Roth and other naturalists framed dingoes as human introductions, the species' classification remained contested. By the early 1900s, legislative efforts to define nativeness were already entangled with agricultural interests. Although dingoes were recognised as unique, they were also viewed as a threat to livestock, leading to widespread bounty systems and government-funded extermination programs.

By the mid-20th century, legal definitions of nativeness had begun to take a more formalised approach, with scientists advocating for criteria based on evolutionary adaptation rather than arrival history. However, ecologist Chris Dickman notes in his 2021 review of the history of the 'dingo dilemma', that throughout this period, the dingo's legal status shifted repeatedly, shaped more by economic and political considerations rather than ecological evidence. The introduction of the EPBC Act in 1999 finally provided a clear legal cutoff for nativeness, which should have resolved the debate. Yet, the dingo's classification remained vulnerable to shifting definitions, particularly where legislation intersects with agricultural and biosecurity policy.

As such, the legal system has long struggled to categorise the dingo, which remains trapped between being a revered cultural figure, an ecological keystone species and a perceived pest.

Even among scientists, there is no consensus. As the Delphi study of dingo experts found (first mentioned in Chapter 1), disagreements around the dingo's status are shaped not only by conflicting evidence, but also by divergent values. Most particularly, these centre on whether purity matters more than function, whether nativeness should be about time of arrival or ecological role and even whether nativeness should matter at all.

These tensions are most evident in the following case studies, where legal definitions of nativeness have been shaped as much by political convenience as by ecological reality.

Western Australia: redefining nativeness for political convenience

In 2018, under the leadership of Premier Mark McGowan, the Western Australian government attempted to strip dingoes of their native fauna status, a move I have personally found to be one of the most striking examples of politically motivated reclassification to exist. Despite dingoes meeting every scientific and legal benchmark for nativeness under the EPBC Act, the McGowan administration sought to have them reclassified as 'non-fauna', a move that would have effectively removed them from the state's conservation framework and permitted unrestricted lethal control.

The decision rested on two core arguments: 1) dingoes were introduced by humans and therefore cannot be native and (2) hybridisation with domestic dogs had blurred the definition of what a dingo is.

These justifications were deeply flawed. As I've previously discussed, the dingo is just as likely to have reached Australia by walking from New Guinea, making human introduction far from a certainty. But even if dingoes were brought to the continent by people, that alone does not disqualify them from being considered native. Many species have been transported by humans and are still protected, recognised as an integral part of their ecosystems. The Channel Island fox, for example, that was mentioned in Chapter 11, was likely moved between islands off California by Indigenous peoples thousands of years ago. Over time, these foxes adapted to their new environments, evolving into distinct subspecies that are now recognised as a unique, native component of the Channel Islands' ecology. Similarly, in Australia, species like tamarind trees (*Tamarindus indica*), introduced by Macassan traders, and yams (*Dioscorea* spp.), cultivated by Aboriginal Australians for thousands of years, remain an accepted part of the landscape. These examples highlight the arbitrary nature of using human involvement as a primary reason to deny a species its native status.

The hybridisation argument is equally misleading. Recent genetic research covered in Chapter 13, has shown that hybridisation with domestic dogs is not widespread across Australia and does not threaten the integrity of most dingo populations. Notably, in Western Australia, modern dog introgression is actually at its lowest, with the state having some of the highest populations of genetically pure dingoes in the country. In no other context do we so readily dismiss a species' ecological function based on the presence of gene flow.

The scientific backlash was immediate. Along with 25 other dingo researchers, I co-authored an open letter condemning the decision, arguing that it was politically driven rather than biologically justified.

Ultimately, the proposal to reclassify dingoes as 'non-fauna' did not proceed. Dingoes remain listed as native fauna under the *Biodiversity Conservation Act (2016)*, but are simultaneously declared pests under the *Biosecurity and Agriculture Management Act (2007)*, allowing continued lethal control across the state.

New South Wales: the split identity of dingoes

New South Wales also enforces a contradictory approach to dingo management. Although dingoes are listed as native fauna under the *Biodiversity Conservation Act 2016*, they are explicitly excluded from protection, even within national parks. At the same time, the *Biosecurity Act 2015* classifies all wild dogs, including dingoes and hybrids, as biosecurity threats. This imposes a legal duty on landholders to control them. The result is widespread lethal control, often without genetic testing or clear distinction between pure dingoes and hybrids. This dual legal framework has triggered growing calls for reform, urging recognition of the dingo's ecological role and cultural significance.

These contradictions are amplified at the local level, where on-the-ground management is largely shaped by landholder groups and regional pest strategies. Local Land Services (LLS) facilitates Wild Dog Management Groups, networks of landholders, council staff and public land managers, that coordinate control methods such as baiting, trapping and shooting. Region-specific pest management plans guide these efforts, listing dingoes as priority pests and applying control measures across both public and private land. Although participation in these groups is technically voluntary, social and economic pressures often compel landholders to conform. In practice, this devolved system means agricultural interests frequently dominate, sidelining ecological and cultural considerations in decisions about dingo control.

Victoria: a native species without true protection

Victoria has been more progressive in recognising the dingo's ecological importance. In 2008, the government formally listed the dingo as a threatened species and in 2024, it paused lethal control in the state's north-west by lifting what's known as an 'unprotection order' – a bureaucratic mechanism that had previously allowed dingoes to be killed in areas where they were otherwise listed for protection. This marked one of the few instances where a state actively reconsidered its approach to dingo management, responding to new evidence that populations in the north-west had fallen to critically low numbers. Yet the move remains fragile and politically sensitive and, crucially, it only applies to one region.

Despite these steps, dingoes can still be legally culled across much of Victoria, including areas where they are listed as threatened. Unprotection orders remain in place for eastern parts of the state and their effect is far-reaching. They override conservation listings, allowing lethal control in agricultural zones, buffer areas and even parts of the public estate, including some state parks. Extensive ground baiting, trapping, and shooting continue with government sanction. The term itself, unprotection, reveals the contradiction at the heart of policy: a threatened species made killable not because of science, but because of where it lives and what economic interests are at stake.

This reflects a broader national pattern: dingoes are recognised when convenient, but persecuted when they interfere with agriculture. Across Australia, dingo conservation is dictated by politics, not science. Their status remains precarious, just one unprotection order away from being treated as invasive.

THE SCIENTIFIC CASE FOR DINGO NATIVENESS

The notion that a species is only native if it arrived without human involvement is both outdated and misleading. It reduces nativeness to a moment of arrival rather than considering what a species does once it has established. After all, some species integrate, evolve and help regulate ecosystems, while others disrupt them.

Conservation ecologist from the University of NSW, Peter Banks, has proposed an eco-evolutionary framework for assessing nativeness, shifting the focus from arrival history to function. This approach evaluates species based on three core criteria:

1. Evolutionary adaptation: have dingoes changed since arriving in Australia?

As has been covered elsewhere (Part 2), at least 5,000 years of natural selection in Australia have shaped dingoes into a distinct wild canid, adapted to this landscape in ways that set them apart from both their ancestors and free-ranging domestic dogs. For all intents and purposes, dingoes' bodies, their behaviour and their reproductive cycles are all akin to those of other wild canids around the world, allowing them to survive independently of humans in Australian unique conditions.

2. Prey recognition: do native species respond to dingoes as natural predators?

Another key test of nativeness is whether native prey recognises a predator as part of their ecosystem. If dingoes were a recent introduction, Australian prey would lack avoidance behaviours, just as they failed to recognise foxes and cats as threats when those species were introduced, leading to devastating declines in native mammals.

Fig. 14.1. Experimental 'cafeteria trials' conducted by Michael Parsons and Daniel Blumstein demonstrating macropod avoidance of dingo scent. The central container, treated with dingo urine, remains untouched while kangaroos and wallabies prefer the untreated containers. This response indicates an evolved predator recognition mechanism, distinguishing dingoes from introduced predators such as foxes and cats. Source: Parsons and Blumstein (2010).

A meta-analysis (i.e. a systematic review of the scientific literature, typically on a topic) by Banks found that 83% of native Australian mammals exhibit anti-predator behaviours when exposed to dingo cues, a response only seen with long-term predator–prey relationships.

For example, in 2010 Michael Parsons and Daniel Blumstein conducted trials exposing macropods (kangaroos and wallabies) to dingo urine and faeces. The results were extremely clear. Kangaroos displayed increased vigilance and avoidance behaviours when exposed to dingo scent. They reduced foraging activity, even when food was available (Fig. 14.1). And, most significantly, they persistently avoided dingo-marked areas over time, showing that their response was not a short-term learned behaviour, but an evolved recognition of dingoes as predators.

This contrasts sharply with invasive predators like foxes and cats, which Australian prey did not instinctively recognise, leading to massive population crashes and extinctions. That kangaroos and other native prey have evolved strategies to avoid dingoes suggests a long shared history.

3. Ecological function: do dingoes act like a native species?

Invasive species disrupt ecosystems. They outcompete native species, cause population crashes and create ecological imbalances. If dingoes were truly invasive, we would expect similar patterns. Instead, as was covered in Part 3, research consistently shows that dingoes fill an essential ecological role that no other species has replaced. With the extinction of the thylacine, dingoes have taken over as the top predator in many Australian ecosystems, controlling prey populations and keeping ecosystems in balance.

Unlike foxes and cats, which drive native species towards extinction, dingoes act as ecosystem regulators, not destroyers. This is the defining distinction between a native species and an introduced pest.

If nativeness is about function, adaptation and belonging, then dingoes have more than earned their place.

Despite the eco-evolutionary framework proposed by Banks providing a compelling and modern case for dingo nativeness, it remains controversial in some scientific circles. Not all researchers agree that function should override arrival history. More traditional views favour chronological or biogeographical definitions of nativeness that prioritise how and when a species arrived.

These differing views were clearly reflected in the Delphi study of dingo experts, which found no consensus on how nativeness should be defined. Some participants supported functional, ecological definitions, while others insisted on more traditional, taxonomic or historical criteria. For them, the question was not how dingoes behave now, but how they got here and what that says about their right to belong.

From this perspective, a native species is one that evolved *in situ*, without human assistance, and forms part of an ecosystem's 'original' pre-human assemblage. Because dingoes did not originate in Australia, and are believed to have arrived with human movements from South-East Asia thousands of years after the continent's first settlement, critics argue they fall outside this definition. Even if dingoes now play a vital ecological role, they are, according to this logic, best considered naturalised or long-established non-natives rather than truly native.

One particularly cautious position comes from some taxonomists who point to the lack of formal taxonomic separation between dingoes and domestic dogs. For these scientists, the ongoing potential for hybridisation muddies the waters, making it difficult to classify dingoes as a distinct native species under conventional definitions. In their view, a native species must not only arrive unassisted, but also remain genetically isolated from its domestic relatives, something the dingo, in practice, cannot fully meet.

A CULTURAL PERSPECTIVE ON NATIVENESS

While the legal and scientific debates over dingo nativeness continue, for Aboriginal Australians, there has never been any question. Dingoes belong to Country. Interestingly, the Delphi study found rare consensus among experts on one point: the cultural significance of the dingo. Even those who disagreed on taxonomy or hybridisation acknowledged that dingoes hold a unique place in Australian society.

This perspective was formally recognised at the First Nations Dingo Forum I described in Chapter 10, where Aboriginal representatives from around Australia met to discuss the dingo and issued the First Nations Dingo Declaration (Fig. 14.2).

The declaration recognises dingoes as a culturally significant, protected native species and describes the dingo as a Creator Being, a cultural icon and an inseparable part of First Nations identity. It states that dingoes formed the lands, the waterways and the constellations and are responsible for mapping ancestral Songlines across the continent. More than just animals, 'The dingo is deeply sacred to Australia's First Nations peoples. They are family.'

Unlike Western legal frameworks, which rely on rigid categories of native versus introduced, Indigenous knowledge recognises species through their relationships with the land. Belonging is not just about when a species arrived, but how it interacts with and sustains the environment. From this perspective, the idea that dingoes could be considered 'not native' is nonsensical.

The declaration also directly opposes the language of dingo misclassification:

> *We do not support the use of the term 'wild dog'. This term diminishes the Dingo. It is a deliberate misrepresentation to justify killing. It disrespects and disregards culture.*

The First Nations Dingo Declaration is a direct challenge to the Western legal and political systems that have long denied dingoes protection. While governments continue to debate the

Fig. 14.2. Elliot Hunter, a Nyangumarta Ranger, signs the First Nations Dingo Declaration at the inaugural First Nations Dingo Forum held in Cairns in 2023. The declaration formally recognises the dingo as a culturally significant, protected native species, rejecting its misclassification as a 'wild dog' and advocating for its legal protection. Photo by Guy Burrows / Girringun Aboriginal Corporation.

dingo's nativeness, Aboriginal Australians have never questioned it. Dingoes are not pests nor are they introduced, but are inseparable from the land and its people.

Despite this deep cultural significance, Indigenous perspectives have largely been ignored in dingo management decisions. Lily van Eeden, a social scientist and leading researcher on public perceptions of dingo management, and her colleagues highlight that dingo management policies in Australia have historically sidelined Aboriginal knowledge systems, even though dingoes have been integral to Aboriginal cultures for thousands of years.

van Eeden further argues that Western conservation frameworks often fail to account for species that exist in cultural liminality – neither fully 'wild' nor fully 'domestic', which is precisely how many Aboriginal communities have historically related to dingoes. Eurocentric rigidity has resulted in policies that prioritise agricultural and political interests over cultural heritage and ecological balance.

The declaration calls for a new model of dingo management, one based on coexistence (something we get to in Chapters 18 and 19), Indigenous knowledge and legal recognition of dingoes as a culturally significant and protected species:

> *We advocate for a different model: caring for Dingoes on Country. This model sees dingo recognised in legislation as a culturally significant and protected native species. This model must use traditional knowledge, and evidence-based practice, focusing on peaceful co-existence between Dingoes and all stakeholders.*

MISCLASSIFICATION AND ITS CONSEQUENCES

Generations of Australians have been conditioned to see dingoes not as a native apex predator, but as a nuisance. A rogue wild dog that exists outside the bounds of protected wildlife. The deliberate use of terms like 'wild dog' by government agencies and industry bodies is a strategic effort to undermine the dingo's status in the public mind.

Language shapes public attitudes to wildlife. 'Dingo', evokes an animal with cultural, ecological and historical significance. 'Wild dog' suggests something invasive, uncontrolled and undeserving of protection. Species labelled as vermin or pests attract little empathy or political support. Part of the success of koala conservation, for example, lies in their powerful cultural identity as unquestionably Australian.

If the public sees dingoes as nothing more than feral dogs, or believes that there are no 'pure' dingoes left in the wild, there is no demand for conservation efforts. Instead, lethal control policies continue largely unchallenged, hidden behind the sanitising language of 'wild dog management'. The more we allow misclassification to persist, the more entrenched this narrative becomes. The harder it is to build public will to protect them.

Lily van Eeden's research demonstrates the power of this framing. In a national survey of 811 Australians, only 19% were aware that dingo control was even occurring. This is largely because it is often concealed under the more politically neutral language of 'wild dog management'. Attitudes towards 'wild dogs' were significantly more negative than towards 'dingoes', and approval for lethal control was much higher when the former term was used.

Even within the scientific community, the terminology is contested. The Delphi study revealed sharp divisions among dingo experts. Some participants defended 'wild dog' as a practical management term acknowledging hybridisation. Others saw its use as a deliberate act of erasure, a way of stripping dingoes of their ecological role and cultural identity.

Without clear legal distinction from domestic dogs, dingoes remain in a precarious limbo. They are native, but not protected. Recognised, but not respected. Their identity shifts with political convenience. Native when useful, pest when not.

As environmental policy researcher Karen Hytten argues, dingoes occupy a space of contradictions. They are 'simultaneously native and feral, protected and persecuted, pure and hybrid'. These dualisms, as she calls them, drive policy far more than science.

Australia risks losing a uniquely adapted apex predator, not because the predator failed to belong, but because our systems failed to see it clearly. Misclassification, left unchallenged, risks rewriting the dingo out of its own country.

DINGOES: A WILD, NATIVE CANID

The collective evidence paints a convincing picture: dingoes are a native species. Legally, they are recognised as such under the EPBC Act. Yet in practice, this status means little. They remain in a precarious limbo, their identity manipulated by politics, their protections eroded by contradictions in legislation. In Western Australia, they are simultaneously classified as native under conservation law yet treated as pests under biosecurity regulations. The fact that the government wanted to take it further and remove their status altogether speaks volumes.

As the Delphi study reminds us, the debate about dingo nativeness is not just a contest of facts, but of frameworks. That is, what kind of knowledge we prioritise, what values we hold and what future we envision for Australia's most controversial carnivore.

But the question of whether dingoes are native perhaps misses the point. What matters is not how they arrived or might have arrived, but how they have shaped this land and how this land has shaped them in return. For thousands of years, dingoes have been integral to Australia's ecosystems, its landscapes and its cultural traditions. Their presence is not one of invasion, but of adaptation, survival and deep connection to Country.

If 'true blue' means something unmistakably Australian, something that belongs, then the dingo is exactly that. Not because of a legal definition, but because of what it is: a wild canid, an apex predator and a keystone species that has earned its place through time, evolution and function. It is not an outsider, yet its fate hinges not on ecological reality, but on political convenience.

The path forward is clear. Australia needs a unified, scientifically grounded taxonomic definition of the dingo that is formally recognised in law and policy. One that recognises its uniqueness – not as a domestic stray, but as a wild, native canid – and aligns with conservation policy, Indigenous knowledge and ecological reality.

Part 5

Dingoes in the crossfire

15
Dangerous liaisons

IT WAS HOT AND SUNNY. THE HEAT WAS DRY AND PRESSING, THE KIND THAT hits you like the blast from an oven when you open the door. This was the Pilbara, the desert in the north of Western Australia, where summer temperatures often soar above 40°C. It's the very definition of 'dog days' – a term Australians use to describe heat so intense that even dogs can do little but sprawl in the shade.

A mining company had brought me in to better understand the dingoes that lived alongside their operation. My job was to assess the population, build a clear picture of how the animals were living in and around the site, and listen to the miners' experiences and attitudes. Then it was to apply this knowledge to help the company develop a plan to manage them. In other words, to find ways for the dingoes to remain without being a nuisance and while keeping staff safe.

The dingoes had made the mine their home, thriving on its resources that included food scraps, water and the shelter of man-made structures such as pipes, buildings and large stores of truck tyres. Their silhouettes became a constant, greeting me at dawn, always watching from a distance.

One day, I set out to explore an area referred to as a topsoil dump. These are mounds of rich dirt the miners had scraped from the ground before they dug holes or built roads, to then use when they eventually rehabilitate the site. The mounds sat in uneven rows, a little like the moguls you see on a winter Olympic ski run. I was drawn by the faint hope of finding active dens. It was August, the season when dingoes have their litter and the signs were there: paw prints in the red dirt, abandoned dig sites and the occasional small hole burrowed into the mounds. Some of the older mounds were already covered with vegetation that had sprouted from the seeds mixed through the topsoil, and together they offered the perfect combination of stability and shade for raising pups.

As I walked through the scrub, the air shimmered with heat. Between the rustling of spinifex grass and my steady footsteps, I heard the unmistakable squeaks and whines of dingo pups, the contact calls young canids use to draw their mother's attention. Scanning the landscape, I suddenly spotted a dingo watching me. Then another appeared and soon six dingoes trailed behind me.

My pulse quickened, but I kept walking. They followed. When I stopped at a small clearing, so did they. Carefully, I placed my bag on the ground, then kneeled and eased out my camera, keeping my movements slow and calm. The dingoes circled me, their movements deliberate and measured. One lunged forward, barking sharply, while another crept behind me, nose already sniffing at my open bag. It felt like a test. A confusing mix of curiosity and caution. The only thing I did know, was the dingo behind me kept trying to sneak up to take something from

Fig. 15.1. A wild dingo from the Great Sandy Desert in Western Australia, interacting with the photographer, not unlike what I experienced at the mine site. Source: Gaz Meredith Images

my bag. They appeared to be taking their time trying to figure me out, assessing what I was, whether I was a threat and, perhaps, whether I was a risk worth taking (Fig. 15.1).

I stayed still, calm and non-threatening. The lead dingo barked again, its tail raised high and the hair on its back bristling. The others mirrored its movements, lunging and darting back and forth. It was intense, like staring down a pack of wild predators, because that's exactly what it was. Yet, as moments stretched into minutes, their energy began to change. Slowly, one by one, they lay down, watching me from a safe distance of around 3 metres.

I took the opportunity to snap a few photos, but after a while, I decided not to push my luck. They clearly didn't really want me around, no doubt wanting to protect the pups being housed nearby. So, I stood, slung my backpack over my shoulders and started walking back towards the gravel road. The dingoes followed, escorting me the way a security detail might walk a trouble-maker out of a nightclub, making sure I left the premises.

When I got back to the car, I felt a rush of relief and exhilaration. I had just shared a moment with these wild animals, a brief window into a world few people ever see. And though I felt safe, perhaps safer than I might around an unpredictable domestic dog, I couldn't ignore the reminder of how these encounters might feel to someone unprepared or unfamiliar with dingo behaviour. The experience gave me a new perspective on how miners, or anyone, in fact, might feel when dingoes approach them unexpectedly during work or leisure.

In most situations like this, dingoes behave not maliciously but calculating, acting on instinct. It was clear that conflicts often arise not because dingoes are inherently dangerous, but because humans misunderstand or mishandle these interactions. I knew what to do to de-escalate the situation, by remaining calm and non-threatening, and never felt it would escalate, but many don't. We often make little effort to understand their behaviour and, in panic, act in ways that only make things worse. There's also a degree of wilful ignorance at play. People are warned not to run around dingoes, yet they still go for jogs on the beach; they're told to stay close to their children, but let them wander out of reach. But more on that later.

The miners enjoyed seeing and interacting with dingoes in the mine, which was otherwise a lonely and remote place. Those who miss their pet dog at home, get to observe a dingo playing or hunting insects nearby while they work or are relaxing. Often, however, the dingoes became a bother, stealing food, chasing people on bikes, chewing up work boots or damaging equipment. Some of the stories you hear are quite humorous, others come at significant costs, especially if interrupting production. To management in city offices far removed from the desert, the dingoes were a liability to be controlled through regular culls. Pest control companies were employed to carry out culls to reduce the population. Over the 2 years I visited the site, they had killed over 100 dingoes. To the senior managers, this seemed necessary. To me, it felt like the easy way out.

This is where the heart of the conflict lies. Dingoes are tolerated as long as they stay out of sight, out of mind and out of trouble. But the moment they step into our world, we struggle to coexist. They are persecuted simply for being dingoes, for doing what any predator would do in a human-dominated landscape where our actions are what attract and keep them there in the first place. We only seem to tolerate nature when it is on our terms: predictable, controlled and compliant.

HISTORY OF HUMAN–DINGO INTERACTIONS

The relationship between humans and dingoes in Australia has undergone several dramatic shifts since European settlement. In the early colonial period of the late 18th and 19th centuries, dingoes were widely regarded as dangerous predators capable of attacking and killing humans, particularly children. This 'deadly dingo' trope emerged from both direct experiences and colonial folklore.

Early newspaper accounts from the *Sydney Gazette* in 1804 document dingoes brazenly entering settlements, killing poultry and domestic dogs. Settlers reported dingoes 'swarming round Sydney' and forcing all animals to be stabled at night to prevent predation. By the mid-19th century, a common belief had taken hold that dingoes would opportunistically attack vulnerable humans, especially lost children, intoxicated adults and lone travellers in the bush.

This perception was reinforced through numerous reported incidents. Adam Brumm, an archaeologist from Griffith University and expert in the people and animals of the Wallacean archipelago (South-East Asia), reviewed historical documentation. He found that between 1804 and 1928, at least 52 dingo attacks were reported in colonial newspapers and records, including 28 alleged fatalities (examples presented in Fig. 15.2). The veracity of individual cases

Fig. 15.2. Early artistic depictions of dingoes attacking humans, as discussed by Brumm (2022). Some illustrations draw on reported historical events, while others are purely fictional. (Top left) A pair of dingoes 'treeing' a woman, published in *The Bulletin* in 1884 and based on a contemporary account from the Dawson River region, where a young female cattle drover was said to have climbed a tree to escape. (Top middle) An illustrated scene from the fictional tale *Warrigal the Wild Dingo*, showing a lost settler's child surrounded by dingoes. (Top right) An anonymous 1920 illustration of a dingo attacking a woman and child, accompanying a fictional story set in colonial New South Wales. (Bottom) Dingoes attacking a lone rider, reportedly in the Para River region in the mid-19th century. Source: remastered and reproduced by Brumm (2022).

is difficult to assess, given the fragmentary nature of historical records, but the frequency and consistency of reports suggest some basis in actual truth. Notably, reported attacks showed clear seasonal patterns coinciding with dingo breeding and whelping seasons, lending credibility to the accounts.

However, during the interwar period of the 1920s and 1930s, public attitudes shifted dramatically. The once-common belief that dingoes posed a direct threat to humans was largely replaced by a new narrative, that dingoes were timid creatures that would flee from even small children. Brumm argues this transformation was linked to several factors: the near elimination of dingoes from settled areas due to widespread poisoning, trapping and shooting campaigns; reduced human–dingo interactions in more urbanised regions; and selection pressures favouring warier dingoes in areas of intensive persecution. Broader cultural shifts also played a role as Australians began to romanticise rural life and its wildlife, reimagining the dingo as a shy inhabitant of the bush rather than a menace.

This perception of harmlessness, however, was shattered in 1980 with the disappearance of 9-week-old Azaria at Uluru (then Ayers Rock). During a family camping trip, Lindy Chamberlain made the horrifying discovery that her baby was missing from the family's tent. Upon discovery, she was famously heard to have cried, 'A dingo took my baby!', a claim met with widespread scepticism. At the time, many Australians found it hard to believe that a dingo could harm a human, let alone take a child. This cultural disbelief, paired with circumstantial evidence, led to Lindy Chamberlain's wrongful conviction for murder. Her exoneration years later, following the discovery of crucial evidence (namely the child's clothes found by a dingo den), providing evidence that a dingo was indeed responsible. This forced a dramatic reassessment of dingo behaviour and potential danger.

The Chamberlain case became a global phenomenon and a media spectacle – on the same scale that the OJ Simpson trial would a few years later. Headlines dominated for years and the tragedy immortalised in the 1988 film *A Cry in the Dark*, in which Meryl Streep portrayed Lindy Chamberlain, and Sam Neill the father. For better or worse, the case embedded the dingo in cultural consciousness, fuelling debates and cementing its controversial image as both predator and scapegoat.

In 2001, another tragedy unfolded on K'gari. Nine-year-old Clinton Gage was fatally attacked by dingoes while playing unsupervised with his friend near an area known as Waddy Point. The boys had wandered from their family on a bush track when a pair of dingoes approached. Clinton's friend managed to escape and raise the alarm, but Clinton sustained fatal injuries (according to later accounts, a bite to his inner leg punctured a major artery, and he died from rapid blood loss before help could arrive). This marked only the second human fatality widely attributed to dingoes since European settlement. The other being the death of Azaria Chamberlain in 1980, which a coroner ultimately ruled was caused by a dingo, although the exact remain uncertain.

The attack reignited public concern about dingo-related risks, but it also highlights their rarity. Despite the millions of tourists who visit dingo-inhabited areas each year, fatal attacks are extraordinarily uncommon. Essentially non existent. The far majority of human–dingo encounters there and elsewhere end without incident.

UNDERSTANDING MODERN CONFLICTS

Although human–dingo conflicts occur across Australia (there are a few urban and regional towns where the occasional dingo takes domestic dogs or chickens), K'gari has emerged as the epicentre in recent decades. Incidents on K'gari dominate media reports of dingo (known as wongari to the Buthchulla people) 'attacks', with the island consistently associated with the highest number of documented negative interactions. With over 400,000 visitors annually to this relatively small World Heritage-listed island, only 123 km long, the potential for conflict is heightened. Of the many thousands of interactions between humans and K'gari's estimated 120 dingoes that live on the island, are benign. On average, only about 10 serious incidents are reported each year. These range from threatening behaviour, such as loitering and nipping, to rare cases where someone (often a child) is seriously bitten or injured. Regardless of the severity of the injury or how overblown the reporting may be, such incidents are terrifying for anyone.

Tourism on K'gari has long been a double-edged sword for dingoes. Being able to see these iconic animals in the wild attracts hordes of visitors to the island each year, but the behaviour of tourists often exacerbates dingo–human conflicts. Many visitors treat dingoes as photo opportunities rather than wild predators, creating unsafe situations that endanger both themselves and the animals. As one Butchulla community member observed, 'You see one poor dingo coming up the beach and you'll see them all pull-up and pile out of their cars, like have 80 people all around the dingoes. It's going to get scared or something, and I reckon they'll end up biting someone, but they'll get their photo.' This highlights a disconnect, where human entertainment often takes precedence over safety and respect for wildlife.

The consequences of such behaviour extend beyond individual incidents. Over time, this normalisation of close human–dingo contact undermines conservation efforts and erodes the wild nature of the dingoes themselves. These encounters condition dingoes to associate humans with food, thereby encouraging dingoes to seek out humans and exacerbating risk of aggression or harm. For the Butchulla people, this behaviour also disrespects the cultural and ecological significance of the dingo, reducing it to a commodity for tourist consumption.

Seasonal variation plays a crucial role in shaping dingo behaviour and influencing interactions with humans. There are two peaks across a year. The first occurs during the breeding season, from March to May. Hormones surge in young male dingoes, driving them to disperse as well as to test boundaries and explore their environment more boldly. The second occurs during the whelping season, from June to August. Here dingoes defend their dens and become more territorial and protective.

These seasonal changes in dingo behaviour also overlap with peak tourism periods, particularly during school holidays in March, April and July. Thus, the influx of visitors to K'gari during these times significantly increases the likelihood of encounters. Geographically, conflicts are disproportionately concentrated along the eastern beaches near popular tourist hubs like Eurong and Hook Point. In these areas, dingo packs, accustomed to human presence, are often responsible for the majority of serious incidents.

Children are particularly vulnerable in dingo–human conflicts. Analysis I have conducted on incident reports over decades with Rob Appleby (2017a) shows that the risk of serious outcomes increases significantly when children are unsupervised or stray more than a few metres from adults during interactions. Both of the recorded fatalities in Australia involved children, and in both case the children did not have an adult nearby. On K'gari, unsupervised children are occasionally bitten or chased and these incidents typically escalate when children run or scream, triggering the dingo's natural predatory instincts. In contrast, adults are rarely the target of serious dingo aggression. Adult visitors occasionally report being nipped or lunged at, but these incidents often occur when people engage in risky behaviours, such as approaching dingoes for photos, feeding them or failing to secure food and waste properly. Of course, incidents also occur where dingoes bother people and steal their belongings (such as shoes) simply because they are curious. And in some cases still, I look at the video of a dingo nipping an adult and I have no clue as to what led to it happening.

Food accessibility remains one of the primary drivers of dingo–human conflict. Dingoes that associate humans with food, whether from deliberate feeding, unsecured rubbish or fishing scraps, are incentivised to approach humans and spend time in human-use areas. As there are rarely negative consequences, dingoes learn to approach tourists or campsites with alarming and growing confidence

Linda Behrendorff, a Senior Ranger with the Queensland Parks and Wildlife Service (QPWS), has worked tirelessly for decades to manage both the dingoes and the conflicts that arise, though, ultimately, this means managing people. She has instigated many strategies and had lots of successes, but emphasises that 'food conditioning is the single biggest factor influencing negative dingo–human interactions on the island'. Even inadvertent feeding, such as leaving food scraps behind or failing to secure rubbish, creates a cycle where dingoes begin to associate humans with easy resources. This association not only alters their natural behaviour, but also sets the scene for potential serious conflict.

Despite ongoing warnings, feeding remains a persistent issue, driven by three main motivations. Some people feed dingoes out of misplaced concern, believing they are starving. Others do it for personal gratification, to capture close-up photos or experience an interaction. The third and most common form is accidental: leaving food waste unsecured, dropping scraps or discarding fish remains. Regardless of intent, the outcome is the same. Dingoes learn that humans are a reliable source of food, reinforcing behaviours that increase the risk of negative encounters. Yet, research has consistently shown that K'gari's dingoes are not struggling to find food. A comprehensive diet study conducted by Behrendorff found that dingoes have access to a diverse range of natural food sources, from insects and bandicoots to fish and even stranded whales. So the issue isn't food scarcity. By feeding dingoes, whether out of concern, for the thrill or by accident, people unwittingly contribute to patterns that place both dingoes and humans at greater risk.

Interestingly, not all dingoes are equally likely to engage in conflict. The same biological and hormonal forces that shape seasonal patterns of conflict also influence which individuals are most often involved. Subadult males, especially juveniles transitioning to independence, are disproportionately involved in serious interactions. These individuals are naturally more curious, energetic and willing to take risks. During the breeding season, male dingoes of all ages tend to display heightened territoriality, which can manifest as behaviours like lunging, circling or barking when humans enter their perceived space. Many of these behaviours, often interpreted as aggression, may actually reflect curiosity, boundary testing or even play, particularly among younger dingoes. However, humans frequently misunderstand these signals and may escalate the situation by running, yelling or attempting to scare the dingoes away, actions that can inadvertently provoke the animals further.

A significant contributing factor to these conflicts consists in knowledge, attitudes and perceptions of dingoes. Surveys conducted on the island over the years reveal a consistent pattern that many visitors and residents fail to respect them as wild predators. The dingoes' dog-like appearance fosters a false sense of security and many people feel no fear when encountering them. Unlike tourists on safari in Africa, who are acutely aware of the dangers posed by lions or leopards, visitors to K'gari often underestimate the risks of interacting with dingoes. This lack of respect or caution frequently leads to behaviour such as approaching, feeding or attempting to pet dingoes, actions that increase the likelihood of a negative encounter.

This combination of factors, seasonal variation, vulnerable groups, food accessibility and human perceptions, creates a complex web of interactions on K'gari. At the heart of these conflicts is the need to understand what dingoes intend when they bark, lunge or approach. Insights that could hold the key to reducing conflict.

DECODING DINGO BEHAVIOUR

Picture yourself walking along the picturesque white sands on the eastern beach of K'gari, camera in hand. You spot a dingo trotting along the tideline. It notices you too and begins moving in your direction, not directly, but in a loose arc. Its ears are pricked forward, tail raised with a slight curve. As it gets closer, its movements become more deliberate, leaping side to side, dipping its head and shifting its weight with an almost exaggerated precision. It even gives a little bark. You wonder what this dingo is trying to tell you. It is hard to tell whether it is being playful, aggressive or something else entirely.

Fig. 15.3. Screenshots captured by Rob Appleby during his research on dingo–human interactions on K'gari. These images depict typical dingo behaviours, highlighting their complex and nuanced interactions with humans. Source: Bradley Smith.

This scene plays out hundreds of times each year on K'gari and interpreting these encounters remains one of the greatest challenges in managing human–dingo interactions. Even with domestic dogs, animals we've lived alongside for millennia, misunderstandings of their signals are common. Wild dingoes, however, speak an even subtler language, one that is more ambiguous and far less familiar. Scientists are only beginning to decode the complex behaviours of these wild canids, uncovering nuances that challenge long-held assumptions about their intentions.

'The complexity of dingo behaviour during these interactions is remarkable', explains Rob Appleby, who spent his PhD studying and analysing dingo–human encounters on K'gari. 'We're seeing behaviours that don't fit neatly into traditional categories of aggression or submission (Fig. 15.3). Sometimes, it feels like the dingoes themselves are unsure how to engage with us.'

One of the most surprising insights from the work by Appleby and me has been the role of play in dingo–human interactions. Far from being the antagonists they are often portrayed as, many dingoes, especially juveniles, engage in encounters out of curiosity or a desire for social play. Exaggerated movements, bouncy approaches and even the manipulation of objects like sticks or seedpods mimic the way young dingoes play with packmates. This playful behaviour, while fascinating, can easily be, and often is, misinterpreted by humans as threatening. Conversely, truly aggressive cues, such as stiff body posture, direct staring, raised hackles and low growling, are sometimes missed by untrained eyes, leading to dangerous situations.

This ambiguity is both fascinating and fraught with consequences. Dingoes often exhibit mixed signals, approaching confidently, only to suddenly avert their gaze or lick their lips, a known calming signal in canid communication. They might offer a play bow, the universal invitation to play, only to retreat and growl softly moments later. These behaviours suggest that dingoes, much like humans, are navigating the unspoken rules of cross-species encounters and they are perhaps as confused by us as we are by them. Unfortunately, how we interpret and respond to these cues often dictates the outcome of such interactions.

Sometimes, dingo motivations during interactions remain unclear. In 2023, for instance, footage of a French tourist sunbathing on K'gari among a group of young holidaymakers made global news. In the video you can see three dingoes casually roaming the beach area, sniffing people and their belongings. One of the dingoes approached the tourist, sniffing her as she lay on the sand. When she realised what was happening, she instinctively got up and lurched forward to move away from the dingo. This sudden movement led to the dingo nipping her on the bottom. Such encounters are sometimes labelled as 'testing', where dingoes are testing their prey to instigate a reaction. However, the incident did not look like predatory behaviour, in the sense that the dingo was not pursuing her as prey and did not continue to bite or pursue. It appeared to be a reactive bite or nip, something akin to 'if in doubt lunge out', as sometimes striking first can be the best defensive strategy. The tourist herself, handled the incident with remarkable understanding and grace. 'The dingoes shouldn't have to change their behaviour because I'm there – I should change my behaviour because they're there.'

Such incidents highlight the importance of understanding and respecting dingoes as wild animals. As one Butchulla K'gari ranger, Tessa Waia put it, 'I know people are on holiday, so

Fig. 15.4. A dingo pursuing children on K'gari, as their father rushes to intervene. The scene illustrates the dangers of leaving children unsupervised in areas where dingoes are present. Source: Bradley Smith.

they're in holiday mode. But keep in mind, this place is the wongari's [dingo] country.' Visitors must act responsibly, respecting both the dingoes' habitat and their wild nature to ensure safety for all.

As we have seen, children are overrepresented in serious incidents involving dingoes. Their small size, high-pitched voices and tendency to run or squeal, traits that mimic aspects of the prey dingoes would naturally hunt, make them especially vulnerable. One such incident occurred while Rob and I were observing a dingo on the beach (Fig. 15.4). A tourist bus had stopped nearby, allowing visitors to get out and enjoy their coffee while watching the animal. Two young children wandered away from the group, unaware of the danger and the dingo, previously indifferent to the crowd, immediately began sprinting towards them from almost 50 metres away. Rob and I, usually passive observers, had no choice but to intervene, shouting to alert the parents. Thankfully, the father acted quickly, scooping up his youngest child and shooing the dingo away. He did so nonchalantly, never letting go of his half-filled coffee cup. This encounter ended without injury, but this is not always the case. Media reports describing such events as 'dingoes hunting small children' may seem sensational, but they are not entirely unfounded. The risks are real and highlight the importance of keeping children supervised – or perhaps reconsidering whether islands with wild predators are suitable family destinations.

The stakes in decoding dingo behaviour are high. Misinterpreting playful or investigative behaviour as aggression often results in dingoes being labelled as dangerous and subjected to lethal control measures. Conversely, failing to recognise legitimate warning signs puts people at serious risk. Dingoes on K'gari that are deemed an unmanageable risk to human safety will often be killed. Although this policy aims to ensure public safety, and this is a classic example of a 'wicked problem' (with competing values, high stakes and no simple solution), at the end of the day it is most often the dingoes that pay the highest price.

MEDIA AND PUBLIC PERCEPTION

Language shapes perception and the media's choice of words profoundly influences how the public views dingoes. Consider the earlier incident involving the French tourist. Despite it being a minor interaction resulting in a small mark, the media sensationalised the story with headlines like 'French Tourist Attacked by Dingo'. This characterisation was far from reality. If dingoes were truly hunting or attacking humans, the injuries would be severe, even catastrophic. Such headlines unfairly and unnecessarily evoke fear and drama, offering clickable content to sell newspapers or drive online engagement. However, they also perpetuate the narrative of dingoes as dangerous predators while ignoring the critical context of the encounter.

The language of conflict dominates media coverage, overshadowing the more mundane reality of coexistence. Most of interactions between people and dingoes are benign or even positive. Many visitors to K'gari hope to see a dingo in its natural environment and the animal is a central feature of the island's tourism marketing. Dingoes navigating human presence is a daily occurrence and they most often do so relatively peacefully, yet these moments rarely make headlines. Positive encounters don't sell newspapers.

Instead, even minor incidents are treated as proof of danger, reinforcing outdated stereotypes. Every incident that occurs makes the news, often nationally. Especially on a slow news day. Yet the contrast with domestic dogs is striking. Each year, over 10,000 Australians seek hospital treatment for dog attacks, many of which result in serious injuries or fatalities. Postal workers alone report an average of 10 negative dog encounters daily, according to Australia Post. The risk of harm from a domestic dog in Australia far outweighs the risk of encountering a dangerous dingo on K'gari or anywhere else.

This careless framing has real consequences. By portraying curiosity or boundary-testing as aggression, the media fuels fear and strengthens calls for extreme measures. Shifting this narrative requires more than factual accuracy; it demands language that reflects the complexities of dingo behaviour and acknowledges the shared space humans and dingoes occupy. It also requires a cultural shift where humans take responsibility for their actions and their dominant role in these conflicts.

A shifting narrative

When I first visited K'gari to assist Appleby with his research several decades ago, dingoes almost always bore the brunt of blame for conflicts with humans. Media reports routinely described calls for culls and arguments that dingoes didn't belong on the island, as people take precedence.

The emergence of social media has amplified the impact of negative dingo stories, with platforms like Facebook and Instagram spreading stories rapidly. Emotive language and graphic imagery heighten public fear and polarised comment sections reveal competing perspectives: some demand lethal control, while others advocate for coexistence. Along with undergraduate research students (thus far unpublished), we have analysed social media commentary and found that fear-based narratives increase public support for extreme management measures like culling, often ignoring the ecological importance of dingoes and the human behaviours that contribute to conflicts.

However, the past decade has also seen a notable and encouraging shift. The spotlight is increasingly turning towards the recognition that it is human behaviour most in need of management. Feeding dingoes, leaving food scraps and failing to supervise children are well-established factors in serious incidents, something that rarely made headlines in the past. This reframing is important. Only by acknowledging our shared responsibility in human–dingo interactions can we reduce conflict and ensure the survival of this iconic species.

I believe that education campaigns have been pivotal in driving this change, teaching visitors to secure food, avoid feeding dingoes and respect their habitats. The messaging increasingly emphasises that coexistence requires more effort and responsibility on the part of humans than it does from dingoes. This shift in perception, though gradual, marks a crucial step forward in balancing safety, respect and conservation.

MANAGEMENT APPROACHES

Managing dingo–human conflict is a delicate balancing act. On one side, there is a need to protect people in high-interaction areas such as tourist and mining sites. On the other, dingoes play a vital ecological and cultural role and their persecution risks disrupting ecosystems and endangering populations. Over the years, various management strategies have been employed, each with its own challenges, degree of effectiveness and ethical implications.

After the fatal attack on K'gari in 2001, for example, 31 dingoes were immediately killed by QPWS rangers, regardless of their involvement in the incident. Today, policy allows for individual dingoes to be destroyed only in cases of repeated high-risk interactions or serious injury. Generally, research shows that lethal control does little to reduce long-term conflict. Removing dingoes often disrupts pack dynamics, leading to increased conflict as younger, inexperienced individuals take over territories. More importantly, it fails to address underlying causes, such as food conditioning, meaning the problems continually occur.

As we discussed, records from decades of data collected by the park rangers reveal that juvenile dingoes are disproportionately involved in incidents. Yet, as these 'teenage' dingoes mature, it is reasonable to expect that they'd typically settle into stable pack roles and cause fewer conflicts. This underscores the importance of patience and non-lethal management during their developmental stages.

Non-lethal approaches are increasingly recognised as effective alternatives to culling, promoting coexistence while maintaining ecological balance. Appleby has been exploring aversive conditioning methods on the island to mitigate or prevent conflicts entirely. In learning theory, 'aversive' doesn't mean harmful or cruel, it simply refers to adding something unpleasant to discourage a behaviour. Techniques such as umbrellas, water pistols, air horns and animal-borne shock collars have shown promise in deterring dingoes from approaching humans. The results are sometimes mixed due to logistical and technical challenges, but these methods demonstrate the potential to modify dingo behaviour without resorting to lethal means. On K'gari, visitor education programs emphasising proper behaviour, such as not feeding dingoes, supervising children and securing food waste, have significantly reduced serious incidents, though compliance remains a challenge. Between 2003 and 2012, over 450 penalty notices were issued for breaches, including deliberate feeding, with fines ranging from hundreds to tens of thousands of dollars.

Physical measures, such as fences and designated 'dingo-free' zones, have also been implemented to reduce interactions in high-risk areas, particularly during the whelping season. These measures can be effective, but are not foolproof and require consistent maintenance, as dingoes often find ways to breach barriers or take advantage of open gates. Moreover, fencing can be contentious, as it may restrict people's access to areas they feel entitled to use or create a sense of exclusion. Despite these challenges, such strategies have contributed to a significant decline in the number of dingoes killed annually for management reasons, making culling the exception rather than the rule.

Public opinion plays a critical role in shaping dingo management policy. The concept of Social Licence to Operate is particularly relevant, as public support, or opposition, can significantly influence management approaches. Growing awareness of dingoes' ecological value and the inherent value of wildlife has led to increased resistance to culling and a preference for non-lethal strategies. However, balancing public safety with dingo conservation remains challenging. Missteps in policy or execution can easily exacerbate tensions, highlighting the need for evidence-based, adaptive management that prioritises both safety and ecological integrity.

HARMONY IN THE WILD

We cannot blame dingoes for behaviours we unwittingly encourage. Killing animals that fail to conform to our expectations is not the answer to resolving the conflict, especially when we are the ones who set the conditions for these behaviours in the first place. Rather than scapegoating dingoes for the consequences of our actions, whether it's leaving food unsecured, wandering too far from safety or underestimating their wildness, we must act responsibly. Simple measures like keeping children close, camping in secure locations, properly storing food and maintaining a respectful distance can prevent most conflicts. For families with small children, visiting a dingo-free destination may be a better option.

I am somewhat relieved to witness public attitudes and management strategies beginning to shift towards coexistence, reflecting a growing understanding of shared responsibility. It's a small but vital step towards a future where humans and dingoes can navigate their shared spaces with respect, safety and harmony. Because in the end, living peacefully with dingoes isn't about expecting dingoes to behave differently, it's about changing how we behave around them.

16
The mad red hour

High on a Warrumbungle scarp, under the stars' cold light,
A red dog lifted a questing cry and flung it to the night.
The startled silence woke and cried along the valley wall
And the broken uplands echoed the wild dog's lonely call.
And far below on the open plain, where the hand of man was set,
The tame dogs rattled their iron chains and knew an age-old fret.
The wild dog snarled on his mountain ledge at the smell on the mountain wind
That brought from the sleeping valley floor a challenge to his kind:
Tang of sheep and milch cow's breath and the hostile scent of man,
Swelling up in the darkness there across the valley's span.
Then down from the Warrumbungle height, a shadow on the hill,
With ears alert and eyes alight the dingo sped to the kill.
By straggly oak and tumbled rocks, running the ridges down,
He dropped in the dark to the last long slope under the mountain's frown;
And things of the night that saw him pass, owls and their silent race,
Froze on the limb or under the rock and held them still a space.
Saliva dropped from his open jaws and his whipped tail made a sweep
As, fresh and strong, on the night there came the urging smell of sheep.
Oh, hot the blood and soft the flesh, as, ewe by ewe, they died.
They fell together, the foolish things; he struck from side to side.
Under the cold, uncaring stars, king for a mad, red hour,
Heart ablaze with the primal flame – the ecstasy of power,
The warrigal raged till, spent at last, all bloody from the kill,
He ate his fill on the dreadful plain, then crept to the waiting hill.
Deep in a Warrumbungle hole, replete and warm, asleep,
The dingo lay as dawn light came to the field of mangled sheep;
Deaf to the curses far below, blind to the hatred there,
Mild and soft in his winter coat, safe in his secret lair.
Fair is the day on the mountain spur, sapphire, purple and brown,
But night is the time of the dingo where the Warrumbungles frown.

THE SCENE IMMORTALISED IN CHARLES SHAW'S POEM *THE DINGO* PUBLISHED in a 1946 edition of *The Bulletin*, transports us to the rugged Warrumbungle mountains of New South Wales, where wilderness converges with farmland (also neatly captured in Fig. 16.1). The poem evokes a chilling image of a primal predator unleashing devastation in the stillness of the night. It is a moment of chaos that represents both survival and destruction.

As dawn breaks, the poetry of the night transforms into harsh reality. It is a sight that has haunted graziers for generations: sheep lying lifeless, their bodies scattered across the paddock, many barely touched by their predator.

Observations of dingoes appearing to kill beyond what they need to survive date back to the earliest days of European settlement (Fig. 16.2). In 1839 for instance, Charles Hamilton Smith described the dingo's pleasure in killing, writing: 'When they attack sheep, their delight

Fig. 16.1. A historical depiction of dingoes howling near a shepherd's hut – a scene that captures the dread felt by early settlers whose livelihoods depended on keeping sheep safe in wild country. To hear a howl in the dark was to feel the edge of the untamed pressing in. Titled 'Dingoes prowling round the sheepfold, Australia from Dr Doyle's sketch book, ca. 1862–1863' by John Thomas Doyle and Samuel Thomas Gill. Source: National Library of Australia.

is to kill as many as they can overtake; and their bite is so severe that few who are wounded recover.'

For farmers, it isn't just the killing itself that stings, it is the sheer scale of destruction. In 2023, Brad Page, principal biosecurity officer for pest animals for the South Australian government, described to the ABC media what this behaviour looks like in practice. He noted that it was not unusual to see a dingo, or pair of dingoes, maul 20–30 sheep in one event. 'The wild dogs aren't actually chasing sheep to eat them', he explained. 'They're chasing the sheep and they'll kill them, but then they'll look up and there are more sheep from the flock running away, and the wild dog will leave the one that they've either half-killed or completely killed, and they'll move on to the next one and then the next one.'

Losing one or two sheep might be an expected cost of coexisting with predators, but the sight of an entire flock slaughtered in a single night understandably sparks outrage. This frustration was captured in 1956 by WP Bluett in the *Canberra Times*: 'It is not the sheep he kills for his two-day meal that worries the grazier so much as those he kills for the fun of it. The dog will come on a sheep camp in the early dawn, and before the muddled sheep know, he is among them, he has killed half-a-dozen.' Such accounts of dingoes wreaking havoc, often described as acts of chaos or cruelty, have all but cemented their reputation as ruthless villains of the bush.

While stories of mass slaughter circulated widely through rural communities, the first systematic documentation came through Neil Macintosh's mid-20th-century survey of

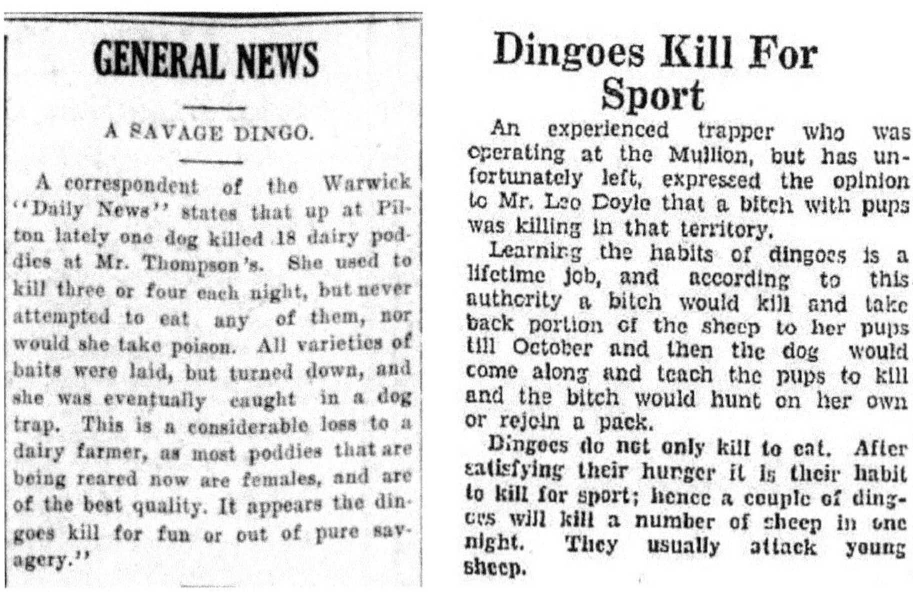

GENERAL NEWS

A SAVAGE DINGO.

A correspondent of the Warwick "Daily News" states that up at Pilton lately one dog killed 18 dairy poddies at Mr. Thompson's. She used to kill three or four each night, but never attempted to eat any of them, nor would she take poison. All varieties of baits were laid, but turned down, and she was eventually caught in a dog trap. This is a considerable loss to a dairy farmer, as most poddies that are being reared now are females, and are of the best quality. It appears the dingoes kill for fun or out of pure savagery."

Dingoes Kill For Sport

An experienced trapper who was operating at the Mullion, but has unfortunately left, expressed the opinion to Mr. Leo Doyle that a bitch with pups was killing in that territory.

Learning the habits of dingoes is a lifetime job, and according to this authority a bitch would kill and take back portion of the sheep to her pups till October and then the dog would come along and teach the pups to kill and the bitch would hunt on her own or rejoin a pack.

Dingoes do not only kill to eat. After satisfying their hunger it is their habit to kill for sport; hence a couple of dingoes will kill a number of sheep in one night. They usually attack young sheep.

Fig. 16.2. Historical newspaper reports of dingoes killing for sport and pleasure. (Left) *Western Star and Roma Advertiser* (QLD), Saturday 19 June 1926. (Right) *Tumut and Adelong Times* (NSW), Tuesday 24 August,1948, page 6. The narrative has remained unchanged. Source: National Library of Australia (Trove).

Australian graziers in relation to dingo behaviour.[1] The results were striking in their consistency. Among the 144 participants, an overwhelming 90% affirmed they had witnessed dingoes killing for sport. The accounts that poured in ranged from occasional incidents to dramatic claims. One Queensland farmer reported having witnessed these killing sprees 'thousands of times', while another documented a case of a single dingo killing 80 sheep in one night.

Yet as these stories accumulated, distinct patterns began to emerge. The behaviour was most frequently attributed to young or inexperienced dingoes, often during the period when adults were teaching pups to hunt. Environmental stressors played a role too, reports peaked during times of drought or in the aftermath of heavy rains, when prey animals were particularly vulnerable. These patterns suggested something more complex than simple bloodlust was at work.

These tales have persisted for centuries, passed down through farming communities and reinforced by repetition. Despite decades of observations, few pause to question whether there might be more to this story than simple bloodlust or sport. It makes you wonder what drives this behaviour. Do dingoes, or any predator that must kill to live, ever kill for reasons other than necessity?

Intrigued by this question, I teamed up with Rob Appleby to dig deeper into this complex topic. Combining his expertise in ecology with my background in psychology, we set out to unravel whether the dingo's reputation as a 'sport killer' is grounded in fact or myth. Our investigation, published in 2018, led us through decades of scientific literature, field studies and behavioural research on dingoes and canids from around the world.

WHAT IS SURPLUS KILLING?

To begin our investigation, we first needed to define exactly what we were looking at. The mass killing of prey that farmers describe has a specific name in the scientific literature: surplus killing. Zoologist Hans Kruuk coined this term in 1972, defining it as situations when a predator kills more prey than it or its social group can immediately consume, despite having free access to the carcasses and the prey being a species the predator normally eats.

Importantly, surplus killing isn't unique to dingoes or even to canids. It's been observed in a wide range of predators, from tiny zooplankton all the way up to large carnivores like bears. This suggests that rather than being an aberration, surplus killing might serve some evolutionary purpose.

To understand why predators might engage in such behaviour, it is necessary to examine the evolutionary, neurological and behavioural mechanisms that drive predatory behaviour.

THE CHALLENGES OF HUNTING

Scientific studies have shed light on just how difficult it is for dingoes to secure a meal. Far from being unstoppable forces of nature, these predators face significant challenges in their hunting efforts.

---

1 The outcomes of Macintosh's survey were never published, but the original data is available from the archives of the Shellshear Museum (University of Sydney). In 2018, Lily van Eeden and I published other outcomes from the survey along with a modern comparison.

One particularly illuminating study comes from Peter Thomson, who conducted extensive observations of dingo hunting behaviour in the Fortescue River region of north-western Australia. Thomson's research provides concrete data that illustrates the challenges dingoes face.

The most striking finding was the low success rate of dingo hunts. Of 272 observed chases of kangaroos by dingoes, only 25 were successful. This translates to a mere 9.2% success rate. In other words, for every 10 hunting attempts on kangaroos, dingoes failed more than 9 times. This statistic alone paints a picture of hunting as a difficult and often fruitless endeavour.

The time and energy investment required for hunting is substantial. Dingoes may spend several hours travelling just to locate potential prey. Even after finding prey, there's no guarantee of a successful capture. This represents a significant investment of time and energy, with no assurance of a return.

Wild prey animals have evolved various defensive strategies that make them formidable opponents. Kangaroos, for instance, are not only fast but also agile. They can use their powerful hindlegs to deliver potentially lethal kicks to pursuing dingoes. This makes every hunt a potentially dangerous affair for the dingo (Fig. 16.3).

Indeed, every hunt carries the risk of injury for the dingo. A well-placed kick from a kangaroo, for example, could seriously injure or even kill a dingo. Such injuries could compromise the dingo's ability to hunt in the future, potentially leading to starvation.

Fig. 16.3. A dingo standing off against a kangaroo. Source: photo by Charles Davis.

Recent advancements in technology have allowed researchers to witness dingo hunting strategies in unprecedented detail. Wildlife cinematographer Daniel Hunter captured remarkable drone footage of dingoes hunting in Namadgi National Park, near Canberra. This footage, later analysed by Tahlia Pollock and colleagues, provided some of the first published video records of wild dingoes in pursuit of eastern grey kangaroos. The drones revealed a diverse hunting repertoire, including cooperative pack strategies where dingoes flushed prey from cover, ambush tactics where they waited for prey to approach and even the use of landscape features like rocky outcrops to slow down fleeing kangaroos. The footage also captured a mother dingo teaching her pups to hunt. These observations, reminiscent of wolf hunting behaviours, reinforce the idea that dingoes are strategic, adaptable predators.

Even when dingoes manage to catch up to their prey, kangaroos can still escape. Peter observed instances where kangaroos successfully outmanoeuvred dingoes, using their agility to evade capture at the last moment. Other researchers, such as Brad Purcell (2010), have reported kangaroos retreating to water, ready to drown any pursuing dingo that follows them in.

These challenges hit home why hunting is such a serious business for dingoes. It's not a game or a sport, but a critical and dangerous activity necessary for survival. It serves as a powerful counterpoint to the notion of dingoes as ruthless, efficient killers – instead showing that their chances of survival are against the odds.

EVOLUTIONARY PERSPECTIVES

From an evolutionary standpoint, any behaviour that persists in a species must provide some advantage, or at least not be significantly detrimental. Surplus killing appears to be an adaptive strategy that allows predators to capitalise on rare opportunities. L. David Mech, a renowned wolf biologist, puts it this way: 'They kill as much as they can when killing is easy because such opportunities are rare, and if undisturbed, such surplus is entirely consumed.'

This makes evolutionary sense when we consider the challenges of hunting as we have just examined, when circumstances tip the balance in their favour, whether through drought weakening prey or deep snow hampering escape, predators must, as the saying goes, 'make hay while the sun shines'.

Research on wolves in North America supports this view: surplus killing events were most common during harsh winters when prey was weakened by deep snow and lack of food. In these conditions, the energy expended in hunting was far outweighed by the potential gain. The wolves would often return to feed on the carcasses over several days or cache parts of their kills for leaner times. These carcasses also served an important ecological function, providing food for scavengers and other wildlife when food is scarce.

This behaviour exists within a broader evolutionary context – a constant 'arms race' between predator and prey. As prey animals develop better strategies to avoid being eaten (like increased speed, better camouflage or more effective group defence behaviours), predators must evolve counter-strategies to keep up. The result is typically a delicate balance that keeps prey populations in check without wiping them out entirely. However, when this balance is disrupted, whether through environmental conditions or human interference, surplus killing can occur. As Philip

Lehner observed in his studies of coyotes, such behaviour is 'not the result of a "blood-thirsty killer," but an efficient predator which finds itself amidst an abundance of easy prey.'

WHEN OPPORTUNITY STRIKES

Most of the controversy around surplus killing by dingoes focuses on livestock predation, but it's crucial to understand that these are rare and exceptions to the normal predator–prey dynamics observed in the wild. Two well-documented cases stand out in understanding surplus killing by dingoes: The Kangaroo Slaughter at Olive Downs, and The Massacre on Swan Island.

In the summer of 1980, at a lonely watering hole in north-west New South Wales, a small group of dingoes carried out a puzzling killing spree. Over the course of just 7 weeks, five dingoes killed at least 83 red kangaroos, most of them juveniles. This was not a case of dingoes simply dispatching weak or sick animals, because detailed autopsies revealed no signs of disease or debilitation that might have hindered the kangaroos' ability to escape.

What made this case particularly notable was the nature of the predation. Despite the staggering number of kills, the dingoes only consumed portions of about half their victims. Some nights, they left carcasses untouched. Neil Shepherd, who documented this event, observed that the behaviour of both predator and prey was unusually relaxed. Large male kangaroos, typically wary, often ignored the dingoes entirely. Even smaller individuals, normally hypervigilant around predators, would sometimes pass within metres of a resting dingo without showing concern. But when the chase began, the tables turned. Shepherd described a scene in which a young kangaroo darted for safety, the dingo effortlessly keeping pace. At one point, the kangaroo veered towards the water, attempting to shake its pursuer. It almost worked, except that, in an instant, the dingo adjusted its stride, reached out and struck.

The answer to why the dingoes killed so many so quickly may lie in the convergence of conditions: a concentrated prey population drawn to a limited water source, young and vulnerable kangaroos still learning the subtleties of evasion and dingoes acting not just out of hunger, but perhaps from deeply ingrained predatory instincts.

In another extraordinary case, surplus killing took place on the water. In 1983, Alan Newsome, Laurie Corbett and their CSIRO colleagues such as Peter Catling documented an event on Swans Island in the Northern Territory, where dingoes launched an assault on a vulnerable population of black swans.

At the time, the swans were undergoing their annual moult, a natural process during which they shed their flight feathers and become temporarily grounded. With no means of escape, they were sitting targets. The dingoes took full advantage. They waded into the shallows, striking down the defenceless birds one after another. By the time it was over, the shoreline was littered with dead and dying swans. Yet, as in the kangaroo case, the dingoes only ate a fraction of their kills.

The explanation is likely the same. An unusual abundance of easy prey and low risk of harm, disrupted the usual balance of predator and prey dynamics. For a brief moment, the normal constraints on predation, difficulty in catching prey and the energy cost of hunting, no longer applied. The dingoes were simply acting on instinct, making the most of an unprecedented opportunity.

BEHAVIOURAL PERSPECTIVES

Predatory behaviour emerges from a sophisticated interplay of innate patterns, learned responses and cognition.

The concept of fixed action patterns, as described by ethologists, offers an intriguing framework for understanding how evolutionary pressures have shaped the very mechanics of predation. These patterns help explain both the precision of normal hunting behaviour and what happens when predators encounter situations beyond their evolutionary experience.

According to this view, hunting behaviour in canids can be broken down into a series of stereotyped, intrinsic motor patterns. Ray Coppinger and Feinstein (2004), prominent canine experts, identify five basic steps in this sequence.

1. Orient: the predator detects and orients towards the prey.
2. Eye: the predator remains motionless, staring fixedly at the prey.
3. Stalk: with its body lowered and head down, the predator moves slowly towards the prey.
4. Chase: the predator transitions into full-speed pursuit.
5. Bite: includes a grabbing bite to disable the prey and a killing bite to dispatch it.

This predatory sequence is not exclusive to wild canids; it has been selectively preserved, modified or suppressed in domestic dog breeds. For example, border collies are known for exhibiting the orient, eye, stalk and chase steps, but they stop short of the bite, allowing them to herd livestock without causing harm. In contrast, livestock guardian dogs have been bred to suppress the entire sequence, prioritising protection over predation.

Coppinger argues that this predatory sequence is intrinsic, automatic and requires little learning (effectively hardwired into their brains), with each step naturally motivating the next. If the sequence is interrupted, the predator may be unable to complete the kill. This perspective offers a potential explanation for surplus killing. If the sequence is a fixed action pattern, triggered by the presence of prey and driven by intrinsic motivation rather than hunger, it might continue even after the predator's immediate nutritional needs are met.

This model provides a compelling framework, but may oversimplify the complexities of real-world hunting behaviour. Critics note that it doesn't fully account for the flexibility and adaptability observed in predators. Hunting, after all, is not a rigid process, as it often involves higher-order mental processes such as planning, learning and responding to environmental cues. That is, hunting behaviour often combines innate drives with learned strategies, allowing predators to adapt to their environment.

Wildlife ecologist Daniel MacNulty and his colleagues demonstrated this complexity through their detailed study of wolves in Yellowstone National Park. After spending over 2,000 hours observing wolf hunting behaviour, they identified six distinct phases in the hunting process,

Table 16.1. Proposed ethogram of the behaviour of wolves hunting ungulate prey

Source: MacNulty *et al.* (2007).

Foraging state	Definition
Search	Travelling without fixating on and moving towards prey
Approach	Fixating on and travelling towards prey
Watch	Fixating on prey while not travelling (e.g. standing, sitting, crouching)
Attack-group	Running after a fleeing group or lunging at a standing group while glancing about at different group members (e.g. scanning)
Attack-individual	Running after or lunging at a solitary individual or a single member of a group while ignoring all other group members
Capture	Biting and restraining prey

from initially searching for prey through to the final capture (Table 16.1). Their observations showed how flexible hunting strategies can be, with predators adjusting their behaviour based on many factors: the type of prey they're hunting, how vulnerable that prey is, whether they're hunting alone or in a group and even their own energy levels and motivation. This adaptability is particularly relevant to surplus killing, where predators may exploit a sudden opportunity or continue killing in response to specific cues, rather than stopping when sated (feeling full).

Planning and strategy

Even though behavioural patterns provide a framework for understanding hunting mechanics, they raise questions about the role of cognition. Here we encounter a long-standing debate in animal cognition relating to how much thought goes into a predator's hunting strategy. Are dingoes simply following instinctive patterns as Coppinger would argue, or are they capable of more complex planning and problem-solving?

Beyond instinct and cognition lies a third powerful force shaping predatory behaviour: learning through experience. Like many intelligent animals, dingoes refine their hunting skills over time, testing strategies, adjusting to outcomes, and gradually improving through trial-and-error learning.

For instance, when a young dingo chases a kangaroo and fails, the failed attempt becomes part of its internal archive of experience. If it later tries a different tactic, such as approaching from a new angle or using terrain to its advantage, and succeeds, the outcome functions as natural reinforcement. Over time, the dingo becomes more efficient, not simply because of instinct or intelligence, but through feedback from the environment.

This form of learning allows dingoes to tailor their approach depending on the prey, the landscape and even the social dynamics within the pack. It may also help explain some of the more innovative hunting strategies observed in the wild. Strategies that go beyond what fixed action patterns would predict, yet don't necessarily require advanced planning or foresight. They are simply learned solutions to recurring problems.

But even though some researchers argue that hunting is largely hard-wired (a behavioural 'rule book', as Coppinger put it), field observations of dingoes using sophisticated strategies suggest their abilities extend beyond fixed instinct. Consider the remarkable observations made by Linda Behrendorff on K'gari. In 2018 she documented dingoes using water as a hunting tool. In one case, a dingo was seen catching an echidna, carrying it to the ocean, digging a shallow hole in the wet sand and drowning it over the course of an hour before eating it. This type of problem-solving demonstrates an understanding of both the prey's defences and how to overcome them.

Even more intriguing are observations from biologist Alan Newsome in the 1960s. He noticed that dingoes would selectively target female wallabies carrying joeys in their pouches. When the mothers were chased, the joeys would often be ejected, providing an easy meal. Newsome described this as a form of 'farming'; the dingoes seemed to recognise that pursuing mothers with young offered a better return on their hunting effort.

Together, these examples show that dingoes are not operating on instinct alone. Their hunting behaviour emerges from a mix of inherited patterns, experience and context. This makes them flexible enough to innovate, without necessarily requiring complex foresight.

The neuroscience of predation

Understanding the cognitive aspects of hunting naturally leads into the neuroscience behind these behaviours. At a neurological level, it is important to recognise that 'fun' is a subjective experience that's difficult to measure, even in humans. When we talk about animals experiencing pleasure, we're really talking about reward mechanisms in the brain.

To understand why dingoes might engage in surplus killing, we need to look at what's happening in their brains during hunting. In mammals, including dingoes, reward mechanisms involve complex networks of neurotransmitters, with dopamine playing a central role. This chemical messenger is associated with pleasure and motivation, but its role is more nuanced than simply signalling 'happiness'.

Recent neuroscience research has revealed three distinct components of reward processing: Liking (the actual pleasure or hedonic impact of a reward); Wanting (the motivation to pursue a reward); and Learning (the ability to form associations and predictions about rewards).

Importantly, dopamine isn't only released when an animal consumes food; it's also released during the pursuit. In other words, the act of hunting itself triggers reward pathways in the brain, independent of the eventual consumption of prey. This system evolved to keep animals motivated even when success isn't guaranteed.

The distinction between liking and wanting is particularly relevant to understanding surplus killing. These two components can operate independently: an animal might experience strong motivation to pursue prey (wanting) even when they're not hungry (liking). This dissociation helps explain why predators might continue hunting even after their immediate nutritional needs are met.

Furthermore, the prediction and anticipation of rewards can be even more powerful motivators than the rewards themselves. This phenomenon, known as reward prediction error, might help explain why dingoes might be especially triggered by situations where prey is unusually easy to catch – the unexpected success could enhance the reward response.

So, in a sense, we could say that aspects of hunting are likely pleasurable for dingoes. The chase, the capture, the kill; all of these might trigger reward pathways in the dingo's brain. But it should not be confused with human notions of 'fun' or 'sport'.

For dingoes, hunting isn't a leisure activity; it's serious business, essential for survival. The pleasure they might derive from it is more akin to the satisfaction we might feel from doing our job well, rather than the carefree enjoyment of a game.

Play versus predation

To further complicate matters, dingoes, like many intelligent animals, engage in play behaviour that can resemble hunting. Young dingoes will chase, pounce on and play-fight with their siblings, behaviours that help develop the skills they'll need for real hunting later in life.

Because of these similarities, some observers have suggested a link between play behaviour and surplus killing. But researchers who study animal behaviour draw clear distinctions between the two. Play is typically restricted to juveniles and is characterised by repetition, exaggeration and incompleteness compared to its real-world counterpart. Crucially, it stops short of causing serious harm.

Surplus killing is very different. It results in actual deaths and occurs within the context of genuine predatory behaviour. While aspects of hunting may activate reward pathways for dingoes, equating surplus killing with play oversimplifies the behaviour and risks anthropomorphising motivation.

THE PERFECT STORM

A clearer picture emerges when we apply this understanding to the case of dingoes killing sheep. The disruption of natural predator–prey dynamics is particularly evident in agricultural settings. Sheep, for instance, have been bred for docility and wool production, not for avoiding predators. They lack the speed, agility and defensive behaviours of prey such as kangaroos.

When dingoes encounter a flock of sheep, they're faced with an unnaturally easy hunting opportunity. The sheep's panic and inability to escape can trigger a heightened predatory response in the dingoes, leading to far more kills than they could possibly eat at the time. This situation is compounded by several factors: sheep tend to bunch together when threatened, which can trigger strong predatory responses in canids; they're confined to paddocks, unable to truly escape; and they present a prey density far higher than what dingoes would typically encounter in nature.

In essence, a paddock full of sheep is an ecological trap, an environment that triggers normal predatory behaviours in dingoes, but with abnormal results due to the sheep's inability to escape or effectively defend themselves. This situation bears little resemblance to the natural predator–prey relationships that shaped dingo evolution.

This dynamic was documented in Peter Thomson's 1992 study of dingoes in north-western Australia, where he observed 61 attacks on sheep, with only eight sheep killed outright. A further 26 sustained visible injuries and some were even fed upon while still alive. This exemplifies just how defenceless sheep are in these encounters. Notably, dingoes often abandoned their attacks not because the sheep escaped, but because the predatory sequence had been triggered in an environment where it could not naturally resolve.

Some have also proposed that dingoes may deliberately or inadvertently incapacitate prey, leaving them alive but wounded, effectively delaying consumption. This strategy could reduce the energy needed to subdue the prey again later. The idea remains speculative, but it reflects the complex and perhaps flexible nature of dingo predation when dealing with prey as unfamiliar and vulnerable as domestic sheep.

This all raises a compelling question: if sheep are such easy targets, why doesn't this kind of killing spree happen more often? The answer may lie in the same mechanisms that drive the behaviour in the first place. Just as the dopamine system can motivate a predator to chase, it doesn't compel them to do so every time. Humans, of course, are also wired for dopamine-driven reward, but not everyone becomes a compulsive gambler. The neurological triggers for reward exist, but so too do limits, learned controls and individual differences in behaviour. The same may apply to dingoes. Despite the opportunity being present, the urge to act on it may vary depending on the dingo's age, experience, social context or even prior learning.

This could help explain why surplus killing is often associated with adolescent dingoes, individuals still learning the rules, both environmental and neurological. Like human teenagers, they may not yet have developed the cognitive control to regulate strong impulses. Additionally, some researchers have suggested that pack disruption caused by baiting or culling may amplify the likelihood of such events. With social structures fragmented and experienced hunters removed, younger or less skilled dingoes may engage in more erratic, opportunistic killing.

In this sense, surplus killing is not just a biological phenomenon, but a behavioural one, shaped by experience, environment and context.

RETHINKING SURPLUS KILLING

At the outset, Rob and I were curious whether surplus killing in dingoes was truly wasteful or if there were deeper ecological and evolutionary explanations behind it. The deeper we delved, the more it became clear that surplus killing is not an act of senseless destruction, but a calculated response to rare opportunities. Rather than the 'mad, red hour' of Shaw's poem, an image of chaotic slaughter, these events reflect the cold logic of survival. When conditions briefly shift in favour of the predator, instinct and experience drive them to seize the moment, securing food in a world where the next meal is never guaranteed. It is not bloodlust that fuels these events, but the same fundamental forces that shape all predator–prey interactions: learning, risk management and the evolutionary rewards of survival.

Dingoes aren't mindless killing machines nor are they engaged in slaughter for enjoyment. They're complex animals navigating a complex world, making decisions based on innate drives, learned behaviours and environmental cues.

This perspective doesn't negate the very real impact on farmers. The loss of multiple animals in a single night is devastating, both economically and emotionally. But attributing human motives like cruelty or sport to dingoes creates dangerous misconceptions. When we label them as wasteful or vindictive, we not only misrepresent their behaviour but risk undermining their ecological role.

Surplus killing, rather than being a sign of imbalance, is often a mechanism through which balance is restored. When prey populations temporarily boom, predators respond in kind, reducing numbers and redistributing nutrients through scavenger networks. In these moments, dingoes are not just hunters but facilitators of equilibrium, regulating species in ways that sustain ecological stability over time. Seen in this light, surplus killing is not a failure of nature's design, but an essential function of it.

In the end, I would argue that only humans kill for sport. Dingoes, like all wild carnivores, kill to live.

17
Wild dog menace

FOR AS LONG AS THERE HAVE BEEN SHEEP IN AUSTRALIA, THERE HAVE BEEN dingoes trying to eat them and farmers trying to stop them (Fig. 17.1). This battle has shaped the landscape, industries, and entire communities, leaving an enduring mark on rural Australia.

One can hardly blame the farmers. A single dingo slipping through the fence one night is bad enough, but when it happens again, and again, it's devastating. Lambs vanish from the flock, some dragged away, others left behind with wounds they won't survive. Traps are set, fences patched, poison laid, but the attacks continue. The routine becomes a slow-motion nightmare, one where the enemy is cunning, elusive and always watching. Some heavily impacted farmers barely sleep, listening for the howls in the dark, their minds haunted by the uncertainty of what they'll wake up to.

The emotional toll is real. Psychologist Sarah Ecker's research found that the stress of dingo predation is comparable to major traumatic life events, akin to the anxiety felt by victims of serious accidents or life-threatening illnesses. The constant vigilance, the sense of helplessness, the sheer exhaustion. It wears people down. Many farmers have abandoned

Fig. 17.1. A dingo caught in a trap, sketched by Richard Wingfield Stuart around 1860. Beneath the drawing, he scrawled: "What I'm waiting to see immediately. The trapping of the 'Dingo'." Source: National Library of Australia.

Table 17.1. Summary of economic and social impacts associated with dingo predation and management

Economic impacts	Social impacts
Direct livestock losses: predation on sheep, goats and calves leads to financial loss for producers	Psychological stress: farmers report persistent anxiety, sleep disruption and emotional fatigue comparable to major life stressors
High lamb and kid mortality: young animals are especially vulnerable, reducing survival rates and breeding success	Social conflict: dingo management sparks tension between farmers, conservationists and policymakers
Reduced wool and meat production: lower stock numbers diminish overall productivity and profitability	Rural decline: the retreat from sheep farming in some areas contributes to economic downturns in regional communities
Shifts from sheep to cattle: farmers switch industries, altering land use and regional economies	Loss of tradition: multigenerational sheep farms close, taking with them cultural knowledge and identity
High cost of control: fencing, baiting, trapping and aerial shooting can exceed the cost of actual stock losses	Legal disputes: predator control leads to lawsuits and political battles over responsibility and regulation
Management complexity: dingoes often shelter in national parks, complicating control efforts and fuelling frustration	Emotional toll: constant vigilance, distressing scenes of predation and ongoing stress affect family life and wellbeing

sheep entirely, switching to cattle, hoping for a reprieve. Some have walked away from the land altogether.

Table 17.1 provides a snapshot of the impacts most commonly cited by industry and rural voices driving for lethal control.

But while dingoes do cause losses, their role as the ultimate threat to the sheep industry is often overstated. Markets shift, droughts bite harder, wildfires tear through grazing land and labour shortages rise. And that's not to mention the impact caused by foxes, feral cats and rabbits. Foxes raid lambing paddocks, feral cats drive native prey to extinction and rabbits strip the land bare, yet it's always the dingo that takes the blame and remains enemy number 1.

Why?

It does not appear to be purely about economics. Understanding the persistence of dingoes as an agricultural adversary requires examining management approaches, policies, propaganda and a deep-seated colonial belief that Australia's landscape could, and should, be controlled.

That belief has been made visible for all to see. Across outback Australia, the war against dingoes has been marked by gruesome displays, where dead dingoes are strung on fences or hung from trees, forming what some call 'dingo trees' (Fig. 17.2). These disturbing markers of course represent frustration or fear, but more than that they carry a deeper symbolism. They are warnings, declarations of defiance and attempts to assert dominance over a landscape that continues to resist control.

THE MACHINERY OF EXTERMINATION

No figure better embodies the early war on dingoes than John S. Bacon, a New England sheep farmer who made dingo killing his life's mission from the late 1800s to the 1940s. His 1946

Fig. 17.2. A stark symbol of conflict: Farmers erect 'dingo trees' to display the carcasses of dingoes they have killed, a practice born of frustration and fear over livestock losses. These grim trophies reveal the deep-rooted tensions between Australia's apex predator and the agricultural industry. Source: Angus Emmott.

book, *The Australian Dingo (The King of the Bush),* captured both the methods and mindset of colonial-era predator control.

> *We had one real pest – the dingo. In the early period of development, dingoes caused a great deal of work and expense, so it was only natural that we decided to hunt them in earnest in an endeavour to clean them right out of our area.*

Bacon wasn't just another farmer trying to protect his flock, he was a man on a mission, turning dingo extermination into an industry.

He achieved notoriety for commanding the highest bounty ever paid for a single dingo – £72/10 – for an infamous sheep-killer that had taken down over 1,000 sheep before Bacon tracked and killed it in just 7 days. His achievement was celebrated including by powerful community figures. PP Abbott (Federal Member for New England), W Stevenson (a local bank manager) and O Sloman (a leading grazier) all hailed his success, exemplifying the rise of the professional dingo hunter.

But Bacon's legacy extends beyond his individual success. His massive arsenal, 'three or four horses, two or three dogs, 52 double spring dingo traps, three rifles, two guns, one revolver, and adequate poison', revealed the industrialisation of killing, while his lifetime tally of over 440 dingoes demonstrated its effectiveness.

He proposed organised dingo-killing groups where trappers would be linked across vast areas to systematically eliminate the species. This vision extended to institutional support, with Bacon arguing that 'The Federal and State Governments must encourage all sheep and cattle owners, doggers, and trappers to organise themselves into dingo association killing groups.' His estimate of dingoes' national economic toll at £6,500,000 – counting dead sheep, stressed wool production and hunting expenses helped justify this escalation.

This crusade was definitely an ideological one. As anthropologist Merryl Parker has shown, the colonial war on dingoes was not just about protecting sheep, but about refusing to accept the natural consequences of introducing prey animals into predator country. Rather than acknowledge that bringing sheep into dingo territory would naturally lead to losses, settlers reframed the relationship, casting the dingo not as a wild animal fulfilling its ecological role, but as a transgressor, a threat to order and civilisation.

This rhetorical shift was significant. By portraying themselves as righteous defenders of progress, settlers absolved themselves of responsibility for disrupting ancient ecological systems. The dingo was no longer just a predator, it was a villain, a symbol of the wildness that settlers believed they had a moral duty to tame.

Violence, then, became not just necessary but noble. Killing dingoes became embedded in the cultural fabric of Australian pastoralism, passed down through generations, and eventually written into laws and policies that still persist today.

THE INDUSTRY OF KILLING

John Bacon may have envisioned a network of organised dingo hunters wiping the species from sheep country, but he could hardly have imagined the scale to which his vision would be realised. Today, killing dingoes is no longer the mission of a few determined farmers, it's written into legislation, supported by public funding, national bodies and embedded in government policy.

Dingo control now operates through a legal and bureaucratic machine that functions across three levels: national, state and local. But at every level, the same interest groups prevail. The National Wild Dog Action Plan (NWDAP), the federal framework guiding dingo management, claims to be a neutral, science-based initiative. In reality, it is led by industry representatives: cattle producers, wool growers and landholders. Very few, if at all, conservation biologists or First Nations voices are represented, and there is little independent oversight.

This framework defaults to lethal management of dingoes, and is rarely encouraged to consider or promote non-lethal or different management approaches. Rather, it provides a recipe for lethal control, such as baiting schedules, trapping protocols and eradication goals. And it ensures that decision-making remains in the hands of those with the most to gain. At the state level, agricultural lobbies have shaped legislation that prioritises killing over coexistence. At the local level, wild dog control groups, often landholder collectives, decide how and where the killing takes place.

It is a self-regulating system built to continue rather than to question, and to keep dingo control firmly at the top of the political agenda.

The EPBC Act recognises the dingo as a native species, but it offers no real protection. It hasn't been listed as threatened, despite multiple applications, and so, regulation is left to the states. Yet this too is steered by the NWDAP, which offers a strong lobby voice and regularly gets a seat at the table. As a result, states end up using similar control methods nationwide, even though the justification for those methods is far less consistent. Government agencies endorse and fund the program, but its agenda is shaped by those already committed to lethal control.

Imagine if mining companies wrote environmental policy. Or if developers set the rules for heritage protection. That's the dynamic at play here. The very people responsible for the killing are the ones deciding what counts as success and setting the rules of the game.

Local wild dog control groups function as the boots on the ground. Their tasks include:

- coordinating baiting programs with 1080 poison across public and private lands
- employing professional 'doggers' to trap and shoot
- collecting bounties and scalps as proof of success
- reporting kill numbers back to state agencies.

But they rarely ask, or are asked, whether what they are doing is working. Their purpose is not evaluation, but elimination. And because the same individuals sit across policy-making, implementation and reporting, there's little room for challenge or accountability. Each year, we hear claims through the media of wild dogs in plague proportions. But when results fall short, no one is held responsible. Few would argue that these are independent agencies. As implied above, they are insiders marking their own homework.

One of the most subtle, yet insidious, tools used to sustain this system is language. Across public messaging and official documents, the term 'wild dog' has replaced 'dingo'.

As discussed in Chapter 12, 'wild dog' conjures an image of something feral, impure, invasive. It strips away the cultural, ecological and legal identity of the dingo. By grouping dingoes with domestic dog hybrids and strays, policymakers sidestep the thorny issue of protection, Most government departments, landholder groups, management plans, media campaigns, all consciously continue using 'wild dog' to justify lethal control, and ignore or dismiss the evidence that this is no longer scientifically accurate.

For the dingo, that linguistic sleight of hand is a death sentence. Dingoes, if named, might trigger environmental protection. Wild dogs, by contrast, are pests. And so, by erasing the name, the system ensures the killing can continue. The industry has won the language war. And with it, carte blanche to wage the real one.

Killing dingoes became the default not necessarily because it is effective, but because it has always been done and because it has become a profitable enterprise. In some ways, it is a policy on autopilot, propped up by industry pressure, bureaucratic inertia and outdated assumptions. I cannot help to see it as a system in which the goal of protecting sheep has become secondary to the killing dingoes.

And so the conflict persists, framed not as a management issue, but as a war. This might seem like rhetorical excess, but government documents and media headlines confirm otherwise. In

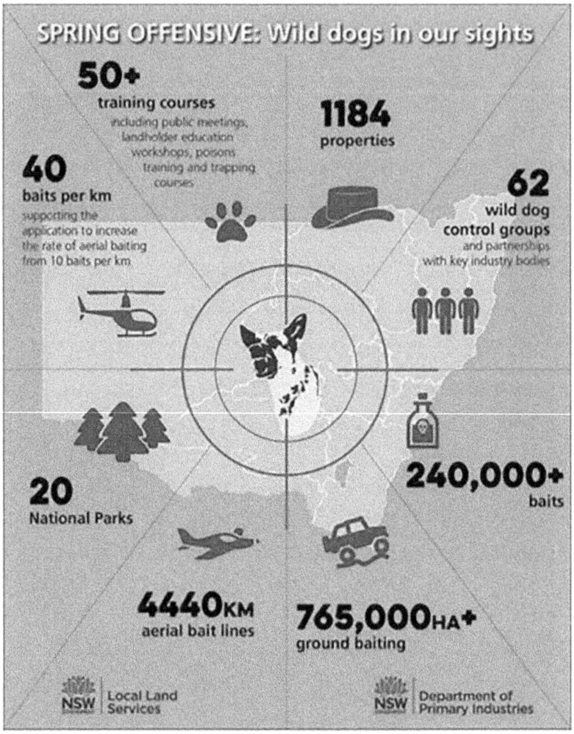

Fig. 17.3. Resource developed by the NSW Government's Department of Primary Industries and Regional Development, and Local Land Services, to engage with landholders and stakeholders with regard to the management of wild dogs. Source: New South Wales State Government.

Queensland, officials tout 'new weapons' in the fight. New South Wales launches its 'Spring Offensive' (Fig. 17.3). In South Australia, the dingo management program is called the 'Bite-back Program'. Wild dog control is branded as defence, as battle, and as victory or defeat.

Dingoes are not being managed. They are being fought. This has since morphed into an entire industry of its own. Poison baiting, bounty programs and professional doggers are all part of a publicly funded machine with its own financial logic. Entire businesses profit from the continuation of lethal control, regardless of whether it achieves its original goal.

THE WAR WITHOUT A SCOREBOARD

Despite the scale and cost of dingo control, meaningful evaluation is almost non-existent. Success is measured not by outcomes, but by inputs: how many dingoes are killed, how many baits are dropped, how many doggers are hired. It is a circular logic that confuses action with achievement.

Government agencies might track the number of 1080 baits deployed, but rarely assess whether these efforts reduce predation. Aerial baiting programs scatter thousands of poisoned carcasses across the landscape, yet there is little follow-up to check if stock losses decline as a result. No long-term monitoring. No comparison sites. No publicly reported cost–benefit analyses. The metric is quantity, not consequence.

The same applies to professional doggers. Paid to trap and shoot dingoes, their performance is judged by body count. Fewer kills might suggest fewer predators, but instead of being seen as a success, it often prompts concern that effort is falling short.

This absence of evaluation is not a minor oversight. Unlike predator control efforts in other parts of the world, Australia has no national program to track the real impacts of dingo removal on livestock predation over time, and individual states also avoid this. Most justifications rest on anecdotal evidence or industry pressure, rather than rigorous data.

By this logic, more killing equals more success, even when livestock losses remain unchanged. It is like measuring the effectiveness of a medicine by the number of pills dispensed, rather than checking whether the patient improves.

And yet, despite this vacuum of evidence, and in some cases, evidence to the contrary, the war against dingoes continues. Without clear evaluation, the assumption of necessity replaces the need for proof. Every report of a dingo sighting, every anecdote of sheep loss, is enough to keep the wheels turning. You can see increases in the number of media stories regarding dingo predation- they are often tied with changes in governments, when programs and funding come to an end, or in response to ideas of reintroduction or stays in dingo control (which I have covered elsewhere).

When pressed, government agencies and industry representatives rarely define what success would actually look like. Is it zero predation? A stable sheep population? A dingo-free region? The lack of a clear endpoint allows the system to perpetuate itself. Without targets, there can be no failure. And without failure, there is no reason to stop.

The result is a system on autopilot, propped up by tradition, funded by taxpayers and governed by institutions with a vested interest in keeping it running. Even when alternative approaches, such as coexistence strategies or targeted non-lethal control, are proposed, they are often sidelined or dismissed.

THE (EXAGGERATED) THREAT TO LIVESTOCK

For an animal so fiercely hunted, so deeply embedded in the pastoral mindset as an enemy of livestock, one might assume that dingoes are responsible for widespread devastation across the sheep industry. That without constant and costly intervention, entire flocks would be wiped out. One might wonder how Australia could have a sheep industry at all.

But the numbers tell a different story. Dingo predation accounts for only a small fraction of total livestock losses. The real pressures on graziers, such as climate variability, market downturns, poor pasture, disease, are rarely given the same attention, but have more impact on stock numbers. As Fiona Probyn-Rapsey argues, the dingo becomes a scapegoat, deflecting blame from deeper structural issues while reinforcing the industry's self-image as protector, not exploiter.

Take Victoria. Home to more than 15 million sheep, it's a major agricultural state. Yet in 2016–17, across all 16 Wild Dog Management Zones, only 288 confirmed attacks on livestock were reported. That's less than 0.001% of the flock. Still, the response was vast:

- 18.75 full-time equivalent wild dog controllers employed by the relevant state government department (at the time of writing, Department of Environment, Land, Water and Planning).
- 57,740 trap-nights
- 2,131 km of baited transects
- aerial baiting across six zones
- 195 landholders laying 37,584 additional baits
- $120 bounties paid per dingo scalp.

The scale of the response dwarfs the scale of the problem.

Estimates claim that dingoes cost Victoria AU$13–18 million each year. But even using the highest sheep prices of 2017, replacing the 288 reported livestock losses would have cost around $111,000, which is a fraction of the millions spent on control. The rest includes the control efforts themselves, creating a circular economy where the expense is used to justify more expense.

Elsewhere, the picture is similar. In New South Wales, one study found average sheep losses to dingoes at just 0.74% across four grazing properties. Nationally, figures swing wildly depending on who is reporting and how they calculate: some put losses under 1%, others as high as 16%. In one year, national losses were estimated at AU$48 million, while another report in the same year claimed $67 million for Queensland alone. These contradictions are rarely questioned. Instead, they are used to demand more funding.

Nationally, lamb deaths average 10 to 25% each year, mostly around birth, while adult ewe mortality sits between 2 and 10% depending on season and management.

Most lamb losses stem from environmental and maternal factors, which account for roughly 45% of deaths, including starvation, exposure, mismothering and poor bonding. Birthing complications contribute a further around 30% in lambs and 20% in ewes through issues such as dystocia, stillbirth and prolapse. In adult ewes, metabolic disorders like pregnancy toxaemia, milk fever and grass tetany make up about 30% of deaths. Infectious diseases and parasites account for 8% of lamb losses and around a quarter of ewe deaths, covering pneumonia, mastitis, scours, footrot, internal parasites and flystrike. Other causes, including accidents, casting and ageing, contribute 7% of lamb deaths and 15% of ewe losses.

Predation, often assumed to be the primary threat, typically represents less than 10% of mortality for both lambs and ewes when averaged nationally. Of this, foxes are responsible for the majority (around 6–8%), with dingoes contributing only 2–4% and wedge-tailed eagles generally less than 1%, though localised 'hot spots' and extensive systems can vary substantially.

These figures fluctuate with rainfall, enterprise type and husbandry practices, but the broad pattern is consistent. Most sheep losses arise from environmental, maternal, metabolic and disease-related factors, not predators, and dingoes represent only a small fraction of the overall picture.

I often think about the comparison between dingo impacts and live-export losses. Each year, more sheep die in transit during live export voyages to the Middle East than are lost to dingo predation. Although mortality rates on export ships are relatively low – typically 0.1–0.3% of the stock shipped in recent years – when millions of animals are shipped that translates into thousands of deaths. In 2020, for example, Australia exported over 777,000 sheep by sea and recorded a mortality of about 1,740 animals (0.224%) and exported over 1 million cattle with about 1,100 deaths (0.11%) during transit.

Sheep stocking rates, and whether running sheep remains profitable for producers, are shaped far more by broad economic and environmental forces than by predators. Seasonal conditions such as weather and feed availability account for around 30% of national flock change, followed by commodity prices at roughly 20%. Enterprise and breed shifts and costs of production each contribute about 15%, while regulation and trade settings and land-use change influence a further 10% each. These figures represent indicative national averages drawn from producer intention surveys and industry analyses, and they vary regionally with rainfall, enterprise type and market cycles. For example, when livestock prices fall, farmers are often forced to reduce herd or flock sizes; when prices rise, they rebuild. In Australia in recent years, this has been starkly visible. Drought and feed scarcity in southern Australia during 2025 prompted a mass sell-off, with many farmers unable to afford feeding animals, leading to an extra 30–35% of calves being sold and one in five ewes being processed even when pregnant. At the same time, industry analysts warned of a 'mountain of supply' in sheep markets after two strong seasons, with inflated inventories depressing prices and encouraging further turnoff.

Yet all these problems receive far less attention than dingoes. Probyn-Rapsey captures the contradiction perfectly in her article 'Eating Dingoes' published in 2017: 'It is not our eating of sheep that is the problem, but their eating of sheep that is the problem.' Dingoes are condemned for doing what graziers do themselves.

Although dingo predation is minimal at the national level it is obvious that impact is not evenly distributed. In high-density grazing regions, a single attack can be traumatic, both financially and emotionally. Sheep are especially vulnerable to surplus killing, where a single dingo may kill several animals in one night (covered in Chapter 16). Even if overall losses are low, that concentrated impact leaves a mark.

But this does not justify a nationwide war. It points to a need for targeted, responsive management. Indiscriminate lethal control treats all dingoes as equally dangerous, ignoring the difference between high-risk zones and areas where coexistence is possible. A smarter approach, one based on evidence rather than fear, would distinguish between genuine threats and background noise.

And then there is the cost. The Dingo Fence alone costs at least AU$10 million each year to maintain. Add baiting, bounties, doggers and exclusion fencing and total spending climbs into the hundreds of millions. These millions might be better targeted to those causes of mortality that have more impact on the industry as a whole.

Perception is what seems to be driving this war. A paddock littered with carcasses is a powerful image. It fuels fear, justifies action and lingers in memory. Slow-burning threats like climate change or market decline are harder to visualise, harder to blame, and arguably harder to fix.

Governments respond to that fear. It is easier and more popular among rural voters to subsidise baiting and doggers than to address the long-term problems facing agriculture. Lethal control offers visible, immediate action, even when it fails to solve the actual problem, and even when the evidence for its effectiveness is unclear or contested.

Worse still, in many places, removing dingoes has made things worse. The ecological fallout of dingo removal and the impact of unchecked populations of herbivores stripping vegetation and competing with stock has been covered earlier in this book. But its relevance here is that predator control is not delivering the results it is routinely claimed to achieve.

SHEEP VERSUS CATTLE

Sheep are small, defenceless and prone to panic. A single dingo can trigger a frenzy, killing multiple animals in a night, some dragged away, others left with wounds they won't survive. Cattle, by contrast, are simply harder to kill. A full-grown cow is too large for a lone dingo to take down. Calves are more vulnerable, but losses tend to be far lower and attacks less frequent.

This difference in vulnerability is evident in the numbers. For sheep, losses to predators can make up close to 10% of total mortality in some regions, though the true national average is lower. For cattle, the picture is very different. Predation usually accounts for less than 5% of deaths, and in many southern and mixed-farming areas it isn't as significant.

Most cattle deaths are caused by calving complications (~25%), diseases and parasites (~20–25%), metabolic disorders (~25%), accidents and misadventure (~20–25%), and poisoning events (~5%). Predation, by contrast, usually sits at only 3–4%.

Yet even when the percentages look similar on paper, they represent very different realities. In sheep enterprises, 10% can mean hundreds or even thousands of animals lost in a single season, especially during lambing. For cattle, that same proportion applies to a much smaller baseline mortality rate, often translating to just a handful of calves on a large property. A comparable percentage does not equate to comparable impact.

This uneven risk has created two very different attitudes. For sheep producers, dingoes are an existential threat. An enemy to be fought with exclusion fencing, baiting, trapping and coordinated campaigns. Cattle producers, by contrast, often take a more pragmatic or ambivalent view. Some use baiting when losses occur, but many see little need for intervention. On large properties with few incidents, dingoes are tolerated or even welcomed. Some graziers believe they help control kangaroos, feral pigs, and goats – species that compete with cattle for grass and water.

Over time, this divide has reshaped the agricultural map. In areas where dingoes are common and hard to control, sheep numbers have steadily declined. Exhausted by the conflict, many farmers have switched to cattle, trading wool and lamb for beef and a quieter life. The shift is especially stark outside the Dingo Fence, where sheep farming has all but vanished and cattle now dominate.

In this way, dingoes have become silent architects of the industry. Their presence, or absence, has dictated where sheep can persist, where cattle take over and how farmers relate to the land. For sheep producers, dingoes can threaten the viability of an enterprise. For cattle producers, they represent a manageable nuisance: sometimes a problem, sometimes a useful ally.

WHAT IF WE STOPPED?

One of the most common arguments for maintaining lethal dingo control is that without it, attacks would skyrocket. Industry groups insist that predator control programs are all that stands between farmers and devastation. And certainly, for individual producers who experience high losses, the impact of dingo predation can feel significant.

But if lethal control works, it doesn't make logical sense to me why it has to be repeated endlessly, and why many claim that lethal control needs to increase despite efforts to control dingoes being higher than any point in history.

Industry responses often fall back on the claim that control efforts fail because they aren't coordinated enough, that some landholders or public lands, like national parks, aren't 'doing their part'. This so-called nil-tenure approach argues that dingoes reinvade from unmanaged areas and, therefore, lethal control must be applied everywhere, all the time, to be effective. But even in well-coordinated programs, results often fail to support the rhetoric.

Take the Brindabella and Wee Jasper Wild Dog/Fox Control Program, which operates along the NSW–ACT border. It's often cited by government and industry as a success story for the nil-tenure approach, where lethal control is applied across both public and private lands in a coordinated effort. In theory, this should represent a best-case scenario. And yet, the program recorded a peak of just 200 sheep killed in a single year across the entire area.

Supporters might argue that this low number shows the program is working, that without coordinated control, the losses would be much higher. But the problem is, we simply don't know. There is no control site, no counterfactual and no long-term monitoring to assess whether these efforts are truly reducing stock losses. In the absence of proper evaluation, programs like these become self-justifying: if sheep are killed, it proves the need for more control; if few are killed, it proves the control is working.

Either way, the killing continues, without ever answering the fundamental question of whether it works.

I will happily concede that we need lethal control of dingoes in some areas, should the evidence be presented. Before then, I remain somewhat sceptical.

A WAR WITHOUT END

More than a century of persecution has not eliminated the dingo, but it has left an indelible mark on Australia's landscapes, policies and psyche. What began as a desperate response to livestock predation has hardened into something else entirely: an institutional cycle of killing, sustained not by results, but by habit, expectation and economic momentum. Governments continue to fund lethal control not because it necessarily works, but because it is politically easier than addressing the deeper challenges of climate volatility, market instability and land degradation.

And so, the war drags on. As cattle farmer Angus Emmott once told me, 'We are in a place where no generation of farmer has farmed without killing dingoes. They know no other way.' And in that, there's a deeper truth: no one really knows what would happen if we stopped. The system has continued for so long, with so little evaluation, that the idea of doing things differently is met with hesitation, if not outright fear. In the absence of evidence, uncertainty fills the gap and the killing habit continues.

I think we need to collectively decide what victory even looks like. If two centuries of culling, fencing and baiting have failed to solve the so-called dingo problem and if many farmers argue it's only getting worse, then what will. What if we have been fighting the wrong war the whole time?

Part 6

Embracing the wild

18
The baiting paradox

IF LETHAL CONTROL METHODS REALLY WORKED, THE DINGO PROBLEM would have been solved long ago. Yet year after year, the same measures are rolled out, as if more poison or more traps will finally deliver victory. The deeper question, as posed in the previous chapter – whether these strategies reduce conflict in the long term, or lead to the best possible outcomes – is rarely asked.

It was this contradiction that I brought to a special dingo forum in 2019, convened by the Royal Zoological Society of New South Wales. There I presented on non-lethal dingo management, a topic I had recently turned my research focus towards. Alongside Rob Appleby, we developed a model synthesising the work of many researchers, mapping out this self-perpetuating cycle of lethal control (Fig. 18.1).

Standing before the audience, I laid out our argument: that killing dingoes doesn't just fail to stop the conflict, it may be the very thing fuelling it. The logic was unsettling but clear, and I was certainly not the first to have said it. Nor is this something unique to dingoes. In Australia, broad-scale and indiscriminate dingo control – poison baiting being the preferred method – makes the problem worse. As I will attempt to show as I unpack the elements, it's a perfect example of how a solution can become part of the problem it was meant to solve.

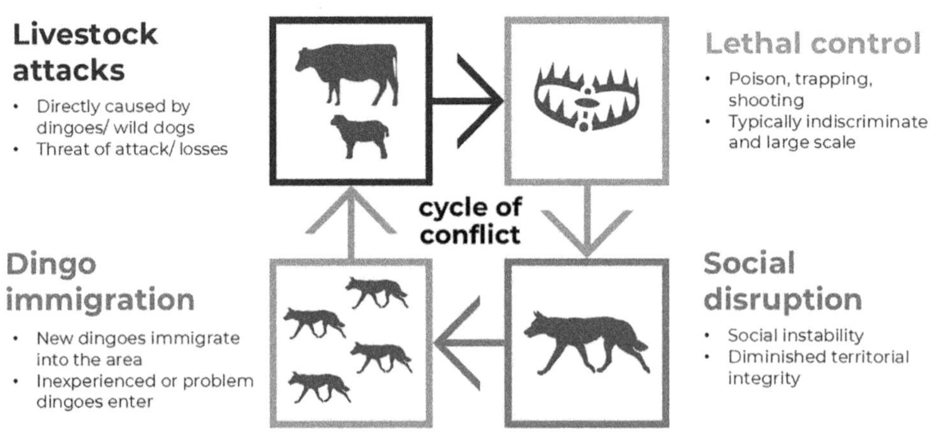

Fig. 18.1. The cycle of conflict in dingo management. Lethal control is meant to reduce livestock attacks, yet it disrupts dingo social structures, leading to instability and an influx of new dingoes. This often increases predation risk, as do management methods that leave livestock vulnerable, triggering yet another round of control, perpetuating an endless cycle of conflict. Source: Bradley Smith.

THE REALITY OF LETHAL CONTROL

It's an equation that seems to add up: fewer dingoes should mean fewer attacks. The reality, however, is far more complex.

When faced with immediate stock losses, the impulse to take direct action is understandable. Community expectations and traditional management approaches play a powerful role in rural areas, where participation in regional baiting programs is often seen as a responsibility to neighbours and often a legal requirement under biosecurity laws. The temporary relief that sometimes follows baiting reinforces the perception that it works, even when longer-term data suggests otherwise.

Research in 2020 led by Guy Ballard from the NSW Government's Department of Primary Industries tested different aerial baiting rates to determine the most effective strategy for reducing dingo numbers. In that region of NSW, the highest rate for maximal impact was found to be 40 poisoned baits laid or dropped every kilometre. On the back of this research, it is a baiting rate adopted across many jurisdictions as best practice.

However, even under these most intensive conditions, the best mortality rate achieved was around 90%. The reasons why 1 in 10 dingoes survived are uncertain. My theory is that some dingoes, particularly those on the cautious end of the bold–shy spectrum, instinctively avoid strange or unfamiliar objects. Others may learn to steer clear through social transmission, especially if a non-lethal dose has made them, or another dingo in their social group, sick.

The temporary relief that sometimes follows baiting reinforces the perception that it works, even when longer-term data suggest otherwise.

This intense level of pressure doesn't eliminate dingoes from the landscape overall, but it does inflict a heavy toll on population size and genetic diversity. In some regions, repeated and extremely intensive baiting does remove dingoes entirely, simply by killing individuals faster than they can reproduce or recolonise. These pockets of high mortality and local eradication come at a high cost. A single aerial baiting campaign can cost upwards of AU$20,000 for a large pastoral property or national park. This figure includes the purchase and/or preparation of the baits, hiring helicopters or aircraft, and labour.

Even where numbers crash, the effect rarely lasts. Dingoes from neighbouring areas soon move in to fill the vacuum, drawn by vacant territory and food. Where recolonisation is slower, numbers typically rebound within a breeding cycle or two. This cycle of removal and recolonisation, sometimes described as a 'boom-and-bust' effect, means any relief for farmers is typically short-lived and requires continual effort to maintain.

Compared to other wildlife management approaches, such as targeted removal of problem individuals (which typically aims for a 5–10% population reduction) or exclusion fencing (which aims for 100% exclusion but rarely achieves it), broadscale baiting sits in an unfortunate middle ground: intensive enough to disrupt social cohesion, but not comprehensive enough to eliminate the species entirely from a region.

The challenges with lethal control don't stop there. Several studies have found that carrying out lethal control can increase stock predation over time, not reduce it as intended. The disruption of predator social structures, rather than just reducing numbers, often plays a key role in shaping predatory behaviour- as is explored in the following section.

In one of the most extensive reviews of dingo management, Lee Allen from the Queensland Government's Department of Natural Resources examined 50 years of historical data to determine how control measures, particularly baiting, impacted livestock losses across Queensland. The analysis tracked dingo bounty records, livestock production data and changes in control strategies over that time period. The findings showed that cattle calf predation increased, rather than decreased, in regions where dingoes were heavily baited.

More alarmingly, the void left by lethal control invited new dingoes into the area, leading to high population turnover and chronic social disruption. The more dingoes were removed, the more unpredictable their presence became, with livestock predation fluctuating wildly rather than declining.

The Allen-led study also found that dingo impacts varied between livestock types. Unlike cattle, which can sometimes defend themselves, sheep are highly vulnerable and often killed in numbers greater than dingoes consume. This has forced sheep producers into more aggressive control measures. Yet, as with cattle, these efforts often made the problem worse, creating population instability and, in some cases, increasing predation rather than preventing it.

The work provided one of the first large-scale empirical demonstrations of the baiting paradox: attempts to exterminate dingoes led to more predation, not less. This isn't unique to wildlife management. We see similar unintended consequences in medicine, where the overuse of antibiotics was meant to eliminate infections, but instead led to something far worse – antibiotic-resistant superbugs. By wiping out bacteria indiscriminately, we removed the weaker strains, allowing the strongest, most resilient ones to thrive. Dingo control doesn't create super-predators or drive genetic resistance in the same way, but the analogy still holds: indiscriminate removal disrupts complex systems. In some cases, it may even favour bait-shy or risk-averse individuals, making control efforts less effective over time.

BAITING FRACTURES SOCIAL STRUCTURES

Dingoes are not solitary wanderers. As already covered, they exist within tightly structured family groups, where cooperation is essential for survival. Within a stable pack, a dominant breeding pair leads, while subordinates assist in raising pups, hunting and defending their territory. This structure enforces order, where territories are strictly patrolled, outsiders are kept at bay and hunting behaviours are passed down through generations.

Ecologist Arian Wallach and her team (2009), in studying the impacts of dingo persecution across Australia, argue that dingo packs function as a form of natural regulation. Stable groups defend their territories, suppress problem individuals and enforce learned hunting traditions focused on wild prey. But when baiting indiscriminately removes dominant adults, this results in consequences that appear through the entire system. With older, experienced dingoes removed, younger, socially unstable individuals are left to fend for themselves. Lacking the cooperative hunting strategies taught by intact packs, these juveniles turn to easier prey: sheep and calves.

Baiting also triggers a surge in reproduction. In a stable dingo pack, breeding is tightly regulated; only the dominant pair typically successfully raises a litter. When those controls are lost, subordinate females begin producing litters, often leading to a spike in dingo numbers rather than a decline. And then there's the issue of territorial collapse. With no dominant dingoes left to defend their range, the vacuum invites newcomers, dispersing juveniles or even feral domestic dogs, which are even less predictable and more likely to target livestock. The landscape becomes fragmented, chaotic and far harder to manage than if stable dingo populations had remained in place.

Wallach's findings align with studies on predator control worldwide. Research on wolves in North America has shown that culling disrupts pack structures, often leading to an increase in livestock attacks. Similar patterns have been observed with African wild dogs, lions and leopards, where the removal of dominant individuals weakens social cohesion and results in more opportunistic predation.

The assumption that fewer predators automatically mean fewer attacks is deeply flawed. A fragmented predator population is far less predictable than one that is stable. In areas where dingoes are left alone, packs regulate their own numbers, defend their territories and suppress unwanted predators, including foxes and feral dogs. When human intervention destabilises these systems, the cycle of conflict continues.

THE COSTLY ILLUSION OF FENCING

At first glance, like lethal control, fencing seems like an obvious fix: keep the dingoes out, protect the livestock. This is the assumption behind cell fencing, where large sections of land are enclosed to exclude predators.

Lee Allen, from the same piece of research mentioned earlier, also examined fencing as a control measure The data suggested that while fencing might provide localised protection, it required extensive upkeep and was only as effective as its maintenance. More recently, a 2025 study by Stuart Dawson and colleagues in Western Australian Government's Primary Industries and Regional Development put the assumption that fencing protects livestock to the test, asking whether cell fencing actually saved graziers money.

Dawson's team analysed the economic viability of fencing projects by modelling thousands of different scenarios, testing the effects of livestock type, predator control costs and potential gains in production. The results showed that in most cases, the cost of fencing was higher than the financial losses caused by dingoes in the first place. Only 37% of cell-fenced scenarios resulted in a profitable outcome. In other words, even with successful dingo exclusion, most pastoral enterprises still struggled to make a profit.

One of the key problems was maintenance costs. Fences require constant repair due to weather damage, wear and tear, and animals breaking through. And when dingoes were excluded, a new problem emerged: kangaroo overpopulation. Without dingoes to keep them in check, kangaroo numbers surged inside the fences, leading to overgrazing and increased competition with livestock for pasture. In some cases, the cost of lost grazing land due to kangaroos was worse than the original dingo problem.

Another issue was that cattle operations gained little benefit from fencing. Sheep enterprises were more vulnerable to dingoes, and had more to gain, but cattle producers saw almost no economic return from fencing, with profits lower than those of unfenced cattle stations. The study suggests that alternative management strategies, such as focusing on cattle in unfenced areas, may be far more viable than sinking money into fencing and lethal control.

Dingo control is often justified as a financial necessity, yet the evidence increasingly shows the opposite: the cost of control is higher than the cost of coexistence. When the price of keeping dingoes out outweighs the damage they might do, the logic behind predator control quickly falls apart, as we'll see throughout this and the next chapter, where alternative strategies often prove more cost-effective.

DINGOES OFTEN PREFER WILD PREY OVER LIVESTOCK

Dingoes have long been cast as the ultimate livestock killer, their very presence in the landscape seen as an inherent threat to sheep and cattle. But the idea that dingoes naturally prefer livestock over native prey could be one of the most enduring misconceptions in dingo management. The reality is that sheep are not, in any meaningful biological sense, a preferred choice.

The diet of a dingo is highly influenced by what it learns to hunt from an early age. Peter Thomson, a researcher with CSIRO who spent the 1970s and 1980s studying predator–prey dynamics in Australia, was among the first to demonstrate that dingoes hunt based on familiarity, not preference. If a young dingo is raised in an area where macropods (kangaroos and wallabies) are abundant, it will continue hunting macropods throughout its life. But if a dingo grows up in a region where sheep are the most available food source, livestock may become its primary target.

This means that most dingoes, particularly those in areas rich with native prey, have no reason to target sheep at all, unless desperate or opportunistically. Their natural prey is faster, more agile and better suited to their evolved hunting techniques. Sheep, by contrast, lack the speed and defensive strategies of kangaroos, making them an easier, but not necessarily preferred option.

Native prey are also far more abundant across most landscapes, while sheep are confined to human-dominated areas. And even when sheep are available, they are not an optimal food source. Compared with kangaroos and wallabies, sheep carry far more fat, and dingoes have evolved on the lean, high-protein meat of macropods. Their digestive system and energy demands are built around that profile. Very fatty meat is harder for them to process and provides a different kind of energy return than the prey they are adapted to hunt. So although sheep are easier to catch, they are not, in any meaningful biological sense, a preferred choice. If dingoes truly preferred livestock over native prey, it would show up more frequently in their diet. Across 73 studies and over 32,000 samples of dingo scats and stomach contents from across the continent, a clear pattern emerges:

- macropods dominate. kangaroos and wallabies are consistently the most frequently consumed prey
- livestock remains are rare. On average, sheep and cattle appear in only ~7% of samples, with most regions reporting much lower rates, often just a few per cent
- much of it is scavenged. In many cases, livestock remains in dingo scats come from carrion, not active kills, meaning dingoes are often cleaning up carcasses rather than creating them.

These findings align with what graziers themselves have observed: in regions where dingoes coexist with healthy macropod populations, stock losses are often minimal. The problem arises when native prey is scarce, either due to drought, habitat destruction or past mismanagement of the landscape. In these cases, dingoes may turn to livestock, but this is an issue of necessity, not preference.

Thomson's work reinforced this point: dingoes are selective hunters. They hunt what they know. Their preference for wild prey is a learned behaviour passed down through generations. This is why indiscriminate baiting and culling can create more livestock predation, not less; by wiping out stable packs that have spent their lives hunting macropods, it opens the door for new, inexperienced dingoes that may be more likely to target stock.

Recent drone footage analysed by Thalia Pollock and colleagues revealed that dingo hunting skills, particularly when targeting difficult prey like kangaroos, are learned, not innate (see Chapter 16). In one sequence, a mother dingo led her three yearling pups in a coordinated hunt. Although the pups managed to corral the kangaroo, they hesitated to engage until their mother arrived and took charge, showing them how to confront and attempt to take down the prey. This suggests that effective macropod hunting is a complex skill passed down socially. When lethal control removes experienced adults from the population, it potentially disrupts this learning process, leaving behind or inviting in younger dingoes that lack those skills, making them more likely to turn to easier targets like sheep.

WHAT IF WE LEFT DINGOES ALONE?

We've already established the paradox of lethal control. If we simply stopped interfering, would it unleash large packs of dingoes across the landscape, terrorising farms and nearby towns?

The answer is likely not.

Dingo society comes with its own checks and balances. Dingoes don't breed unchecked or spiral out of control when left undisturbed. Unlike some other predators, their pack structure enforces natural population limits that reflect the available resources (native prey) of the land, just as their home-range size adjusts to resource availability. Only the dominant breeding pair in a pack typically reproduces and subordinates help raise pups rather than having litters of their own. Young animals will disperse once they reach maturity. In this sense, dingoes regulate themselves

Perhaps most importantly, a stable dingo pack defends its territory, keeping outsiders, whether dispersing wild dingoes or roaming domestic dogs, at bay. This territorial behaviour protects livestock more effectively than many human interventions because it prevents the arrival of unfamiliar, inexperienced dingoes that are far more likely to target sheep or calves.

I am not naïve. Some rise in livestock attacks is to be expected if lethal control is removed, especially in areas where native prey are scarce or farms sit right on the edge of wild country. But the increases are unlikely to be as widespread or dramatic as many imagine or fear, and they would vary greatly from region to region.

The places most exposed are already the most marginal: bushland, remote boundaries, and properties where sheep are grazed in open, extensive systems. In those areas, coexistence isn't impossible, but does require planning and a combination of management options that are covered later in Chapter 19.

Part of this also requires adjusting our expectations. Some level of loss is inevitable when farming alongside a native apex predator. It's often not feasible to eliminate it entirely, but to keep losses within an acceptable range and manage them in ways that are humane, effective and ecologically responsible (as I explore in Chapter 20).

NOT KILLING DINGOES CAN BE MORE PROFITABLE

For decades, graziers have been told that killing dingoes is essential for profitability. What if the opposite were true? A groundbreaking study by Thomas Prowse, Corey Bradshaw and their team from Adelaide University modelled the financial impact of dingoes on cattle stations and their findings turn conventional wisdom on its head.

Using a sophisticated economic model, the researchers simulated real-world conditions on cattle properties, factoring in dingo predation on calves, the role of dingoes in suppressing kangaroo numbers and how this influenced pasture availability and cattle productivity. Their results showed that on stations where dingoes were left alone, graziers consistently saw higher profits than those who spent money on lethal control. The reason came down to dingoes keeping kangaroo numbers down. With fewer kangaroos competing for pasture, cattle had more feed, leading to better weight gain, improved fertility and healthier stock. Over time, this translated to stronger profit margins.

The financial benefits didn't stop there. Farmers who didn't bait or shoot dingoes also saved on pest control costs. Dingoes naturally suppressed mesopredators like foxes and feral cats, which meant fewer lamb and small livestock losses, as well as improved survival rates for native wildlife. The land itself benefited too. Without unchecked kangaroo grazing, pastures recovered, water retention improved and the long-term productivity of the land increased.

Perhaps the most striking takeaway from the study was that, even accounting for calf losses to dingo predation, the overall economic benefits of dingoes outweighed the costs. The model showed that a 100,000-ha cattle station could increase profit margins by as much as AU$83,000 per year simply by allowing dingoes to remain in the landscape. At moderate stocking densities, the presence of dingoes had little impact on overall cattle numbers, while the benefits of healthier pastures and better grazing conditions made up for any losses.

This challenges a core belief in the grazing industry that dingoes are a direct financial threat. In reality, the real financial risk isn't dingoes, it's the cost of trying to eradicate them. Graziers who spend their time, along with tens of thousands of dollars each year on baiting, shooting and exclusion fencing, may actually be undercutting their own profits. As discussed in Chapter 17, the effort to kill dingoes has in some cases become an industry in itself, one that continues regardless of whether the benefits outweigh the costs.

If financial security is the priority, in some regions keeping dingoes could be one of the smartest economic decisions a cattle grazier can make. Rather than being seen as a liability, dingoes should be recognised as a natural, cost-free means of reducing grazing pressure, controlling pests and improving land health. The numbers speak for themselves.

THE BIGGER PICTURE

The financial burden of lethal dingo control is not just an issue for individual graziers; it's an issue for taxpayers and land managers across the country. The millions spent on baiting and fencing would be justified if they produced clear, measurable benefits. But as I have shown, the evidence suggests otherwise. Lethal control is not only a recurring expense rather than a long-term solution, but also exacerbates environmental problems that graziers must then manage separately.

CASE STUDY: ANGUS EMMOTT'S CATTLE ENTERPRISE

The dust billowed behind my tyres as I drove my little silver rental car towards Angus and Karen Emmott's property near Longreach in Queensland's majestic channel country. The late afternoon sun casting long shadows over the open country. Out here, the land stretched in every direction, vast and untamed. This was cattle country, the kind of big-sky, open-range landscape that Americans would instantly compare to Texas.

Before I even reached the homestead, I heard the low hum of a quad bike approaching from the distance. A few seconds later, a familiar figure crested the rise. A man, solidly built, his face covered in a white beard and shaded by a weathered Akubra, his dog Banjo perched like a co-pilot on the back of the bike. He rolled to a stop, Banjo hopping down before Angus swung one leg off the quad, dusting off his jeans with the easy confidence of someone who had spent a lifetime on this land.

Angus is an intergenerational farmer, a naturalist and a scientist in his own right. His knowledge of the land and its critters is encyclopaedic, not just in the way of a cattleman but in the way of a scholar. His house is lined with bookshelves. His library of natural history books is so vast it takes up an entire section of his home. He speaks deliberately, carefully weighing his words, a man who understands that nuance is often lost in debates over dingoes.

'I don't think I've got all the answers', he said, leaning against the quad, Banjo still at his feet. 'But I do know this. People think dingoes are a problem, but the real problem is the damage we've done trying to get rid of them.'

That was the heart of it. Unlike most graziers, Angus doesn't bait, shoot or fence out dingoes. Instead, he lets them exist as part of the land and in return, they regulate the ecosystem in ways that benefit both the landscape and his cattle.

Fig. 18.2. Two dingoes living on Angus's property facing off against a kangaroo. Source: Angus Emmott.

'We stopped the baiting, and after a couple of years, we noticed the packs were stable and they weren't interested in the cattle anymore', Angus explained as we walked towards the homestead. 'The real problem before was the young, rogue dingoes that moved in after we wiped out the resident pack.'

We spent the rest of the afternoon driving across parts of his expansive property – much of it not possible to traverse due to swelling creeks and rivers. I could see the evidence all around us. His cattle, well-conditioned and calm, grazed among healthy native grasses. Without the unchecked kangaroo populations that plague many dingo-free areas, including some of his neighbours, his pastures remained intact.

'Look there', Angus pointed towards a shady area by a small dam (same areas shown in Fig. 18.2). 'That's where I often see the alpha female. She's been here for years now. Knows these cattle aren't worth the trouble.'

Angus' approach isn't just a hunch. What he has observed on his land is exactly what research scientists have been showing for years. His experience challenges the dominant mindset of predator control in Australia. And he is not alone in his thinking.

FARMERS SPEAKING OUT FOR DINGOES

For generations, graziers have followed lethal control strategies because they were inherited from those before them. As Emmott, with a cheeky grin, said to me during my visit, 'Tradition is just peer pressure from dead people.'

He is now part of a growing number of graziers who are starting to challenge this inherited wisdom because they're realising that killing dingoes may be costing them more than it's saving them. Some are even speaking out publicly, advocating for the protection of dingoes on their land.

One of the most significant signs of this shift is the emergence of Landholders for Dingoes, a grassroots movement of farmers who reject indiscriminate lethal control and see dingoes as an integral part of the landscape. These landholders aren't idealists or outsiders; they are cattle station owners, graziers and land managers, people who, like Angus Emmott and David Pollock (whom we met earlier), have seen firsthand that dingoes aren't always the problem they've been made out to be.

Groups like Landholders for Dingoes argue that traditional dingo control measures are expensive, ineffective and often counterproductive. Instead of reducing livestock losses, baiting campaigns disrupt predator–prey dynamics, worsen land degradation and ultimately make grazing conditions harder, not easier.

And perhaps most significantly, governments and industry groups are being forced to take notice. Landholders for Dingoes, alongside ecologists and advocacy groups, are pushing for non-lethal management strategies to be taken more seriously, shifting the conversation from 'how do we kill dingoes?' to 'how do we coexist with them?'

For a long time, it seemed unthinkable that farmers would advocate for protecting dingoes. Yet, here we are. The resistance to baiting isn't just coming from environmentalists and urban Australians anymore, it's coming from those who work directly on the land.

THE GROWING RESISTANCE TO LETHAL CONTROL

This shift is part of a broader cultural change taking place. For much of Australia's history, dingo management remained largely unchallenged by those outside the agricultural industry. But over the past two decades researching dingoes, I have noticed a shift in public and scientific discourse on predator control.

Today, growing numbers of Australians oppose the widespread use of poison baits, aerial culling and trapping. Previously baiting programs were carried out with little scrutiny, but they now face increasing pushback from the public. Surveys reveal a changing tide: younger generations and urban populations are particularly opposed to lethal control, with many questioning the ethics, efficacy and long-term consequences of these measures.

Lily van Eeden has documented this emerging shift. Her national surveys have consistently found that while public awareness of dingo control is low, with only 19% of Australians knowing that baiting programs targeted dingoes, there is strong support for more humane, non-lethal alternatives once people are informed. In one study, 85% of respondents preferred methods like exclusion fencing and livestock guardian animals over lethal options such as baiting or shooting. Crucially, van Eeden's work highlights widespread public discomfort with 1080 poison, due to its cruelty, slow-acting nature and the risk it poses to non-target wildlife. As more Australians learn about these practices, particularly younger and urban populations, the legitimacy of widespread lethal control is being actively questioned.

This growing resistance reflects a broader erosion of the Social Licence to Operate (first mentioned in Chapter 15). Social licence refers to the ongoing level of acceptance or approval granted by the public to practices that impact the environment, animals or communities. In one of the most comprehensive studies to date, Audrey Pearson and Mia Cobb from the University of Melbourne surveyed 990 Australians in 2024 about their attitudes towards dingo management. They found that 53% opposed lethal dingo control and only 39% supported it, a clear majority now against practices like baiting and shooting. Support was especially low among younger Australians, urban residents and dog owners. Crucially, those who held more positive attitudes towards Australian wildlife were significantly less likely to support lethal control. These findings point to a generational shift in public values and a growing disconnect between traditional control methods and community expectations. In an era where public opinion can shape policy, the dingo's social licence to be protected is rising, while the licence to kill is wearing thin.

So, what is driving the shift in public attitudes towards dingoes?

It's part of a broader shift in how Australians view the natural world. As awareness of climate change, biodiversity loss and mass extinctions grows, so too does an appreciation for ecological complexity. The staggering number of species lost in Australia has heightened the desire to protect native wildlife, including the dingo. No longer seen solely as a pest, the dingo is increasingly recognised as an integral part of Australia's natural heritage and one that plays a vital role in maintaining ecological stability. This cultural awakening extends beyond dingoes to a greater recognition of the ecological importance of predators. Australians, like people in other parts of the world, are beginning to see that predators are not just threats to livestock, but essential to maintaining balance in ecosystems. In North America, for example, attitudes towards wolves have undergone a dramatic transformation, with increasing opposition to lethal control. A similar shift may now be unfolding for dingoes.

Increased positive media attention has certainly helped. Graphic reports of widespread baiting campaigns, including those conducted on public land and inside national parks have sparked public outrage. The idea that native wildlife is being poisoned en masse in supposed conservation areas doesn't sit well with many Australians, even those who don't typically engage with environmental issues. And more generally, media portrayals are more positive and are challenging the long-standing image of the dingo as a sheep and baby killer, instead highlighting their uniqueness, ecological role and cultural significance.

Science, too, has played a pivotal role in shifting perceptions. Over the past 25 years, research into the dingo has expanded dramatically, spanning multiple disciplines. Studies have revealed that dingoes are highly intelligent, socially complex and critical for maintaining ecosystem health as apex predators. This stands in stark contrast to earlier decades when dingo research was largely funded by the livestock industry and focused on eradication rather than understanding. The broader research base has reshaped public perception, with dingoes increasingly seen not as pests, but as remarkable and ecologically valuable animals.

This shift in public sentiment has real consequences. Governments and land managers are under increasing pressure to justify their lethal control programs or, at least, explore alternatives. What was once seen as a default practice is becoming a contested issue.

The challenge lies in reconciling these shared objectives with the reality of dingo predation. Instead of repeating the mistakes of the past, it's time to explore new strategies. Rather than fighting nature, perhaps it's time we learned to work with it.

19
Breaking the cycle

SO, IF KILLING DINGOES DOESN'T BREAK THE CYCLE, WHAT DOES? THE answer, increasingly, is coexistence.

Coexistence is a fundamental shift in how we see the natural world, one that goes far beyond individual management practices. It means recognising that humans are not the sole architects of the landscape and that other species have an inherent right to exist, regardless of their perceived value or benefit to us. This perspective rejects the utilitarian mindset that evaluates animals solely through the lens of human interests, whether they're 'useful' or 'harmful.'

At its core, coexistence is about living alongside wildlife rather than constantly working against it. It begins with tolerance, requiring us to see predators not as adversaries but as integral parts of the ecosystems we share. This doesn't mean abandoning livestock protection or ignoring the realities of predation. Rather, it means finding practical ways to reduce conflict that don't rely on killing or eradication. Methods that acknowledge the presence of dingoes as natural rather than as a problem to be solved. Instead of seeking dominance over nature, coexistence strives for balance and mutual flourishing, appreciating both the intrinsic worth of all living beings and the ecological roles that predators like dingoes play in maintaining healthy landscapes.

The first people to truly coexist with dingoes were Aboriginal Australians, who have lived alongside dingoes for thousands of years (see Chapter 3). Their relationship with dingoes was neither one of rigid control nor passive indifference, but of deep understanding and adaptation. In some regions, Aboriginal people developed practical methods to minimise conflict, recognising that dingoes, like all predators, respond to incentives.

Storing food on raised platforms, keeping it out of reach of dingoes and other scavengers is one such method. In some cases, these platforms were built within settlements, acting as designated food storage areas that discouraged dingoes from foraging in human spaces (Fig. 3.3). Rather than attempting to eliminate dingoes, this strategy worked with their natural instincts, directing their behaviour in ways that reduced competition with humans.

Other Aboriginal groups integrated dingoes into their daily lives, raising pups, using them as hunting companions and forging strong connections. Dingoes were valued not only for their utility but as part of the broader cultural and ecological landscape.

These Indigenous principles of adaptation and coexistence offer valuable insights that can be useful for transforming modern farming and wildlife management. By understanding that predators respond to behaviour, incentives and ecological conditions rather than simply to force, we can develop approaches that protect both livestock and wildlife.

FROM CONFLICT TO COEXISTENCE: A NEW MODEL FOR DINGO MANAGEMENT

In the last chapter, I introduced you to our model of the cycle of conflict, a self-reinforcing system where lethal control often creates more problems than it solves. As we saw, killing dingoes disrupts their social structures, which paradoxically can lead to increased predation and perpetuate an endless cycle of conflict.

Here I will focus on how we break that cycle by utilising non-lethal alternatives. Rather than a circle of escalating conflict, this model creates a positive feedback loop of coexistence. The proposed model is shown in Fig. 19.1.

Where the conflict cycle begins with lethal control that fractures dingo family groups, this new model maintains stable dingo territories and social structures. Instead of dispersing young dingoes that may lack hunting skills for wild prey, stable family groups teach their young to hunt natural prey and avoid livestock. Rather than employing lethal methods when predation occurs, this model advocates deploying targeted non-lethal strategies and conditioning techniques that address specific problem behaviours without disrupting the entire social system.

The conflict cycle leads to increased livestock losses and declining biodiversity, whereas the alternative approach typically reduces predation over time while supporting healthy ecosystem functions. And where traditional management methods create an endless loop of killing and conflict, this model includes subsidising the uptake of non-lethal approaches and

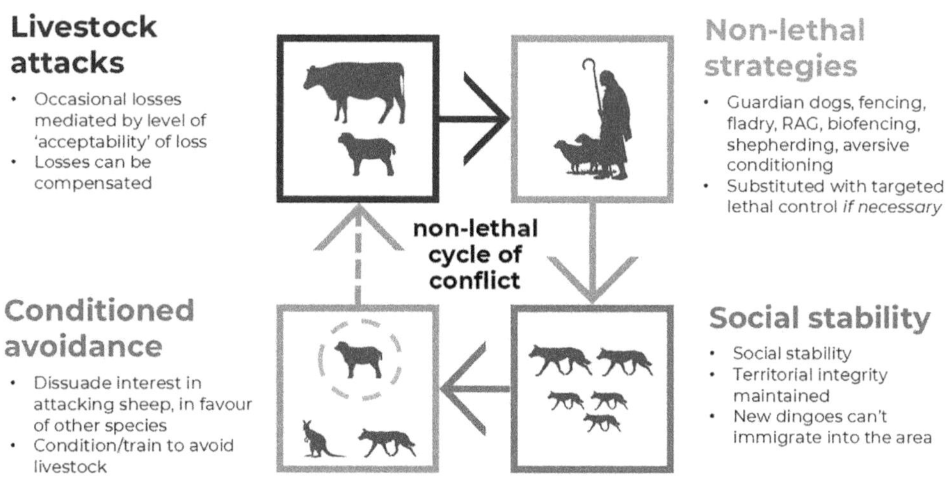

Fig. 19.1. Traditional lethal control methods often backfire by disrupting dingo family groups, leading to more livestock losses. This model outlines a different path: protecting livestock through better husbandry and deterrents and maintaining stable dingo packs that keep out transient individuals. Where needed, specific problem animals are managed using targeted conditioning. With support for farmers during transitions (e.g. compensation for losses), this approach can reduce conflict, support biodiversity and foster long-term coexistence. Source: Bradley Smith

compensation mechanisms for the occasional losses that might still occur, acknowledging that perfect prevention is impossible, but supporting farmers through the transition.

The key difference between these models is their fundamental approach: one reacts to predation after it occurs, often applying a band-aid solution that may trigger more problems than it solves, while the other prevents conflict proactively by working with dingo behaviour. The conflict cycle treats every dingo as a potential threat to be eliminated; the coexistence model recognises that stable dingo populations create predictable, manageable relationships with the landscape and the humans with whom it is shared.

PREDATOR-SMART FARMING

A promising new approach to managing dingoes on farmland is emerging under the banner of 'predator-smart farming'. This represents a management philosophy that seeks to protect livestock while maintaining the integrity of predator populations. This approach acknowledges their ecological role and works with, instead of against, their natural behaviours.

Predator-smart farming is built around proactive deterrence rather than reactive killing. It combines multiple non-lethal strategies: guardian animals, strategic fencing, behavioural deterrents and targeted livestock husbandry practices designed to reduce vulnerability to predation. These include ensuring adequate health and nutrition, encouraging livestock to bunch overnight using salt or mineral licks, increased human presence during high-risk periods like lambing and calving and deploying non-lethal tools when stock are most at risk of attack.

Crucially, this approach capitalises on the social and territorial nature of dingoes. Stable dingo family groups defend their home ranges against intruders, often limiting the movement of transient dingoes that are more likely to target livestock. When these social units remain intact, they act as natural buffers, helping to regulate predation pressure across the landscape.

Social scientist Louise Boronyak has been instrumental in advancing this model and is credited with coining the term predator-smart farming, a reflection of the principle of working smarter, not harder. Her research shows that while non-lethal methods may require upfront investment, they often pay off within a few years, especially when paired with financial incentives, subsidies or technical support. More importantly, predator-smart farming offers a way out of the costly, self-perpetuating cycle of baiting, trapping and culling, a cycle that as previously argued may inadvertently increase livestock losses.

Boronyak emphasises that successful implementation also requires broader institutional support, policy reform and a cultural shift in how dingoes are perceived. Field trials of non-lethal tools, farmer training and the development of peer-support networks are essential to tailor approaches to different regions and livestock systems.

As more farmers embrace this model, there will hopefully be growing momentum for policy changes that move away from lethal control and towards coexistence-based strategies.

WHAT MAKES A GOOD NON-LETHAL DETERRENT?

For all its promise, coexistence with dingoes is not easy. If it were, it would already be the norm. The idea of predator-friendly farming, of managing rather than eliminating dingoes, makes sense in theory. Yet, in practice, very few landholders have fully embraced it because it is fraught with challenges. So what makes living with predators, especially intelligent ones like dingoes, so difficult?

One of the biggest obstacles to effective non-lethal deterrents is habituation. Dingoes are highly adaptable and tenacious. They learn quickly, problem-solve and adjust their behaviour based on experience. A deterrent that works today may be ignored tomorrow. Fences can be climbed or dug under, traps can be avoided and even human presence, once an effective deterrent, can lose its impact over time. For coexistence to succeed, deterrents need to be resilient to habituation and deployed with a deep understanding of dingo biology and behaviour. That includes striking the right balance between predictability and variation, creating strong, memorable associations between behaviours and aversive experiences, without being so consistent that the deterrent becomes background noise.

Around the world, nature provides a blueprint for non-lethal deterrence. Just as bright colours signal danger in venomous animals or thorny plants discourage browsing, we can apply similar principles to dingo management, targeting their instincts and behaviours.

Importantly, effective deterrents must be unpredictable and adapted to dingo intelligence. Dingoes are naturally cautious, which makes them avoid novel stimuli at first. Over time, however, they learn what is dangerous and what isn't. As such, static deterrents quickly lose their effectiveness. A scare device that never changes is likely to soon be ignored as dingoes learn it poses no real threat.

Scarecrows are a perfect example of a non-lethal deterrent with limited effectiveness for intelligent predators. The static figure mimics a human, which seeks to scare off birds. But for dingoes (and intelligent birds like crows), such static deterrents don't last long.

There is also anecdotal evidence that dingoes can learn to avoid poisoned meat baits. When an old or partially degraded bait delivers a sub-lethal dose that makes an animal sick, it can create a powerful aversion to that smell or taste in the future. Social animals like dingoes don't learn in isolation, either; individuals that have had a bad experience can influence the behaviour of their pack mates. Similar avoidance learning has been demonstrated experimentally in species such as rats, suggesting that dingoes may be capable of the same kind of conditioned bait shyness.

The best deterrents are those that either exploit a dingo's natural behaviours, tapping into instincts like territoriality, social structure and their natural fear responses or are resilient to habituation by being unpredictable in the type of stimulus encountered or by reliably creating a strong association between an unwanted behaviour (such as chasing livestock) and an aversive outcome (like being chased off by a livestock guardian dog). This kind of learning is further reinforced when dingoes are presented with clear, low-risk alternatives, such as hunting native prey, which offers no negative consequence.

Historically, shepherding, the classic image of a shepherd with a crook on a hillside alongside the sheep, with sheep bells that would clang upon movement, was the most effective deterrent. When humans actively stayed with their flocks, attacks were rare or at least could be minimised to stop instances such as surplus killing we explored earlier. This has been the best way throughout history, and is still practised in many cultures today, with many in western North America and South Africa returning to the practice of herding. It works because the emphasis is on protecting the sheep, staying with them and fending off would-be attackers. The focus is not on killing the predators. In fact, many cultures adopt this method because the sheep are so valuable and for religious or spiritual reasons they may wish not to harm local predators.

But this labour-intensive practice all but disappeared with the rise of extensive modern grazing in the Western world, replaced by a reliance on passive methods of control. Instead of staying with livestock, farmers shoot, trap and poison dingoes, often relying on government-led predator control programs, expecting the state to keep dingoes at bay. This shift, from hands-on management to outsourced lethal control, left livestock increasingly vulnerable when lethal control fails.

THE NON-LETHAL 'TOOL KIT'

When it comes to reducing predator–livestock conflict, good farm management is by far the most effective and reliable approach. Guardian animals, deterrents and fencing can play important roles, but they are most successful when built on a foundation of sound farm management practices. These proactive measures react to the presence of predators, but importantly help prevent problems from arising in the first place.

Effective farm management

A global review of predator control strategies by van Eeden and colleagues in 2018, found that guardian animals ranked among the best non-lethal methods, but the most consistent success comes from effective property management. Simple yet strategic changes can make a farm far less attractive to predators, reducing the need for other interventions, and thus should be the first and most important step in reducing predation. Key practices can include:

- using temporary or mobile enclosures (such as corrals) during vulnerable periods (e.g. lambing season or harsh weather) to limit exposure to predators
- prompt removal of carcasses to prevent dingoes from associating the property with food
- strategic grazing practices, such as keeping livestock away from high-risk areas during peak predator activity periods
- mixing livestock species (e.g. running sheep with cattle), as larger animals can deter predators
- maintaining optimal stocking rates to prevent animals from being scattered and vulnerable
- regular human presence, such as patrols or shepherding, which significantly reduces predator incursions.

These low-cost, high-reward measures have been shown to significantly reduce predation risk, making them the cornerstone of any non-lethal predator management strategy.

With farm management forming the backbone of predator prevention, other tools can enhance its effectiveness in certain situations.

Livestock guardian animals

Livestock guardian animals are highly effective when properly managed. Livestock guardian dogs (LGD), in particular, not only physically protect stock, but also establish a 'landscape of fear' (see Chapter 9), where predators alter their movement patterns to avoid confrontation. In doing so, they mimic and leverage the territorial behaviour of dingoes themselves, effectively replacing wild territorial cues with those of the guardian dogs. However, they require proper training and ongoing management to be successful.

LGDs are the most widely used, but other species can be just as effective in certain contexts. Donkeys and llamas, for example, are naturally aggressive towards canids, making them a good choice for smaller properties where a pack of guardian dogs may not be practical. Each species has its strengths. For example, dogs work best in large, open environments where they can patrol, whereas donkeys and llamas require less training and ongoing management.

Linda van Bommel has spent years researching LGDs in Australian grazing systems, providing some of the most compelling evidence for non-lethal predator management available today (see her multiple research papers in further reading). Her work confirms what many farmers have been slow to recognise: LGDs are among the most effective tools for protecting livestock from predators like dingoes, feral domestic dogs and foxes, without the ecological disruptions of lethal control.

In one of her landmark studies, which surveyed 150 Australian livestock producers (in 2012 and follow up in 2024), 66% reported that predation on their stock ceased entirely after introducing LGDs and another 30% saw a significant reduction in losses. These figures make it clear that when properly managed, livestock guardian dogs are a long-term, sustainable solution that fundamentally changes how predators interact with livestock.

How do LGDs achieve this? van Bommel's research highlights two primary mechanisms: active deterrence and territorial enforcement. Unlike pet dogs, LGDs bond with livestock from an early age and live among them full-time. When predators approach, the dogs respond immediately – barking, chasing, and, if necessary, confronting the intruder. But their power goes beyond these direct interactions.

In controlled experiments using scent and audio playbacks simulating dingo incursions (2014), LGDs were observed leaving their livestock and travelling up to 570 m to challenge potential threats, vocalising aggressively and marking their presence. Dingoes, highly territorial by nature, responded by avoiding these occupied areas altogether. The sheer presence of LGDs shapes predator movement patterns, creating a powerful psychological barrier that keeps livestock safe.

Importantly, LGDs must be properly bonded, trained and managed to be effective. Simply placing a dog in a paddock is not enough without the right socialisation and oversight; they may roam, harass livestock or fail to deter predators effectively. Large properties often require multiple dogs, as a single animal cannot cover vast areas alone.

Despite their success, some farmers have abandoned LGDs due to challenges in training or unwanted behaviours like excessive roaming. However, van Bommel's research suggests that most of these issues can be mitigated with better training programs, genetic selection and support for farmers new to the system. Many traditional LGD breeds originate from European climates, raising additional challenges in adapting them to Australian conditions. However, for those willing to invest the time in proper management, LGDs remain one of the most cost-effective, ecologically responsible and sustainable ways to protect livestock. To ensure long-term success, there is a clear need for a dedicated organisation or coordinated program to provide ongoing training, resources and support for landholders adopting this method.

Fencing and barriers

Fencing has long been a frontline defence in protecting livestock from dingoes, offering a physical barrier that restricts their access. Electrified fencing, overhanging wires and reinforced netting are among the most commonly used designs, effectively reducing incursions when properly maintained. One large-scale approach in Australia is cluster fencing, where multiple landholders enclose vast areas within a continuous predator-proof fence, pooling resources to protect their stock. Despite being effective, fencing presents challenges. It can be costly to install and requires ongoing maintenance to remain functional. Dingoes, known for their intelligence and adaptability, can find ways under, over or through poorly maintained sections. More concerning is the ecological impact, where barriers can disrupt wildlife movement, alter predator–prey dynamics and create unintended environmental consequences. Despite these concerns, fencing remains one of the most widely used non-lethal strategies and is most effective when combined with other deterrents.

Large-scale predator fencing aims to exclude dingoes across expansive properties, whereas smaller-scale corralling offers a more targeted and flexible approach. These portable or semi-permanent enclosures, often electrified, can be deployed during key periods like lambing or overnight to protect vulnerable stock. Corralling doesn't rely on fencing an entire property, making it more feasible for smaller farms or intensively managed areas. Due to the labour and oversight required, it's less practical for vast, extensively grazed properties. In these remote areas, where constant monitoring is difficult, the suitability of livestock species becomes critical, with cattle generally preferred over sheep due to their lower vulnerability to predation.

Aversive conditioning

Training dingoes to avoid livestock through negative reinforcement is another method that has been trialled with varying success. Aversive conditioning works by associating livestock with an unpleasant experience, discouraging future attacks.

Shock collars (like the ones people give to dogs to stop them barking) have been trialled with wolves and coyotes in the United States. In Wisconsin, Shawn Rossler and team (2012) fitted wolves with remotely activated collars and delivered a mild shock whenever an animal entered a designated 'danger zone' near livestock. The result was that collared wolves reduced their visits to these areas by up to 90% and often avoided them for weeks or months afterwards. Similar work by William Andelt and colleagues (1999) in Colorado used shock collars on problem coyotes and reported short-term reductions in sheep attacks, with some coyotes abandoning predatory behaviour altogether.

Shock collars have been tested on K'gari to deter dingoes from approaching humans and campsites. Effective in some cases, their use is controversial and difficult to implement on a large scale, at least at present. As technology improves, however, this might become a more realistic option, as dingoes can be simultaneously monitored (via GPS and radio transmitters) helping to pre-emptively warn farmers and managers.

But these tools are difficult to use at scale. They require capturing and collaring predators, closely monitoring their movements, and applying the shock at precisely the right moment. They also carry clear ethical and welfare challenges.

Another, potentially promising technique is conditioned taste aversion (CTA), where livestock carcasses are treated with a non-lethal but nausea-inducing substance, such as lithium chloride. When a dingo consumes the treated meat, it experiences temporary illness and may associate that discomfort with eating livestock, discouraging future predation.

Again in North America, CTA has been used successfully on coyotes, significantly reducing their attacks on sheep. In a well-known series of studies in the 1970s and 1980s led by American behavioural ecologist Carl Gustavson (1974), they trained coyotes to avoid specific prey by pairing its taste with nausea. In one experiment, coyotes were fed black rabbits treated with lithium chloride, a harmless chemical that induces temporary illness. After a single episode of nausea, the coyotes refused to hunt black rabbits again, yet continued to chase white rabbits that had not been associated with sickness. The effect lasted for months, showing just how strong and durable these learned aversions can be.

More advanced approaches involve embedding emetic chemicals in bait packets mounted to the necks of sheep, designed to rupture upon being bitten, creating an immediate negative reaction.

However, research relating to CTA has historically been mixed. Some dingoes fail to make the association and new individuals entering the area may not inherit the learned aversion. Newer approaches that incorporate secondary stimuli such as consistent smells with emetic agents (those that induce vomiting or nausea) can help to improve results. Nevertheless, for certain high-conflict areas, it provides a valuable tool, and often last resort, for modifying predator behaviour.

Scent-based deterrents

Dingoes, like many canids, rely heavily on scent marking for communication, using chemical signals to establish territories and warn off intruders. Research into artificial scent boundaries by Ben Walker and Neil Jordan from the University of New South Wales suggests that the urine of dominant male dingoes contains special scent molecules that act like a 'keep out' sign, discouraging other dingoes from entering marked areas.

The practical application is straightforward: by applying dingo urine, or synthetic equivalents, to fence posts, livestock or key entry points, farmers could make an area appear occupied by a dominant dingo, potentially deterring new individuals from encroaching. This plays on dingoes' natural aversion to conflict with stronger rivals. However, the effectiveness of scent-based deterrents remains under study. Initial trials suggest dingoes react to these scents, but they may habituate over time, learning to ignore them if they are not reinforced with an actual territorial threat. This means the markers need to be reapplied frequently or combined with other deterrents for long-term success, unless, of course, scientists can design a synthetic potent urine mix!

At the dingo sanctuary, we once had a group of Buddhist monks regularly visit. Not to see the dingoes, but to collect buckets of dingo poo. They used to take it back and spread it around their monastery to keep possums away. Whether the possums interpreted it as a territorial warning or were simply put off by the smell, the monks swore by its effectiveness. It was an unexpected but fascinating example of how dingo scent, whether from urine or faeces, could be used as a natural deterrent, not just for dingoes but for other animals as well.

Radioactivated guards

Technology has also entered the realm of dingo deterrence, with radioactivated guard (RAG) systems offering a dynamic, responsive approach. These systems use motion sensors or GPS-collared livestock to detect when a dingo approaches a protected area, triggering deterrents such as flashing lights, sirens or alarms. Unlike static deterrents, which dingoes can quickly learn to ignore, RAGs offer the advantage of timing and relevance, activating only when a threat is present, making the deterrent more biologically meaningful. This targeted response helps form a consistent association between the dingo's behaviour and an aversive outcome, while also reducing unnecessary disturbance to livestock and other wildlife. However, RAGs come with challenges. They are expensive to install and maintain, and their effectiveness can be reduced by false triggers, whether from non-target wildlife, environmental factors or malfunctioning sensors. Despite these drawbacks, they represent a promising step towards more intelligent, adaptive predator management.

Scare and sensory deterrents

Exploiting the natural wariness of dingoes, scare devices introduce visual or auditory deterrents designed to startle and repel them. Motion-triggered lights, such as Fox Lights (which mimic a person holding a torch and waving it around), have been widely used to deter nocturnal predators, their unpredictable flashing patterns making livestock areas appear risky to approach.

In my own research, I set out to determine which scare tactics might work best on dingoes. Before any potential method could be tested on a working farm, where failure means real livestock losses, I think it is best to trial them in a controlled environment. Captive dingoes provide the perfect test subjects. Here, I can observe their reactions up close, particularly individuals over repeated trials, and refine the methods.

One of the first techniques I tested was fladry (Fig. 19.2), a method that has been highly successful in deterring wolves in North America. It involves suspending small red flags tied along a fence line, their fluttering motion creating an unpredictable visual barrier that predators supposedly hesitate to cross. Wolves see fladry as an unsettling obstacle. I wanted to know if dingoes did too.

Fig. 19.2. (Top) Fred-a-scare as featured in Smith *et al.* (2020). (Bottom) A captive dingo approaching fladry during my trials. Source: Bradley Smith.

The answer? Not really. When I set up fladry in the dingo enclosure, their reaction was, at best, underwhelming. Initially, they paused, ears pricked, watching the fluttering strips as if something seemed slightly off. But within minutes, about half of them walked straight through to reach the bowl of food placed in the centre of the 'exclusion' zone. A few of them refused to enter, even while watching their mate enter with no consequences. This highlighted that this didn't have the same impact as fladry does for wolves and also that there is a wide range of individual differences for how dingoes respond to novelty.

With fladry ruled out (unless paired with electric wires, which it often called 'turbo fladry'), I needed something more biologically relevant. Something that moved erratically enough to make dingoes genuinely wary. Enter Freddy.

Freddy, or 'Fred-a-scare', as we nicknamed him, was a wacky inflatable scare device (Fig. 19.2). The kind you see at car dealerships, arms flailing wildly in all directions. Unlike fladry, which remained static most of the time, Freddy was pure chaos and much, much taller. The moment he powered on, he exploded to life, whipping and twisting in a completely unpredictable frenzy. This time, the dingoes' reactions were immediate. They bolted. Even the boldest individuals, the ones who had walked straight through the fladry without a second glance, kept their distance. The sheer unpredictability of Freddy's movement seemed to tap into something instinctual.

With every scare tactic, there's always the same looming question: how long before they figure out it might not actually be harmful? Repeated trials showed that most dingoes, when re-entering the enclosure the following few days, did not venture anywhere near the area of the fenced paddock where Freddy stood. A few of the dingoes I tested nervously approached and ate the food positioned near the fence. But this was the strongest deterrent I had tested.

So while a wavy inflatable man might send a dingo running the first time, the real challenge is ensuring they never get comfortable enough to stop running. The key to success isn't just one tactic: it's rotation, variation, and an element of surprise. That's the only way to stay one step ahead of one of Australia's most intelligent predators.

The importance of a multilayered approach

It is probably clear to you by now that no single method will work in isolation. The most effective strategy is a layered approach, starting with proactive farm management, reducing attractants, improving livestock husbandry and increasing human presence. These foundational practices significantly lower predation risk and often make additional interventions unnecessary.

However, when guardian animals, fencing and deterrents are strategically combined with good farm practices, they become far more effective. Guardian animals create a landscape of fear, fencing physically excludes predators and deterrents add unpredictability to disrupt dingo behaviour. Because dingoes quickly habituate to static threats, rotating sensory deterrents, such as motion-triggered lights, sirens and inflatable devices, keep them guessing and maintains their effectiveness.

Ultimately, prevention is always better than reaction. The farms that experience the least conflict are those that manage their land, livestock and risks wisely, reinforcing non-lethal strategies with a flexible, multilayered approach.

DESIGNING FOR DINGO COEXISTENCE

What if we could take everything we've learned about non-lethal management and use it to design a property that embodies those principles? One that makes livestock genuinely difficult for dingoes to predate on livestock. This was the idea behind a model I developed and published in 2021 of a hypothetical farm that promoted coexistence by stacking every evidence-based strategy we have. The goal was to make predation so difficult, so exhausting, risky and so unpredictable that dingoes would choose to hunt elsewhere.

Another key to success (besides a motivated farmer) is timing. Dingoes aren't always a threat and deploying deterrents indiscriminately increases the risk of them learning to ignore them. Instead, smart farms use targeted deterrence and ramp up protection only when it's needed most, such as during lambing season.

So, what does this farm look like? At first glance, it doesn't seem that different from any other livestock property. Paddocks stretch out towards the horizon, stock graze in the open, and windmills spin lazily above the water troughs. But in this hypothetical property, the paddocks near the homestead are smaller and reinforced. These protect the most vulnerable animals, the lambs and pregnant ewes, kept close where they can be watched. The perimeter of these paddocks are lined with electric fencing, forming a psychological barrier as much as a physical one.

Further out, livestock guardian animals roam among the flock. A pair of Maremma sheepdogs move slowly through the mob, appearing indifferent but always alert, scanning the horizon for movement. Elsewhere on the property, a group of guardian donkeys graze alongside cattle. Their territorial nature makes them unpredictable and aggressive towards canines and many a dingo has learned the hard way that approaching the wrong donkey can lead to a swift, skull-cracking kick.

Beyond the paddocks, the farm has been designed to keep dingoes hunting where they prefer, in the bush. Large areas of native habitat have been left intact, full of kangaroos and small mammals. A well-fed dingo with plenty of natural prey is less likely to take the risk of approaching a farm full of barking dogs, electric fences and aggressive donkeys.

Another strategy in play here is mixing livestock species. Sheep are highly vulnerable to predation, but running them alongside cattle changes the dynamic. Cattle are much harder for dingoes to hunt and their presence can help deter attacks on sheep.

For the few that do decide to test their luck, there are further deterrents. Along the fence lines, motion-activated lights and alarms create unpredictable bursts of noise and movement, disrupting any dingo that tries to sneak in. Some areas even employ bioacoustic deterrents, playing recordings of rival dingo howls to create the illusion that this land is already occupied. In more advanced setups, artificial intelligence software can be used to automatically identify dingoes from motion-triggered cameras, based on their shape, gait or other visual cues, and then activate targeted deterrents in real time. These systems, such as a vertebrate pest detect-and-deter device, represent a new generation of smart deterrents that link recognition with immediate response, offering a proactive approach to reducing encounters before conflict occurs

On this property, carcass removal is a routine practice. Leaving dead livestock in the paddock not only attracts dingoes, but encourages them to see sheep as food especially when times are tough. Regularly patrolling and removing carcasses ensures that the farm doesn't inadvertently train dingoes to seek easy meals.

Additionally, the farm employs real-time monitoring technology. GPS-tracked guardian dogs and occasionally the dingoes themselves, coupled with remote cameras linked to an app provide real-time alerts if a dingo is detected near the property. In some areas, thermal and motion-activated detectors or drones trigger alerts, allowing farmers to respond proactively rather than reactively.

This farm isn't 'dingo-proof'. No farm ever will be. But what it does is turn the odds in the livestock's favour. It focuses on protecting livestock rather than removing dingoes. By making every aspect of predation risky, time-consuming and exhausting, it gives dingoes an easy choice: hunt elsewhere. Crucially, farmers also don't expend effort removing kangaroos or other native prey species that dingoes prefer to hunt. This not only saves time and money, but ensures that dingoes have access to abundant natural food sources, making livestock the less attractive option.

Across Australia, a small but growing number of farmers are moving away from lethal control, At present, it's often the same few voices we hear advocating for coexistence, but it's likely there are many more who quietly share these views. Farmers who, for various reasons, don't speak up, perhaps out of fear of backlash or simply a desire to avoid conflict within their communities.

My dream is that, one day, farms like this won't be the exception, they'll be the new normal. Not just in how they operate, but in how farmers think, talk and share. Coexistence should be something farmers feel proud to talk about, not something to keep quiet for fear of backlash. Cultural change often hinges on a tipping point, when enough people start doing things differently, it suddenly becomes the norm. All it takes is a shift in momentum and the quiet supporters become a visible majority.

THE BARRIERS TO NON-LETHAL MANAGEMENT

If non-lethal deterrents can work, why aren't they more widely used? Why do so many farmers still default to lethal control? The answers aren't as simple as cost or convenience, though both play a role. Instead, the barriers to coexistence are deeply rooted in economic structures, social pressures and institutional inertia.

One of the biggest challenges is the initial investment required for non-lethal methods. These approaches can be cost-effective over time, but the upfront costs can be substantial compared to the seemingly inexpensive option of lethal control (especially when baiting is subsidised, and non-lethal requires research and development).

I've encountered this economic argument firsthand in my academic work. When I applied for a research grant to develop and evaluate non-lethal management solutions, the feedback I received from an independent reviewer was rather blunt:

> *'Why should we spend thousands of dollars on deterrents when a bullet only costs 50 cents?'*

This reviewer wasn't personally advocating for lethal control or being deliberately nasty. Rather, they were playing devil's advocate, highlighting the economic reality I needed to address – that for many farmers operating on tight margins, the immediate cost comparison between non-lethal and lethal methods is a significant barrier. Their feedback made clear that for my research to have practical impact, I needed to demonstrate not just that non-lethal methods work, but that they represent a worthwhile economic investment for farmers despite their higher upfront costs.

Beyond financial concerns, social pressure is a major hurdle. Many farmers have spent their entire lives hearing that dingoes are the enemy. The idea of allowing them to remain on the land, even in a controlled way, is often met with scepticism or outright hostility. In some communities, using non-lethal strategies can be seen as a sign of weakness, or worse, as siding with environmentalists over fellow landholders. These attitudes are often reinforced when dingoes are involved in surplus killing events, which can appear to confirm fears about their unpredictability or aggression. Yet ironically, such events are frequently exacerbated by the very lethal control practices used to prevent them.

This reluctance is reinforced by a lack of education and training. While some farmers have successfully adopted guardian animals or specialised fencing, many others simply don't have access to the information or support needed to implement these methods effectively. A poorly trained guardian dog, for example, can become more of a liability than an asset. Without proper guidance, landholders trying non-lethal control for the first time may abandon it at the first sign of failure, reinforcing the belief that only lethal measures work.

And then there's the psychological weight of uncertainty. Lethal control, for all its flaws, offers the illusion of control. Shooting a dingo provides immediate relief; it's a tangible action with an immediate result. Non-lethal strategies, on the other hand, require patience. They require trust in the idea that preventing conflict is better than reacting to it. And in an industry where livelihoods are at stake, trust is hard to come by.

LIVING WITH PREDATORS

If coexistence is the future of dingo management, then the challenge is no longer just about hard evidence that it works (although that remains crucial). It is about making coexistence accessible, viable and scalable for farmers and landholders across Australia. Right now, non-lethal strategies exist, but they remain underutilised, underfunded and underevaluated.

Another challenge lies in turning isolated success stories into widespread adoption. A key part of this shift depends on policy and financial incentives. For decades, governments have funded lethal control measures, subsidising baiting programs, offering bounty payments and maintaining the infrastructure needed for large-scale dingo removal. If these same resources (even a mere 1% of what is spent on lethal control would help!) were redirected towards supporting predator-friendly farming practices, the transition to coexistence would be far easier. Such initiatives could include:

- subsidies for livestock guardian animals and improved fencing, helping farmers invest in non-lethal solutions without the heavy upfront costs
- funding for research and pilot programs, testing new deterrents in real-world conditions and refining best practices

- training and education programs, giving landholders the knowledge and skills to implement effective coexistence strategies
- legal and regulatory support, ensuring that farmers who choose coexistence are not penalised by laws that still mandate lethal control in certain regions.

Although the potential benefits of non-lethal approaches are clear, the transition is not without its challenges. Many landholders face economic constraints, policy limitations and ingrained industry norms that make non-lethal management practices difficult to adopt. These barriers, from financial concerns to social resistance, must be addressed with practical, well-supported solutions if coexistence is to become a viable mainstream approach.

Of course, non-lethal strategies are not perfect. They will not reduce losses to zero. But neither does lethal control, as we have seen. The goal is not to eliminate all risk, but to manage it more intelligently. We need to shift away from the unrealistic expectation that losses must be zero. Non-lethal tools should not be judged by impossible standards. They do not need to be miracle workers or magic bullets to be worth using. Instead, they should be evaluated based on their overall effectiveness, sustainability, and ecological impact.

20
Pathways to coexistence

IN *KING DINGO* BY IVAN SMITH, THE DINGO IS FOUND DEAD AND 'THE HARD, crystallised eyes look out | across the rough skin of Australia'. These words appear alongside a painting of a dead dingo lying on a red desert plain. A confronting image because it is so ordinary. Across much of Australia, this is still how many people encounter dingoes – dead, or not at all. The same poem also tells us, then: 'This is his country. | He is still king of it.'

Dingoes are Australia's wolves. The southern hemisphere's wolf in everything but name. They pair for life. They raise their young together. They hunt cooperatively as a pack and teach their pups the skills they need to survive. In every meaningful way, their behaviour aligns with wolves, coyotes, dholes far more than with domestic, feral or village dogs. They are intelligent, social and they were shaped by Australia- its climate, prey base, social pressures, landscapes and isolation. There is no other canid like them on Earth, and losing them would mean losing something truly irreplaceable.

Recently, I have been spending time in Ngarkat Conservation Park (pronounced *Narr-Kat*), a remote patch of conservation land in South Australia's south-east, as part of a research project to better understand the dingoes that live there. This is classic mallee scrub, gnarled, wind-swept and full of hidden life, with sandy soils and twisted eucalypts stretching in every direction. It's a wild and rugged place and incredibly, despite South Australia's harsh zero-tolerance stance, the only area south of the Dingo Fence where dingoes still persist. But they're barely hanging on.

Ngarkat, along with the adjoining Victorian parks of Big Desert and Wyperfeld, forms an isolated pocket of conservation land surrounded by farmland and sliced by highways. These are ecological islands, cut off from gene flow, and the effects are showing. Dingoes in these parks are under relentless pressure: baited up to four times a year with poisoned meat laid by government-funded rangers, hunted by professional doggers and targeted by neighbouring landholders. The population in Ngarkat has been devastated. In Big Desert and Wyperfeld, the remaining dingoes are highly inbred and genetically imperilled. The new changes to dingo management in this region of north-west Victoria might have come too late.

The real cost of ongoing persecution of the dingo is very real and very present. Whole populations, like those in the Mallee country but equally in areas of human-occupied Australia, are being pushed to the brink, if not already lost. This is not the result of natural forces, but policy choices. This includes in the very places set aside to preserve nature such as national parks; we're actively removing one of its most essential players. Dingoes survive simply because they are resilient.

Ngarkat is special, not just ecologically but culturally. The Aboriginal people who once lived here were known as the Dingo People, a name that speaks to the deep and enduring bond between dingoes and First Nations communities across the continent. That history makes their treatment today all the more painful to witness.

Sometimes I catch myself wondering: what would Ngarkat look like if dingoes were truly allowed to flourish? What kind of country could this be if we chose to share the land with wildlife as equals – if we saw their survival as bound to our own?

I've seen glimpses of it in places like the Great Sandy Desert in Western Australia, where I've been working alongside Aboriginal rangers on their Country (in the form of Indigenous Protected Areas). There, dingoes are respected, admired and protected. Country feels different. The presence of a top predator changes everything, from the behaviour of kangaroos to the presence of threatened species. And it shifts the way people relate to the land, too. There is a sense of balance. A sense of care. These contrasting landscapes, Ngarkat and the desert IPA, symbolise the crossroads we now face.

Throughout this book, I've tried to cut through the myths, the half-truths and the political spin. I've argued, based on years of research and observation of my own, and that of many others, that dingoes are not 'wild dogs'. They're not feral pests. And they're not the hybridised caricatures that outdated science has long claimed. He is 'King dingo'.

Anti dingo narratives persist, but they don't reflect the evidence. They reflect attitudes. But attitudes can change.

Some people will insist there are 'two sides' to the dingo story (exemplified perfectly in Fig. 20.1). That they're dangerous. That they're invasive. That they don't belong here. But these claims often come from vested interests rather than based on the science. From people and industries that benefit from keeping dingoes unprotected and unloved. Without those interests, there would be no real debate. The evidence is strong. The cultural significance is clear. The ecological need is urgent. We must learn to recognise when a 'debate' exists only because someone is trying to preserve power, profit or the status quo. And when that happens, we need to look past the rhetoric and return to the evidence.

Understanding why dingoes deserve protection is one of the most important conversations we need to be having.

I was reminded of this while hosting the Melbourne-based artist Matthew Dunne during a visit to Ngarkat. He'd come to see dingo research in action, to ask questions, observe the landscape and better understand the species through both science and story. One afternoon, we were on a short hike looking for a soak called 'Dingo Well', marked on an old map we'd excitedly discovered that morning at the ranger station. As we walked, he peppered me with question after question, genuinely seeking the underlying truth. He listened closely, challenged assumptions and wanted to get to the bottom of the dingo debate, so that he could represent the truth in his art. One of his most simple questions stood out: 'If you had to explain to someone why dingoes are worth protecting, what would you say?'

Sometimes, it's easy to forget the core reasons why dingoes matter. Being asked to articulate them out loud reminded me how important it is that these reasons drive conversations and that

Fig. 20.1. A sign for Clifton Station, situated in south-western Queensland's Channel Country, captures both sides of the dingo debate. Source: Angus Emmott

we're able to express them clearly and emotionally, in plain terms. Here is a summary of my response, outlining the top three.

Ecological role: Dingoes are native apex predators. They belong here. For thousands of years, they have shaped Australian ecosystems by controlling overabundant herbivores, and more recently, suppressing invasive predators like foxes and cats, and indirectly protecting smaller, threatened animals. In a continent defined by long ecological relationships, dingoes are a key part of how the land works.

Cultural significance: Dingoes are culturally significant. To Aboriginal Australians, dingoes are kin. They are embedded in Dreaming stories, law, ceremony and identity. In many communities, they are family, companions, protectors and spiritual beings. Their tracks are part of songlines that cross the continent, connecting people to Country and to each other. The relationship is deep and enduring, shaped by thousands of years of shared life and meaning.

More broadly, dingoes have long been part of Australia's story. They appear in early exploration journals, settler accounts and bush poetry. They were companions, threats and symbols that were feared, admired, misunderstood. From goldfields to cattle stations, from desert tracks to campfires, the dingo has walked through the pages of Australian history and culture. Their presence has shaped the way Australians have understood the land, its dangers and its wildness.

Unique identity: Dingoes are Australian (or southern hemisphere) wolves. Not in name, but in nature. They pair for life. They raise their young together. They teach their pups to hunt. In every meaningful way, their behaviour aligns with wild canids such as wolves, coyotes, dholes, more than it does to dogs (domestic, feral or village). They are intelligent, social and deeply connected to the land. In his poem *King Dingo*, Ivan Smith captures this when he writes that dingoes are 'not a dog, not a wolf'. There is no other canid like them on Earth. They belong to Australia and to lose them would be to lose something irreplaceable.

It's strange, then, that such an extraordinary animal isn't more widely celebrated. I remember receiving an email many years ago from Peter Savolainen, a canine geneticist from Sweden, who had been reading through the scientific literature on dingoes. He was baffled by the tone and wrote to ask: 'Why does Australia hate the dingo so much? They're so unique, something to be proud of.' He had a point. Australia is a continent largely devoid of predators, and dingoes are the largest and (arguably!) most impressive. You would think they'd be woven into our childhood imagination. The kind of creature that ends up in picture books and toy boxes, the hero or the hunter in the little dramas children act out on their bedroom floors. Instead, it takes an international scientist to remind us just how remarkable dingoes are and how little pride we seem to take in them.

The truth is, we've spent more time trying to get rid of dingoes than trying to understand them, and all it has delivered is a fractured, dysfunctional system. Landscapes are suffering and farmers still don't have the certainty they were told lethal control would bring. Dingoes are protected in one valley and poisoned in the next. We've let emotion and industry interests dictate decisions that should be guided by science, ethics and the broader public good.

At its core, our treatment of dingoes reflects a persistent colonial mindset. One that sees land as something to be tamed and controlled, animals as either resources or threats and ecological balance as something we impose rather than understand. This worldview has already cost Australia dearly in lost species, degraded ecosystems and fractured relationships with Country. Dingoes, as apex predators, fundamentally challenge this control-based approach; they resist being managed like machines, forcing us to confront a profound question: do we continue down this path of suppression, clinging to outdated fears and inherited attitudes? Or do we finally embrace what both science and Indigenous knowledge have long affirmed: that dingoes belong here and our future is immeasurably richer with them in it.

I recognise that for farmers and pastoralists, concerns about livestock losses are real and significant. A family's livelihood can be deeply affected by predation and these economic impacts shouldn't be dismissed. However, the evidence increasingly shows that coexistence strategies can be both effective and economically viable, at least in many scenarios. Guardian animals, like Maremma sheepdogs, the evidence tells us, have been shown to be remarkably successful in protecting livestock in dingo country. Strategic fencing around lambing areas rather than entire properties can provide protection during vulnerable periods. In parts of Europe and North America, similar approaches have allowed wolves and livestock to share landscapes in ways once thought impossible. A handful of progressive graziers in Australia have already begun implementing these methods, finding that working with the landscape, including its predators, can lead to more resilient agricultural systems. The challenge is to develop and support smarter approaches that serve both ecological and economic needs.

Underpinning all of this is the need to recognise that we all have skin in the game. The decisions we make about dingoes affect not just individual landholders or government departments, but the health of the entire Australian landscape.

The dingo has survived ice ages, isolation and invasion. In doing so, this species has kept the future open – not just for itself, but for the landscapes that depend on it. But it won't survive indifference. If there's one thing I hope this book has shown, it's that we have a choice. Whether we accept a landscape without dingoes, or choose a future where we live alongside them. If we are guided by the evidence – and we should be – the path is clear.

COEXISTENCE

Coexistence is not radical. It's a return to relationships that existed long before colonisation carved the land into paddocks and declared war on predators. For thousands of years, First Nations people lived alongside dingoes. Today, that kind of coexistence is no longer just a moral ideal, but a practical, ecological, cultural and economic necessity.

For many farmers, coexistence still feels impossible. Livestock and predators are seen as locked in eternal conflict. Yet it is important to remember that this is belief is not a law of nature, but rather a perspective. True coexistence means recognising dingoes as allies in land management, not threats to be removed. It goes further than simply tolerance.

That shift begins with education, policy reform and public storytelling. Farmers will not adopt non-lethal methods if they do not believe they will work. The public will not support change if dingoes remain cast as villains.

WHY HASN'T COEXISTENCE BEEN FULLY EMBRACED?

As I have tried to argue in the last few chapters, lethal control is not a long-term solution, not for protecting livestock and not for maintaining healthy ecosystems. Baiting, culling and exclusion fences disrupt dingo pack structures, fuel mesopredator outbreaks and can even lead to greater livestock losses.

If science alone guided policy, lethal control would have ended years ago. But the real barriers to coexistence are political, economic and deeply cultural.

Louise Boronyak, through interviews with landholders, found that most do not see lethal control as ideal. But they feel stuck. They have inherited a system in which dingoes are seen as pests, the government supplies poison baits and coexistence sounds like a fantasy dreamed up in the city. Many do not want to manage the issue themselves nor do they have the knowledge of how dingoes behave and what deters them. They just want it taken away.

This expectation has shaped the way funding and responsibility are distributed. Farmers who want to kill dingoes are often subsidised to do so. Those who want to adopt non-lethal strategies are expected to pay themselves. The imbalance is clear: if a landholder sets out baits or shoots dingoes, the government provides resources or bounties. If they want to invest in non-lethal alternatives, they're on their own. Worse, the social cost can be high: landholders who choose non-lethal paths are often judged, second-guessed, or quietly frozen out by the community. Many farmers feel they are on their own, caught in a system that assumes conflict with predators is inevitable.

Government control programs have become self-sustaining. Millions of dollars are poured into baiting, shooting and fencing. Not because the evidence shows that these programs work, but because that is what has always been done. Success is often measured by how many dingoes are killed, not by whether livestock losses are reduced. Boronyak refers to this as 'policy lock-in', where lethal control is not just encouraged. It is built into the system.

In our research in 2021, led by ecologist Helen Waudby, we argued that the persistence of lethal control has less to do with effectiveness and more to do with institutional momentum. Dingo management is shaped by policies, social norms and land tenure systems that continue to favour killing over adaptation. Until we tackle the social, political and institutional conditions that favour killing, no amount of evidence will shift behaviour. As ecologist Thomas Newsome put it, 'We risk finding ourselves in 2040 still arguing about the same things: hybridisation, ecological effects, dingo status, while the killing continues.' Unless we break this cycle, the debates will keep looping and dingoes will keep dying, regardless of how much evidence we have.

Breaking this cycle will require dismantling the structures that make killing the easiest, cheapest and most supported option.

WHO PAYS FOR COEXISTENCE?

If Australia is serious about shifting away from conflict and towards coexistence, then the funding model must change.

Lethal control is cheap for landholders because the public pays for it. Poison baits are supplied by governments. Aerial shooting programs are publicly funded. In some regions, bounties are offered for every dead dingo. These incentives keep lethal control as the default. If farmers are expected to shift their practices in the interest of broader environmental and ethical goals, then they should be supported to do so. This is no different from the way governments subsidise water-saving infrastructure, renewable energy or land regeneration. Coexistence with dingoes is a public good. It should come with public investment.

Where public funds support either lethal or non-lethal strategies, they should be tied to clear, transparent evidence of success. If we are spending millions of dollars managing predators, the public deserves to know whether it is working.

There's also potential for market-driven change. Ethical food and fashion labels have transformed parts of the livestock industry (e.g. free-range, organic, grass-fed meats; responsible wool standards). Why not predator-friendly? Consumers increasingly want to support farms that prioritise animal welfare and environmental responsibility. A certified label or incentive scheme for predator-smart farming could reward landholders who choose coexistence, not control.

The bottom line is that coexistence should not be a personal burden. It should be supported as a matter of national policy. That means shifting incentives, closing legal loopholes and designing support systems that help farmers adapt. Without that shift, coexistence will remain the harder path, even when it's the right one.

PUBLIC PERCEPTIONS AND THE POWER OF LANGUAGE: FROM WILD DOG TO DINGO

For more than a century, dingoes have been trapped in a war of words that has little to do with the dingoes themselves. They are rarely seen for what they are: intelligent, social, ecologically important canids, but instead cast in simplistic terms that strip away their complexity. As environmental social scientist Karen Hytten (2009) has argued, dingoes are often framed in binary ways: native or feral, pest or protector, good or bad. These categories are not neutral. They reflect deep cultural narratives, rooted in settler anxieties and control-based thinking, which have long portrayed dingoes as untrustworthy, dangerous or morally inferior.

This shapes how people relate to dingoes and whether they believe they should be protected or removed. A binary frame helps sustain the idea that dingoes do not belong, that they are a threat to be managed rather than an animal to be understood.

Reforming how dingoes are classified in law is also essential. In many jurisdictions, dingoes are still labelled as 'wild dogs' and treated as biosecurity threats. This legal framing allows governments to justify funding lethal programs under pest control law, while non-lethal strategies fall through the cracks. If dingoes were formally recognised and protected as native apex predators, funding priorities would need to reflect that reality.

I remain hopeful that we can reclaim the name 'dingo', shifting the conversation away from control and toward coexistence. Only then can we break free from the war that has defined Australia's relationship with its top-order predator.

THE NEED FOR CONSISTENT (LEGAL) PROTECTION

As first mentioned in Chapter 1, across Australia, dingoes are treated differently depending on where they are. Even in jurisdictions that recognise their conservation status, contradictory policies persist. In Victoria, dingoes are listed as vulnerable under the *Flora and Fauna Guarantee Act*, yet the state continues to issue lethal 'unprotection orders'. In 2024, one such order was extended to 2028, authorising poison baiting, shooting and trapping across public and private land (and excluding the north-west as we covered in Chapter 10). This decision is now being challenged in the Victorian Supreme Court by Animals Australia, an organisation that argues that it undermines the dingo's listed status and lacks proper transparency.

The island tourist destination K'gari reveals a different kind of contradiction. Despite its World Heritage status and the formal protection of dingoes, lethal control continues under the guise of public safety. There is still no dedicated, up to date conservation plan for dingoes on the island and population monitoring remains limited. The protection that exists is fragile, conditional and easily overridden, highlighting how even recognised conservation areas fail to offer real security for dingoes.

At the local level, some councils are pushing back. In 2023, Byron Shire in New South Wales committed to non-lethal dingo management and introduced a moratorium on lethal control across all council-managed lands. The council publicly recognised the dingo as a native species with ecological importance. Such local reforms are promising, but they highlight the broader problem: dingo protection depends entirely on jurisdiction. There is no unified national framework, and without one, there can be no consistent path to coexistence.

Cultural rights and legal recognition

As legal scholars Rebecca Hoare and Juliette King pointed out in 2023, even though dingoes are a cultural keystone species, there are no formal legal rights for Indigenous communities to be involved in decisions about how dingoes are managed. In Queensland, for example, there's a law on the books that was meant to allow the cultural use of protected animals, but it's never been put into effect. And dingoes aren't even officially recognised as protected in that state to begin with.

In recent years, efforts have been made to change this. The *Dingo Cultural Heritage and Protection Bill*, introduced in the New South Wales Parliament in 2022, proposed formal recognition of the dingo as a species of cultural and ecological significance. The Bill aimed to establish a framework for protecting dingoes as part of Aboriginal cultural heritage, including provisions for Indigenous-led management, limits on lethal control and requirements for consultation and planning processes. Although the Bill did not pass, it offered a clear legislative model that would embed both cultural and ecological values into dingo policy.

The 2023 National First Nations Dingo Forum (mentioned in Chapter 10) reinforced these calls for change. The resulting Dingo Declaration rejected the 'wild dog' label and asserted the right of Aboriginal communities to protect and manage dingoes. These statements reflect internationally recognised rights and present a clear legal and policy direction for reform.

If the law is to reflect the full significance of dingoes, Aboriginal voices must be included. That means formal roles in policy development, funding for Indigenous-led conservation, and legal recognition of dingoes as culturally significant fauna.

Building the legal foundations for coexistence

Several reforms are essential if dingo management is to shift from conflict to coexistence.

- End the 'wild dog' classification. Dingoes should be recognised in law as native fauna in all jurisdictions. Removing the wild dog label, and ensuring their classification is not the same as the domestic dog (*Canis familiaris*) is a necessary first step towards consistent protection.
- Appoint an independent expert panel. To resolve ongoing classification/taxonomy disputes, an independent panel should be established, similar to those used for red and

Mexican wolves in North America, to formally determine the dingo's taxonomic and conservation status.
- Repeal mandatory culling laws. Dingoes should not be seen as a biosecurity threat and landholders should not be legally required to destroy dingoes. Coexistence must be a legally supported option, not an exception.
- Amend the *Environment Protection and Biodiversity Conservation Act.* Dingoes should be formally listed under the EPBC Act as a threatened native species or a species of national environmental significance. This would trigger federal oversight, require the development of national conservation and recovery plans, and establish a consistent baseline for legal protection and management across jurisdictions.
- Mandate non-lethal funding. Currently, government funding overwhelmingly supports lethal control programs. A portion of this investment should be legally directed towards non-lethal strategies, including guardian animal programs, targeted fencing, and community-based coexistence initiatives. Public funds must support coexistence, not just lethal strategies.
- Embed Indigenous leadership in law and policy. Their knowledge, authority and rights must be formally recognised in legislation, with mandated involvement in all decision-making and management relating to dingoes. Populations identified as having particular cultural significance should receive specific legal protection.

In addition, Australia needs a National Dingo Conservation Strategy. Without a consistent national framework, dingo policy will remain fragmented, reactive and dominated by short-term political priorities. This strategy must ensure that all key stakeholders, including Indigenous communities, conservation scientists, landholders and government agencies, have equal say and decision-making rights. Coexistence cannot succeed unless it is built on shared responsibility, mutual respect and genuine collaboration.

There is no technical reason these laws cannot change. It is a question of values, incentives and public interest. If legal protections for dingoes are to improve, those shifts will need to happen not only in legislation, but in the broader conversations that shape what is seen as necessary, acceptable and possible.

A NETWORK OF POSSIBILITIES

The path to coexistence is not linear. It is more like a web: a network of shifting norms, old stories, new tools and the relationships that bind people to place. Policy and law are vital, but they are only one thread in a much larger weave.

Seeds of coexistence

Some of the most powerful transformations happen through conversations: between neighbours, across kitchen tables or at the boundary fence. Farmer networks and peer-to-peer learning break through the isolation that many graziers feel when facing management of predators. When landholders hear how others are succeeding with guardian animals, predator-smart farming or adaptive grazing, coexistence stops being a theory and starts becoming a possibility. These stories, practical and honest, from people who have taken the risk, often carry more weight than any government brochure or academic paper.

Better science is equally critical. Much of what drives lethal control today is based on outdated assumptions or flawed data. Dingo population estimates remain patchy. The long-term effects of culling on pack stability, mesopredator release and ecosystem function are rarely tracked. Non-lethal methods, though promising, are still under-researched. We need robust studies that examine not just predator behaviour, but also the economic outcomes of coexistence: how guardian animals affect long-term livestock productivity, how targeted fencing influences stress and injury rates and what the true cost–benefit profiles of these strategies look like. As such, we need public investment in science that asks different questions. Questions about resilience, regeneration and the long-term sustainability of coexisting with apex predators.

Consumer pressure adds another layer. As ethical food markets grow, there is increasing demand for products that are humane and ecologically responsible. Farmers who adopt coexistence strategies are well positioned to tap into these values, with opportunities for niche markets, export access and brand differentiation. Predator-friendly certification, if implemented credibly, could help shift both perception and practice. These consumers are shaping the future of agriculture.

Coexistence can also represent opportunity. In parts of Australia, there is growing interest in ecotourism centred on dingoes. Visitors are drawn to wild, intact landscapes. The chance to see Australia's native apex predator behaving naturally can be a powerful attraction (as it already is at K'gari). Sanctuaries, cultural tours and educational programs could all build economic value around dingoes, not in opposition to them. This kind of conservation tourism not only brings revenue to regional communities, it also reinforces the importance of preserving natural predator–prey dynamics on Country.

These shifts will depend on something intangible yet powerful: social licence. Industries, whether farming, conservation, or government, only maintain legitimacy if they reflect public values. When the public learns that dingoes are being killed in national parks or that taxpayer dollars fund programs that may do more harm than good, support begins to erode. The case for coexistence is a question of trust, legitimacy and public expectation, just as it is scientific and economic. And the social licence for the broadscale killing of dingoes is wearing thin.

From change to transformation

In 2021, conservation scientist Lily van Eeden and the now familiar names from University of Sydney and University of New South Wales (Chris Dickman, Matthew Crowther and Thomas Newsome) proposed a Theory of Change to shift dingo management away from conflict and towards coexistence. Their approach argues that lasting change is not just about better tools or more evidence. It depends on understanding the mix of values, identities and social pressures that shape how people respond to dingoes.

They identified four key groups: farmers, policymakers, Aboriginal communities and the broader public, and outlined specific strategies for each. For farmers, this included peer-to-peer learning and showing how non-lethal tools can work in real-world conditions. For policymakers, the focus was on improving transparency, embedding cultural perspectives and shifting incentives away from lethal control. The public, meanwhile, plays a role in shaping the social licence for dingo management through awareness, consumer choices and the stories told in the media.

While this framework focuses on changing behaviour and beliefs within different groups, other researchers have looked at how to shift the system itself.

Building on this, Louise Boronyak and Brent Jacobs from the University of Technology Sydney (2023) mapped out seven interconnected pathways for change, ranging from legal and institutional reform to shifts in culture, markets and knowledge systems (Fig. 20.2). One of their proposals was to reframe predator losses as a shared problem rather than an individual burden, which would open the door to public investment in coexistence strategies and support for landholders who adopt non-lethal tools. Their work highlights the need for coordinated action across sectors, including new policies, as well as a deeper transformation in how dingoes are valued, understood and lived with.

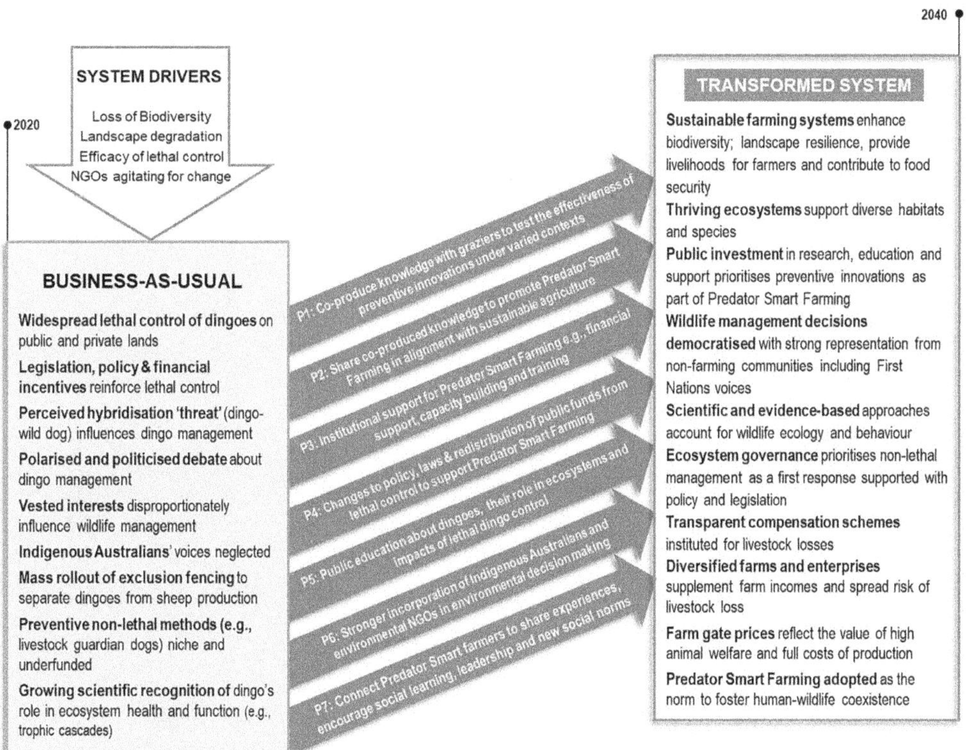

Fig. 20.2. Seven pathways to help graziers and dingoes coexist. This model shows how different changes – better policies, public awareness, Indigenous leadership and support for farmers – can work together to move away from lethal control and towards more sustainable, predator-smart farming. Source: Boronyak and Jacobs (2023).

At its core, both frameworks recognise that dingo management is as much about social and political context as it is about the environment. Changing attitudes and underlying beliefs is slow work, but it may be the most important shift of all.

FINAL THOUGHTS

If you've made it this far, I hope your appreciation for the dingo has grown. For some, that might mean a deeper respect or a few new insights to add to what you already knew. For others, perhaps it means seeing dingoes in a new light – not as villains, but as vital players in one of the world's most remarkable ecosystems. Either way, my hope is the same: that you now see them as rightful inhabitants of this land, deserving of our respect and protection.

To lose the dingo through mismanagement, misinformation, policy failure, ignorance or indifference would be an astonishing tragedy. We've been down this road before. Australia has already wiped out one of its remarkable apex predators: the thylacine, and is rapidly losing more native species each year. We cannot afford to keep making the same mistakes.

Coexistence is possible. It is necessary. And it starts with understanding the dingo, the land and our place within it.

If this book has moved you to action, there are ways to help. Engage with local conservation groups working to protect native predators. Write to your representatives to advocate for stronger dingo protections in national conservation policies. Visit sanctuaries to learn about dingoes first-hand. Support Indigenous ranger programs that embed traditional knowledge into dingo management.

Share what you've learned. Challenge misconceptions when you hear them. Question the narratives. Critically examine the statistics and media coverage. Look behind the headlines and ask: who benefits from anti-dingo sentiment?

Each of these actions, however small, contributes to a future where dingoes are valued, not vilified. Australia's apex predator deserves nothing less.

Acknowledgements

The dingo is a rich and complicated subject, scientifically, culturally and politically, and doing it justice required careful thought and persistence. One of the biggest challenges was finding a way to structure the book because the dingo story isn't linear: its origins, its biology, its difference from dogs and even the laws that govern it are all deeply entangled. Genetics alone weaves through multiple chapters, touching on ancestry, taxonomy, domestication and policy. Pulling apart these threads without losing sight of the whole often felt like trying to map a living system. Fortunately, I had the support, insight and generosity of others to help make sense of it all. I'm deeply grateful to the many people who lent their wisdom, challenged my thinking and helped give shape to these pages.

Particular thanks to Rob Appleby, now also a doctor of dingoes after the longest PhD known to science. He has been a generous source of knowledge and a steady presence since we first met on K'gari in 2008, when he gave me my first encounter with wild dingoes. We'd sit in the sand for hours, scanning the dunes and swapping stories. I'd pass the time learning to make his port-infused tobacco rollies, a fond memory from those early days. And to Kylie Cairns, who has patiently helped me make sense of more genetics papers than I care to admit. Both Rob and Kylie have been constant sources of insight, always up for talking dingoes, and I'm glad to count them as both collaborators and friends.

My heartfelt thanks to those who generously took the time to read and review this book in its various stages: Daniela Osiander (who read early drafts while driving around Australia with her partner Shannon and their two dingoes Jarrah and Bindi), Jenny Morton, Rob Appleby, Kylie Cairns and Stacey Smith. Your feedback, insight and encouragement were invaluable and helped ensure this book stayed grounded in both evidence and clarity.

Every argument made within these pages has been underpinned by rigorous research and for that, I owe a deep debt to the extraordinary community of researchers I've been privileged to know, collaborate with or be inspired by. If I've quoted you, cited you or disagreed with you, know that your work has helped shape this story.

To the dingo researchers and field ecologists who have advanced our understanding of dingoes' behaviour, ecology and place in the landscape: Euan Ritchie, Mike Letnic, Matthew Crowther, Thomas Newsome, Chris Dickman, Chris Johnson, Dale Nimmo, Katherine Moseby, Linda Behrendorff, Linda van Bommel, Arian Wallach, Eloïse Déaux, Tim Doherty, Al Glen, Aaron Greenville, Rob Appleby, Neil Jordan, Ben Pitcher, Damian Morrant, William Parr, Peter Banks, Corey Bradshaw, Daniel Hunter, Brad Purcell, Stephen Fleming, Guy Ballard, Tracey Keplins, Lee Allen, Ben Allen, as well as the 'original' CSIRO dingo scientists including Peter Thomson, Peter Catling, Laurie Corbett and Alan Newsome.

To those working in genetics, taxonomy and origins, whose insights have shaped the foundational questions of what a dingo is and where it came from: Kylie Cairns, Peter

Savolainen, Yassine Souilmi, Danielle Stephens, Andrew Weeks, Gabrielle Conroy, Justin Adams, Stephen Jackson, Kathryn Lord, Clive Wynne, Melanie Fillios, Bill Ballard and the late NWG Macintosh, Klim Gollan, Ray Coppinger, Alan Wilton and Janice Koler-Matznick.

To scholars and advocates exploring the social, cultural and ethical dimensions of human–dingo relationships: Lily van Eeden, Louise Boronyak, Loukas Koungoulos, Adam Brumm, Sonya Takau (Jirrbal, Girringun Nation), Stuart Harradine (Wotjobaluk Nation, Barengi Gadjin Land Council), Fiona Probyn-Rapsey, Justine Philip, Merryl Parker, Paul Taçon, Fred Cahir, Ian Clark, Jane Balme, Sue O'Connor and Clare Archer-Lean, as well as the many anthropologists that covered the dingo, including but not limited to Deborah Rose-Bird, Annette Hamilton, Norman Tindale and Mervyn Meggitt.

Of course, our work is far from finished. As with all science, our understanding of dingoes continues to grow. New discoveries will refine, challenge and reshape parts of what we think we know, and I look forward to seeing the research progress in the years ahead.

To those who speak up for dingoes, in small ways or on big stages, thank you. Your voices matter. These include Roland Breckwoldt, Lyn and Peter Watson, Bernice Walters, Gill Ryhorchuk, Bruce Jacobs, Lionel Hudson, Barry Oakman, Leigh Mullan and Darryl Griffiths, who have either written books and/or set up various organisations to help dingo advocacy and education.

To all the volunteers, past, present and future, thank you. Your dedication, time and passion have made an incredible difference, whether behind the scenes or out in the field. Special thanks to Lyn Watson and Mel Browning from the Australian Dingo Foundation and to the tireless crew at the Dingo Discovery Sanctuary and Research Centre, such as Kevin Newman, Favel Parrett, Gale Day, and Paul Webb (and many others past and present) who devote themselves every day to protecting and advocating for this remarkable animal. May you, and others at similar organisations and facilities, keep up the fight. Your work is more important than ever and while progress is slow, it is being made.

To the team at CSIRO Publishing, Briana Melideo, George Knott, Tracey Kudis and Mark Hamilton. Thank you for not only giving me the opportunity to pursue this project, but for your ongoing support in helping me bring it to life, and for the patience and flexibility you showed as the manuscript continued to be revisited and reshaped over time.

Finally, my deepest thanks go to my family, Stacey, Hunter, Aria and Aster. You've put up with countless dingo-related conversations, travel schedules and long stretches of writing. Thank you for your patience, your love and for reminding me of what really matters. And of course, to Maya, our four-legged friend, who kept me company in my office and distracted me with long walks that helped to clear my head.

These pages carry the weight of many conversations, challenges and moments of shared insight I have had over many years. I might be the one who wrote it down, but this book is the result of years of collective effort, and the curiosity and care of many others who believe the dingo's story is worth telling.

Chapter sources and further reading

INTRODUCTION

Beilby W (1897) *The Dog in Australasia.* George Robinson and Company, Melbourne.

Commonwealth of Australia House of Representatives (1973) Parliamentary Debates. (Official Hansard No. 47). November 20, 1973, p. 3489. https://parlinfo.aph.gov.au/

Department of Primary Industries and Regions (2023) *South Australian Wild Dog Management Strategy 2023–2033.* Government of South Australia, Adelaide. https://www.pir.sa.gov.au/__data/assets/pdf_file/0006/438513/sa-wild-dog-management-strategy.pdf.

Dickman CR, Newsome TM, van Eeden LM (2021) The dingo dilemma: a brief history of debate. *Australian Zoologist* **41**, 298–321. doi:10.7882/AZ.2020.029

Donfrancesco V, Allen B, Appleby R, Behrendorff L, Conroy G, Crowther M, Dickman C, Doherty T, Fancourt B, Gordon C, Jackson S, Johnson C, Kennedy M, Koungoulos L, Letnic M, Leung L, Mitchell K, Nesbitt B, Newsome T, Pacioni C, Phillip J, Purcell B, Ritchie E, Smith B, Stephens D, Tatler J, van Eeden L, Cairns K (2023) Understanding conflict among experts working on controversial species: a case study on the Australian dingo. *Conservation Science and Practice* **5**, e12900. doi:10.1111/csp2.12900

Hudson L (1974) *Dingoes Don't Bark.* Seal Books, Adelaide.

Kreplins TL, Gaynor A, Kennedy MS, Baudains CM, Adams P, Bateman PW, Fleming PA (2018) What to call a dog? A review of the common names for Australian free-ranging dogs. *Pacific Conservation Biology* **25**, 124–134. doi:10.1071/PC18018

Newsome AE, Corbett LK, Best LW, Green B (1972) The dingo. *Australian Meat Research Committee Review* **14**, 1–11.

Newsome TM, Ballard GA, Crowther MS, Dellinger JA, Fleming PJ, Glen AS, Greenville AC, Johnson CN, Letnic M, Moseby KE, Nommo DG, Nelson MP, Read JL, Ripple WJ, Ritchie EG, Shores CR, Wallach AD, Wirsing AJ, Dickman CR (2015) Resolving the value of the dingo in ecological restoration. *Restoration Ecology* **23**, 201–208. doi:10.1111/rec.12186

Parker M (2007) The cunning dingo. *Society & Animals* **15**, 69–78. doi:10.1163/156853007X169351

Pringle EL (1990) The Dingo baby case. *De Rebus* **272**, 570–579.

Probyn-Rapsey F, Lennox R (2022) Feral violence: the Pelorus experiment. *Environment and Planning. E, Nature and Space* **5**, 362–380. doi:10.1177/2514848620976959

Twain M (1897) *Following the Equator - A Journey Around the World.* American Publishing Co. and Doubleday & McLure Co, Hartford and New York.

PART 1: A NEW ARRIVAL

Ardalan A, Oskarsson M, Natanaelsson C, Wilton AN, Ahmadian A, Savolainen P (2012) Narrow genetic basis for the Australian dingo confirmed through analysis of paternal ancestry. *Genetica* **140**, 65–73. doi:10.1007/s10709-012-9658-5

Arendt M, Cairns KM, Ballard JWO, Savolainen P, Axelsson E (2016) Diet adaptation in dog reflects spread of prehistoric agriculture. *Heredity* **117**(5), 301–306. doi:10.1038/hdy.2016.48

Attard MRG, Chamoli U, Ferrara TL, Rogers TL, Wroe S (2011) Skull mechanics and implications for feeding behaviour in a large marsupial carnivore guild: the thylacine, Tasmanian devil and spotted-tailed quoll. *Journal of Zoology* **285**, 292–300. doi:10.1111/j.1469-7998.2011.00844.x

Ballard JWO, Field MA, Edwards RJ, Wilson LA, Koungoulos LG, Rosen BD, Chernoff B, Dudchenko O, Omer A, Keilwagen J, Skvortsova K, Bogdanovic O, Chan E, Zammit R, Hayes V, Aiden EL (2023) The Australasian dingo archetype: de novo chromosome-length genome assembly, DNA methylome, and cranial morphology. *GigaScience* **12**, giad018. doi:10.1093/gigascience/giad018

Balme J, O'Connor S (2016) Dingoes and Aboriginal social organization in Holocene Australia. *Journal of Archaeological Science, Reports* **7**, 775–781. doi:10.1016/j.jasrep.2015.08.015

Balme J, O'Connor S (2021) The coming of the dingo In: *The Oxford Handbook of the Archaeology of Indigenous Australia and New Guinea*. (Eds C Smith and K Smith) pp. 361–380. Oxford University Press.

Balme J, O'Connor S, Fallon S (2018) New dates on dingo bones from Madura Cave provide oldest firm evidence for arrival of the species in Australia. *Scientific Reports* **8**, 9933. doi:10.1038/s41598-018-28324-x

Bergström A, Frantz L, Schmidt R, Ersmark E, Lebrasseur O, Girdland-Flink L, *et al.* (2020) Origins and genetic legacy of prehistoric dogs. *Science* **370**, 557–564. doi:10.1126/science.aba9572

Brumm A, Koungoulos L (2022) The role of socialisation in the taming and management of wild dingoes by Australian Aboriginal people. *Animals (Basel)* **12**, 2285. doi:10.3390/ani12172285

Cahir F, Clark I (2013) The historic importance of the dingo in Aboriginal society in Victoria (Australia): a reconsideration of the archival record. *Anthrozoos* **26**, 185–198. doi:10.2752/175303713X13636846944088

Cairns KM (2021) What is a dingo: origins hybridisation and identity. *Australian Zoologist* **41**, 322–337. doi:10.7882/AZ.2021.004

Cairns KM, Wilton AN (2016) New insights on the history of canids in Oceania based on mitochondrial and nuclear data. *Genetica* **144**, 553–565. doi:10.1007/s10709-016-9924-z

Cairns KM, Shannon LM, Koler-Matznick J, Ballard JWO, Boyko AR (2018) Elucidating biogeographical patterns in Australian native canids using genome wide SNPs. *PLoS One* **13**, e0198754. doi:10.1371/journal.pone.0198754

Cairns KM, Nesbitt BJ, Laffan SW, Letnic M, Crowther MS (2020) Geographic hot spots of dingo genetic ancestry in southeastern Australia despite hybridisation with domestic dogs. *Conservation Genetics* **21**, 77–90. doi:10.1007/s10592-019-01230-z

Cairns KM, Crowther MS, Parker HG, Ostrander EA, Letnic M (2023) Genome-wide variant analyses reveal new patterns of admixture and population structure in Australian dingoes. *Molecular Ecology* **32**, 4133–4150. doi:10.1111/mec.16998

Corbett LK (1985) Morphological comparisons of Australian and Thai dingoes: a reappraisal of dingo status, distribution and ancestry. *Proceedings of the Ecological Society of Australia* **13**, 277–291.

Corbett LK (2001) *The Dingo: In Australia and Asia*. 2nd edn. J.B. Books, Marleston, South Australia.

Costello O, Webster N, Morgan D (2021) A statement on the cultural importance of the dingo. *Australian Zoologist* **41**, 296–297. doi:10.7882/AZ.2021.028

Feigin CY, Newton AH, Doronina L, Schmitz J, Hipsley CA, Mitchell KJ, Gower G, Llamas B, Soubrier J, Heider TN, Menzies BR, Cooper A, O'Neill RJ, Pask AJ (2018) Genome of the Tasmanian tiger provides insights into the evolution and demography of an extinct marsupial carnivore. *Nature Ecology & Evolution* **2**, 182–192. doi:10.1038/s41559-017-0417-y

Field MA, Yadav S, Dudchenko O, Esvaran M, Rosen BD, Skvortsova K, Edwards RJ, Keilwagen J, Cochran BJ, Manandhar B, Bustamante S, Rasmussen JA, Melvin RG, Chernoff B, Omer A, Colaric Z, Chan EKF, Minoche AE, Smith TPL, Gilbert MTP, Bogdanovic O, Zammit RA, Thomas T, Aiden EL, Ballard JWO (2022) The Australian dingo is an early offshoot of modern breed dogs. *Science Advances* **8**, eabm5944. doi:10.1126/sciadv.abm5944

Fillios MA, Taçon PS (2016) Who let the dogs in: a review of the recent genetic evidence for the introduction of the dingo to Australia and implications for the movement of people. *Journal of Archaeological Science, Reports* **7**, 782–792. doi:10.1016/j.jasrep.2016.03.001

Fillios M, Crowther MS, Letnic M (2012) The impact of the dingo on the thylacine in Holocene Australia. *World Archaeology* **44**, 118–134. doi:10.1080/00438243.2012.646112

Flannery TF (1994) *The Future Eaters: An Ecological History of the Australasian Lands and People*. Reed Books, Sydney, Australia.

Johnson C (2006) *Australia's Mammal Extinctions: A 50,000-year History*. Cambridge University Press, Port Melbourne.

Johnson CN, Letnic M (2014) Introducing a new top predator, the dingo. In *Invasion Biology and Ecological Theory: Insights from a Continent in Transformation*. (Eds HHT Prins and IJ Gordon) pp. 414–428. Cambridge University Press, Cambridge.

Johnson CN, Wroe S (2003) Causes of extinction of vertebrates during the Holocene of mainland Australia: arrival of the dingo, or human impact? *The Holocene* **13**, 941–948. doi:10.1191/0959683603hl682fa

Hamilton A (1972) Aboriginal man's best friend? *The Australian Journal of Anthropology* **8**, 287–295. doi:10.1111/j.1835-9310.1972.tb00449.x

Koungoulos L (2017) *Canis dingo* and the Australian smaller-fauna trend: a new explanatory model integrating ecological data. *Journal of Archaeological Science, Reports* **14**, 38–45. doi:10.1016/j.jasrep.2017.05.035

Koungoulos L (2020) Old dogs new tricks: 3D geometric analysis of cranial morphology supports ancient population substructure in the Australian dingo. *Zoomorphology* **139**, 263–275. doi:10.1007/s00435-019-00475-z

Koungoulos L (2021) Domestication through dingo eyes: an Australian perspective on human-canid interactions leading to the earliest dogs. *Human Ecology* **49**, 691–705. doi:10.1007/s10745-021-00262-9

Koungoulos L, Contos P (2019) Global expansion of the Australian biting louse *Heterodoxus spiniger* facilitated by human transport of dog (*Canis familiaris*) and implications for prehistoric cultural interaction in Australasia. *Environmental Archaeology* **26**, 281–294. doi:10.1080/14614103.2019.1653619

Koungoulos L, Fillios M (2020) Hunting dogs down under? On the Aboriginal use of tame dingoes in dietary game acquisition and its relevance to Australian prehistory. *Journal of Anthropological Archaeology* **58**, 101146. doi:10.1016/j.jaa.2020.101146

Koungoulos LG, Balme J, O'Connor S (2023) Dingoes, companions in life and death: the significance of archaeological canid burial practices in Australia. *PLoS One* **18**, e0286576. doi:10.1371/journal.pone.0286576

Koungoulos L, Hulme-Beaman A, Fillios M, Willandra Lakes Region World Heritage Aboriginal Advisory Group (2024) Phenotypic diversity in early Australian dingoes revealed by traditional and 3D geometric morphometric analysis. *Scientific Reports* **14**, 21228. doi:10.1038/s41598-024-65729-3

Kumar M, Conroy G, Ogbourne S, Cairns KM, Borburgh L, Subramanian S (2023) Genomic signatures of bottleneck and founder effects in dingoes. *Ecology and Evolution* **13**, e10525. doi:10.1002/ece3.10525

Letnic M, Fillios M, Crowther MS (2012) Could direct killing by larger dingoes have caused the extinction of the thylacine from mainland Australia? *PLoS One* **7**, e34877. doi:10.1371/journal.pone.0034877

Letnic M, Fillios M, Crowther MS (2014) The arrival and impacts of the dingo. In *Carnivores of Australia: Past, Present and Future.* (Eds A Glen and CR Dickman) pp. 53–67. CSIRO Publishing, Melbourne.

Luomala K (1990) A history of the binomial classification of the Polynesian Native Dog. *Pacific Science* **14**, 193–223.

Ma GC, Ford J, Lucas L, Norris JM, Spencer J, Withers AM, Ward MP (2020) "They reckon they're man's best friend and I believe that." Understanding relationships with dogs in Australian Aboriginal communities to inform effective dog population management. *Animals (Basel)* **10**, 810. doi:10.3390/ani10050810

Macintosh NGW (1975) The origin of the dingo: an enigma In *The Wild Canids: Their Systematics Behavioural Ecology and Evolution*. (Ed. MW Fox) pp 87–106. Van Nostrand Reinhold Company, New York.

Musharbash Y (2017) Telling Warlpiri dog stories. *Anthropological Forum* **27**, 95–113. doi:10.1080/00664677.2017.1303603

Oskarsson MC, Klütsch CF, Boonyaprakob U, Wilton A, Tanabe Y, Savolainen P (2012) Mitochondrial DNA data indicate an introduction through Mainland Southeast Asia for Australian dingoes and Polynesian domestic dogs. *Proceedings. Biological Sciences* **279**, 967–974. doi:10.1098/rspb.2011.1395

Parker MA (2006) Bringing the dingo home: discursive representations of the dingo by Aboriginal, colonial and contemporary Australians. PhD thesis, University of Tasmania.

Philip J (2021) The waterfinders: a cultural history of the Australian dingo. *Australian Zoologist* **41**, 593–607. doi:10.7882/AZ.2020.034

Philip J, Garden D (2016) Walking the thylacine: records of Indigenous companion animals in Australian narrative and photographic history. *Society & Animals* **24**, 34–62. doi:10.1163/15685306-12341386

Prowse TA, Johnson CN, Bradshaw CJ, Brook BW (2014) An ecological regime shift resulting from disrupted predator–prey interactions in Holocene Australia. *Ecology* **95**, 693–702. doi:10.1890/13-0746.1

Rose DB (2000) *Dingo Makes Us Human: Life and Land in an Australian Aboriginal Culture*. CUP Archive.

Roughsey D (1973) *The Giant Devil-Dingo*. Collins, Sydney.

Rovinsky DS, Evans AR, Adams JW (2021) Functional ecological convergence between the thylacine and small prey-focused canids. *BMC Ecology and Evolution* **21**, 58. doi:10.1186/s12862-021-01788-8

Savolainen P, Leitner T, Wilton AN, Matisoo-Smith E, Lundeberg J (2004) A detailed picture of the origin of the Australian dingo obtained from the study of mitochondrial DNA. *Proceedings of the National Academy of Sciences of the United States of America* **101**, 12387–12390. doi:10.1073/pnas.0401814101

Smith BP, Litchfield CA (2009) A review of the relationship between Indigenous Australians, dingoes (*Canis dingo*) and domestic dogs (*Canis familiaris*). *Anthrozoos* **22**, 111–128. doi:10.2752/175303709X434149

Souilmi Y, Wasef S, Williams MP, Conroy G, Bar I, Bover P, Dann J, Heiniger H, Llamas B, Ogbourne S, Archer M, Ballard JWO, Reed E, Tobler R, Koungoulos L, Walshe K, Wright JL, Balme J, O'Connor S, Cooper A, Mitchell KJ (2024) Ancient genomes reveal over two thousand years of dingo population structure. *Proceedings of the National Academy of Sciences of the United States of America* **121**, e2407584121. doi:10.1073/pnas.2407584121

Taçon P, Pardoe C (2002) Dogs make us human. *Nature Australia* **27**, 52–62.

Wilton A (2001) DNA methods of assessing Australian dingo purity. In, A symposium on the dingo. 8 May 1999, Australian Museum, Sydney. (Eds CR Dickman and D Lunney) pp. 49–55. Royal Zoological Society of New South Wales.

Wilton AN, Steward DJ, Zafiris K (1999) Microsatellite variation in the Australian dingo. *The Journal of Heredity* **90**, 108–111. doi:10.1093/jhered/90.1.108

Wroe S, Clausen P, McHenry C, Moreno K, Cunningham E (2007) Computer simulation of feeding behaviour in the thylacine and dingo as a novel test for convergence and niche overlap. *Proceedings Biological Sciences* **274**, 2819–2828. doi:10.1098/rspb.2007.0906

Zhang SJ, Wang GD, Ma P, Zhang LL, Yin TT, Liu YH, Otecko NO, Wang M, Ma Y-P, Wang L, Mao B, Savolainen P, Zhang Y-P (2020) Genomic regions under selection in the feralization of the dingoes. *Nature Communications* **11**, 671. doi:10.1038/s41467-020-14515-6

PART 2: THE DINGO

Appleby R, Smith BP, Jones D (2013) Observations of a free-ranging adult female dingo (*Canis dingo*) and littermates' responses to the death of a pup. *Behavioural Processes* **96**, 42–46. doi:10.1016/j.beproc.2013.02.016

Ballard JWO, Gardner C, Ellem L, Yadav S, Kemp RI (2022) Eye contact and sociability data suggests that Australian dingoes were never domesticated. *Current Zoology* **68**, 423–432. doi:10.1093/cz/zoab024

Ballard JWO, Field MA, Edwards RJ, Wilson LA, Koungoulos LG, Rosen BD, Chernoff B, Dudchenko O, Omer A, Keilwagen J, Skvortsova K, Bogdanovic O, Chan E, Zammit R, Hayes V, Aiden EL (2023) The Australasian dingo archetype: de novo chromosome-length genome assembly, DNA methylome, and cranial morphology. *GigaScience* **12**, giad018. doi:10.1093/gigascience/giad018

Breckwoldt R (1988) *The Dingo: A Very Elegant Animal*. Angus and Robertson, North Ryde, NSW.

Cairns KM (2021) What is a dingo – origins, hybridisation and identity. *Australian Zoologist* **41**, 322–337. doi:10.7882/AZ.2021.004

Corbett LK (1988) Social dynamics of a captive dingo pack: population regulation by dominant female infanticide. *Ethology* **78**, 177–198. doi:10.1111/j.1439-0310.1988.tb00229.x

Corbett LK (2001) The dingo in Australia and Asia (2nd edn). J.B. Books, Marleston.

Crowther MS, Fillios M, Colman N, Letnic M (2014) An updated description of the Australian dingo (*Canis dingo* Meyer, 1793). *Journal of Zoology* **293**, 192–203. doi:10.1111/jzo.12134

Déaux ÉC, Clarke JA (2013) Dingo (*Canis lupus dingo*) acoustic repertoire: form and contexts. *Behaviour* **150**, 75–101. doi:10.1163/1568539X-00003038

Field MA, Yadav S, Dudchenko O, Esvaran M, Rosen BD, Skvortsova K, Edwards RJ, Keilwagen J, Cochran BJ, Manandhar B, Bustamante S, Rasmussen JA, Melvin RG, Chernoff B, Omer A, Colaric Z, Chan EKF, Minoche AE, Smith TPL, Gilbert MTP, Bogdanovic O, Zammit RA, Thomas T, Aiden EL, Ballard JWO (2022) The Australian dingo is an early offshoot of modern breed dogs. *Science Advances* **8**, eabm5944. doi:10.1126/sciadv.abm5944

Hudson R, Rödel HG, Elizalde MT, Kennedy GA, Smith BP (2019) Diurnal pattern of preweaning den visits and nursing in breeding pairs of captive dingoes (*Canis dingo*). *Mammalian Biology* **94**, 86–91. doi:10.1016/j.mambio.2018.07.002

Johnston AM, Turrin C, Watson L, Arre AM, Santos LR (2017) Uncovering the origins of dog–human eye contact: dingoes establish eye contact more than wolves, but less than dogs. *Animal Behaviour* **133**, 123–129. doi:10.1016/j.anbehav.2017.09.002

Lord K, Feinstein M, Smith BP, Coppinger R (2013) Variation in reproductive traits of members of the genus *Canis* with special attention to the domestic dog (*Canis familiaris*). *Behavioural Processes* **92**, 131–142. doi:10.1016/j.beproc.2012.10.009

Miklósi Á, Kubinyi E, Topál J, Gácsi M, Virányi Z, Csányi V (2003) A simple reason for a big difference: wolves do not look back at humans, but dogs do. *Current Biology* **13**, 763–766. doi:10.1016/S0960-9822(03)00263-X

Mouton A, Bird DJ, Li G, Craven BA, Levine JM, Morselli M, Pellegrini M, Van Valkenburgh B, Wayne RK, Murphy WJ (2025) Genetic and anatomical determinants of olfaction in dogs and wild canids. *Molecular Biology and Evolution* **42**, msaf035. doi:org/10.1093/molbev/msaf035

Parr WCH, Wilson LAB, Wroe S, Colman NJ, Crowther MS, Letnic M (2016) Cranial shape and the modularity of hybridization in dingoes and dogs; hybridization does not spell the end for native morphology. *Evolutionary Biology* **43**, 171–187. doi:10.1007/s11692-016-9371-x

Smith B (2014) Living with wild dogs: personality dimensions in captive dingoes (*Canis dingo*) and implications for ownership. *Anthrozoos* **27**, 423–433. doi:10.2752/175303714X14023922797869

Smith B (2015) *The Dingo Debate*. CSIRO Publishing, Clayton South.

Smith B, Litchfield C (2010a) Dingoes (*Canis dingo*) can use human social cues to locate hidden food. *Animal Cognition* **13**, 367–376. doi:10.1007/s10071-009-0287-z

Smith B, Litchfield C (2010b) How well do dingoes (*Canis dingo*) perform on the detour task. *Animal Behaviour* **80**, 155–162. doi:10.1016/j.anbehav.2010.04.017

Smith B, Litchfield C (2013) Looking back at 'looking back': operationalizing referential gaze for dingoes in an unsolvable task. *Animal Cognition* **16**, 961–971. doi:10.1007/s10071-013-0629-8

Smith B, Browne M, Serpell J (2017) Owner-reported behavioural characteristics of dingoes (*Canis dingo*) living as companion animals: a comparison to 'modern' and 'ancient' dog breeds. *Applied Animal Behaviour Science* **187**, 77–84. doi:10.1016/j.applanim.2016.11.010

Smith B, Cairns K, Adams J, Newsome T, Fillios M, Déaux É, Parr W, Letnic M, van Eeden L, Appleby R, Bradshaw C, Savolainen P, Ritchie E, Nimmo D, Archer-Lean C, Greenville A, Dickman C, Watson L, Moseby K, Doherty T, Wallach A, Morrant D, Crowther M (2019) Taxonomic status of the Australian dingo: the case for *Canis dingo* Meyer, 1793. *Zootaxa* **4564**, 173–197. doi:10.11646/zootaxa.4564.1.6

Smith B, Appleby RG, Cairns KM (2023) Reimagining the dingo: the 'Australian wolf' or just a feral dog? In *The Wolf: Culture, Nature, Heritage*. (Eds I Convery, O Nevin, E van Maanen, P Davis and K Lloyd) pp. 161–178. The Boydell Press, Suffolk, UK.

Smith BP (2015) Biology and behaviour of the dingo. In *The Dingo Debate: Origins, Behaviour and Conservation*. (Ed. BP Smith) pp. 25–53. CSIRO Publishing, Clayton South.

Smith BP, Vague A-L (2017) The denning behaviour of dingoes (*Canis dingo*) living in a human-modified environment. *Australian Mammalogy* **39**, 161–168. doi:10.1071/AM16027

Smith BP, Appleby R, Litchfield C (2012) Spontaneous tool-use: An observation of a dingo (*Canis dingo*) using a table to access an out-of-reach food reward. *Behavioural Processes* **89**, 219–224. doi:10.1016/j.beproc.2011.11.004

Smith BP, Lucas T, Norris R, Henneberg M (2018) Brain size/body weight in the dingo (*Canis dingo*): Comparisons with domestic and wild canids. *Australian Journal of Zoology* **65**, 292–301. doi:10.1071/ZO17040

Thomson PC (1992) The behavioural ecology of dingoes in north-western Australia. IV. Social and spatial organisation, and movements. *Wildlife Research* **19**, 543–563. doi:10.1071/WR9920543

Wigmore SM, Nankervis SA, Watson L, Bean DC (2024) Staphylococcal carriage among captive dingoes (*Canis dingo*) in Victoria, Australia. *Veterinary Microbiology* **294**, 110104. doi:10.1016/j.vetmic.2024.110104

Yadav S, Pickford R, Zammit RA, Ballard JWO (2021) Metabolomics shows the Australian dingo has a unique plasma profile. *Scientific Reports* **11**, 5245. doi:10.1038/s41598-021-84411-6

Zhang SJ, Wang GD, Ma P, Zhang LL, Yin TT, Liu YH, Otecko NO, Wang M, Ma Y-P, Wang L, Mao B, Savolainen P, Zhang Y-P (2020) Genomic regions under selection in the feralization of the dingoes. *Nature Communications* **11**, 671. doi:10.1038/s41467-020-14515-6

PART 3: DINGOES IN THE LANDSCAPE

Allen BL, Engeman RM, Allen LR (2011) Wild dogma: An examination of recent "evidence" for dingo regulation of invasive mesopredator release in Australia. *Current Zoology* **57**, 568–583. doi:10.1093/czoolo/57.5.568

Allen BL, Allen LR, Engeman RM, Leung LK-P (2013) Intraguild relationships between sympatric predators exposed to lethal control: predator manipulation experiments. *Frontiers in Zoology* **10**, 39. doi:10.1186/1742-9994-10-39

Allen BL, Fleming PJ, Allen LR, Engeman RM, Ballard G, Leung LK (2013) As clear as mud: a critical review of evidence for the ecological roles of Australian dingoes. *Biological Conservation* **159**, 158–174. doi:10.1016/j.biocon.2012.12.004

ANU Reporter. 'Cats kill more than 1.5 billion native animals per year.' 9 July 2019. https://reporter.anu.edu.au/all-stories/cats-kill-more-than-1-5-billion-native-animals-per-year?utm_source=chatgpt.com (accessed 13 October 2025).

Brook LA, Johnson CN, Ritchie EG (2012) Effects of predator control on behaviour of an apex predator and indirect consequences for mesopredator suppression. *Journal of Applied Ecology* **49**, 1278–1286. doi:10.1111/j.1365-2664.2012.02207.x

Campbell G, Emmott A, Pollock D, Traill BJ (2022) Can dingoes increase graziers' profits and help maintain Australia's rangelands? *The Rangeland Journal* **44**, 129–135. doi:10.1071/RJ22002

Castle G, Kennedy MS, Allen BL (2023) Stuck in the mud: persistent failure of 'the science to provide reliable information on the ecological roles of Australian dingoes. *Biological Conservation* **285**, 110234. doi:10.1016/j.biocon.2023.110234

Choquenot D, Forsyth DM (2013) Exploitation ecosystems and trophic cascades in non-equilibrium systems: pasture–red kangaroo–dingo interactions in arid Australia. *Oikos* **122**, 1292–1306. doi:10.1111/j.1600-0706.2012.20976.x

Claridge AW, Hunt R (2008) Evaluating the role of the dingo as a trophic regulator: additional practical suggestions. *Ecological Management & Restoration* **9**, 116–119. doi:10.1111/j.1442-8903.2008.00402.x

Contos P, Letnic M (2019) Top-down effects of a large mammalian carnivore in arid Australia extend to epigeic arthropod assemblages. *Journal of Arid Environments* **165**, 16–27. doi:10.1016/j.jaridenv.2019.03.002

Costello O, Webster N, Morgan D (2021) A statement on the cultural importance of the dingo. *Australian Zoologist* **41**(3), 296–297. doi:10.7882/AZ.2021.028

Dickman CR, Greenville AC, Newsome TM (2019) Carnivore conservation: the importance of carnivores to the ecosystem, and the value of reintroductions. In *Saving the Tasmanian Devil: Recovery through Science-based Management*. (Eds CJ Hogg, S Fox, D Pemberton and K Belov) pp. 1–10. CSIRO Publishing, Clayton South.

Doherty TS, Davis NE, Dickman CR, Forsyth DM, Letnic M, Nimmo DG, Palmer R, Ritchie EG, Benshemesh J, Edwards G, Lawrence J (2019) Continental patterns in the diet of a top predator: Australia's dingo. *Mammal Review* **49**, 31–44. doi:10.1111/mam.12139

Elliott TF, Paine CT, Ballard GA, Milne H, Van der Eyk J, Elliott K, Meek P, Bruhl JJ, Vernes K (2023) The dingo (*Canis familiaris*) as a secondary disperser of mycorrhizal fungal spores. *Wildlife Research* **51**, WR22057. doi:10.1071/WR22057

Fisher AG, Mills CH, Lyons M, Cornwell WK, Letnic M (2021) Remote sensing of trophic cascades: multi-temporal Landsat imagery reveals vegetation change driven by the removal of an apex predator. *Landscape Ecology* **36**, 1341–1358. doi:10.1007/s10980-021-01206-w

Fleming PJS, Allen BL, Ballard GA (2012) Seven considerations about dingoes as biodiversity engineers: the socioecological niches of dogs in Australia. *Australian Mammalogy* **34**, 119–131. doi:10.1071/AM11012

Glen AS, Dickman CR, Soule ME, Mackey BG (2007) Evaluating the role of the dingo as a trophic regulator in Australian ecosystems. *Austral Ecology* **32**, 492–501. doi:10.1111/j.1442-9993.2007.01721.x

Gordon CE, Moore BD, Letnic M (2014) Dingoes and foxes: complex interactions between native predators and invasive mesopredators. In *Carnivores of Australia: Past, Present and Future*. (Eds AS Glen and CR Dickman) pp. 183–197. CSIRO Publishing, Melbourne.

Greenville A, Wardle G, Ritchie E, Newsome T (2019) Demise of the dingo. *Austral Ecology* **44**, 555–560. doi:10.1111/aec.12731

Johnson C, Isaac J, Fisher D (2007) Rarity of a top predator triggers continent-wide collapse of mammal prey: dingoes and marsupials in Australia. *Proceedings. Biological Sciences* **274**, 341–346. doi:10.1098/rspb.2006.3711

Johnson CN, Ritchie EG (2013) The dingo and biodiversity conservation: response to Fleming *et al.* (2012). *Australian Mammalogy* **35**, 8–14. doi:10.1071/AM12005

Johnson CN, VanDerWal J (2009) Evidence that dingoes limit abundance of a mesopredator in eastern Australian forests. *Journal of Applied Ecology* **46**, 641–646. doi:10.1111/j.1365-2664.2009.01650.x

Johnson CN, Crowther MS, Dickman CR, Letnic MI, Newsome TM, Nimmo DG, Ritchie EG, Wallach AD (2014) Experiments in no-impact control of dingoes: comment on Allen *et al.* 2013. *Frontiers in Zoology* **11**, 17. doi:10.1186/1742-9994-11-17

Kennedy M, Phillips BL, Legge S, Murphy SA, Faulkner RA (2012) Do dingoes suppress the activity of feral cats in northern Australia? *Austral Ecology* **37**, 134–139. doi:10.1111/j.1442-9993.2011.02256.x

Kennedy MS, Phillips BL, Legge S, Murphy SA (2011) Mesopredators suppress small mammals in Australian desert landscapes: evidence from a large-scale experiment. *Ecology* **92**, 1828–1835.

Letnic M, Crowther MS (2013) Patterns in the abundance of kangaroo populations in arid Australia are consistent with the exploitation ecosystems hypothesis. *Oikos* **122**, 761–769. doi:10.1111/j.1600-0706.2012.20425.x

Letnic M, Dworjanyn S (2011) Does a top predator reduce the predatory impact of an invasive mesopredator on an endangered rodent? *Ecography* **34**, 827–835. doi:10.1111/j.1600-0587.2010.06516.x

Letnic M, Koch F (2010) Are dingoes a trophic regulator in arid Australia? A comparison of mammal communities on either side of the dingo fence. *Austral Ecology* **35**, 167–175. doi:10.1111/j.1442-9993.2009.02022.x

Letnic M, Crowther MS, Koch F (2009a) Does a top-predator provide an endangered rodent with refuge from an invasive mesopredator? *Animal Conservation* **12**, 302–312. doi:10.1111/j.1469-1795.2009.00250.x

Letnic M, Koch F, Gordon C, Crowther MS, Dickman CR (2009b) Keystone effects of an alien top-predator stem extinctions of native mammals. *Proceedings. Biological Sciences* **276**, 3249–3256. doi:10.1098/rspb.2009.0574

Letnic M, Greenville A, Denny E, Dickman CR, Tischler M, Gordon C, Koch F (2011) Does a top predator suppress the abundance of an invasive mesopredator at a continental scale? *Global Ecology and Biogeography* **20**, 343–353. doi:10.1111/j.1466-8238.2010.00600.x

Letnic M, Ritchie E, Dickman C (2012) Top predators as biodiversity regulators: the dingo *Canis lupus dingo* as a case study. *Biological Reviews of the Cambridge Philosophical Society* **87**, 390–413. doi:10.1111/j.1469-185X.2011.00203.x

Letnic M, Baker L, Nesbitt B (2013) Ecologically functional landscapes and the role of dingoes as trophic regulators in south-eastern Australia and other habitats. *Ecological Management & Restoration* **14**, 101–105. doi:10.1111/emr.12035

Mitchell DR, Cairns SC, Körtner G, Bradshaw CJ, Saltré F, Weisbecker V (2023) Differential developmental rates and demographics in Red Kangaroo (*Osphranter rufus*) populations separated by the dingo barrier fence. *Journal of Mammalogy* **104**, 929–940. doi:10.1093/jmammal/gyad053

Morris T, Letnic M (2017) Removal of an apex predator initiates a trophic cascade that extends from herbivores to vegetation and the soil nutrient pool. *Proceedings of the Royal Society B: Biological Sciences* **284**(1854), 20170111.

Moseby KE, Read JL, Paton DC, Copley P, Hill BM, Crisp HA (2011) Predation determines the outcome of 10 reintroduction attempts in arid South Australia. *Biological Conservation* **144**, 2863–2872. doi:10.1016/j.biocon.2011.08.003

Moseby KE, Neilly H, Read JL, Crisp HA (2012) Interactions between a top order predator and exotic mesopredators in the Australian rangelands. *International Journal of Ecology* **2012**, 250352. doi:10.1155/2012/250352

Murphy BP, Woolley LA, Geyle HM, Legge SM, Palmer R, Dickman CR, Woinarski JCZ (2017) Introduced cats (*Felis catus*) eating a continental fauna: the number of mammals killed in Australia. *Biological Conservation* **214**, 76–81.

Murphy SA, Austin JJ, Murphy RK, Silcock J, Joseph L, Garnett ST, Watson DM (2020) Observations on breeding Night Parrots (*Pezoporus occidentalis*) in western Queensland. *Emu - Austral Ornithology* **120**, 84–88.

Newsome TM, Ballard GA, Crowther MS, Dellinger JA, Fleming PJ, Glen AS, Greenville AC, Johnson CN, Letnic M, Moseby KE, Nimmo DG (2015) Resolving the value of the dingo in ecological restoration. *Restoration Ecology* **23**, 201–208. doi:10.1111/rec.12186

Newsome TM, Greenville AC, Letnic M, Ritchie EG, Dickman CR (2017) The case for a dingo reintroduction in Australia remains strong: a reply to Morgan *et al.*, 2016. *Food Webs* **10**, 39–41. doi:10.1016/j.fooweb.2017.02.001

Pollock D (2019). *The Wooleen Way: Renewing an Australian Resource*. Scribe Publications, Brunswick, Vic.

Pollock D (2021a) Managing the unmanageable: reinstating the dingo for pastoral sustainability in Australian rangelands. *Proceedings of the Royal Society of Victoria* **133**, 27–31. doi:10.1071/RS21005

Pollock D (2021b) Rediscovering a role for dingoes on Wooleen Station. *Australian Zoologist* **41**, 480–486. doi:10.7882/AZ.2021.006

Ritchie E (2012) Conservation needs more bite. *Australasian Science* **33**, 26–28.

Ritchie E, Smith B, Cairns K, Takau S, Rassip W (2023) 'The boss of Country', not wild dogs to kill: living with dingoes can unite communities. The Conversation. October 2, 2023. https://theconversation.com/the-boss-of-country-not-wild-dogs-to-kill-living-with-dingoes-can-unite-communities-214212

Ritchie EG, Johnson CN (2009) Predator interactions, mesopredator release and biodiversity conservation. *Ecology Letters* **12**, 982–998. doi:10.1111/j.1461-0248.2009.01347.x

Ritchie EG, Elmhagen B, Glen AS, Letnic M, Ludwig G, McDonald RA (2012) Ecosystem restoration with teeth: what role for predators? *Trends in Ecology & Evolution* **27**, 265–271. doi:10.1016/j.tree.2012.01.001

Ryan-Schofield NL, Bannister HL, Ross A, Moseby KE (2025) A comparison of field and genetic methods for identifying feral cat predation events during wildlife translocations. *Australian Mammalogy* **47**, AM24031. doi:10.1071/AM24031

Smith BP, Loughridge J, Rangers N, Wright C, Badal A, Rose NM, Hunter E, Kalpers J (2024) Traditional owner-led wartaji (dingo) research in Pirra Country (Great Sandy Desert): a case study from the Nyangumarta Warrarn Indigenous Protected Area. *Wildlife Research* **51**, WR24082. doi:10.1071/WR24082

Spencer E, Newsome T (2021) Dingoes dining with death. *Australian Zoologist* **41**, 433–451. doi:10.7882/AZ.2021.008

Sweeney OF, Turnbull J, Jones M, Letnic M, Newsome TM, Sharp A (2019) An Australian perspective on rewilding. *Conservation Biology* **33**, 812–820. doi:10.1111/cobi.13280

Wallach AD, Murray BR, O'Neill AJ (2009a) Can threatened species survive where the top predator is absent? *Biological Conservation* **142**, 43–52. doi:10.1016/j.biocon.2008.09.021

Wallach AD, Ritchie EG, Read J, O'Neill AJ (2009b) More than mere numbers: the impact of lethal control on the social stability of a top-order predator. *PLoS One* **4**, e6861. doi:10.1371/journal.pone.0006861

Wallach AD, Johnson CN, Ritchie EG, O'Neill AJ (2010) Predator control promotes invasive dominated ecological states. *Ecology Letters* **13**, 1008–1018. doi:10.1111/j.1461-0248.2010.01492.x

Wallach AD, Izhaki I, Toms JD, Ripple WJ, Shanas U (2015) What is an apex predator? *Oikos* **124**, 1453–1461. doi:10.1111/oik.01977

Wallach AD, Ramp D, O'Neill AJ (2017) Causes of cattle mortality on a predator friendly station in central Australia. *Journal of Mammalogy* **98**, 45–52. doi:10.1093/jmammal/gyw156

Wang Y, Fisher DO (2012) Dingoes affect activity of feral cats, but do not exclude them from the habitat of an endangered macropod. *Wildlife Research* **39**, 611–620. doi:10.1071/WR11210

Woinarski JCZ, Burbidge AA, Harrison PL (2015) Ongoing unraveling of a continental fauna: decline and extinction of Australian mammals since European settlement. *Proceedings of the National Academy of Sciences of the United States of America* **112**, 4531–4540. doi:10.1073/pnas.1417301112

Woinarski J, Legge S, Dickman C (2019) *Cats in Australia: Companion and Killer*. CSIRO Publishing, Melbourne.

PART 4: AN IDENTITY IN CRISIS

Ballard JWO, Wilson LA (2019) The Australian dingo: untamed or feral? *Frontiers in Zoology* **16**, 2. doi:10.1186/s12983-019-0300-6

Banks PB (2021) An eco-evolutionary rationale to distinguish alien and native status: why the dingo is a native species on mainland Australia. *Australian Zoologist* **41**, 358–366. doi:10.7882/AZ.2021.039

Brumm A (2021) Dingoes and domestication. *Archaeology in Oceania* **56**, 17–31. doi:10.1002/arco.5226

Cairns KM (2021) What is a dingo – origins, hybridisation and identity. *Australian Zoologist* **41**, 322–337. doi:10.7882/AZ.2021.004

Cairns KM, Nesbitt BJ, Laffan SW, Letnic M, Crowther MS (2020) Geographic hot spots of dingo genetic ancestry in southeastern Australia despite hybridisation with domestic dogs. *Conservation Genetics* **21**, 77–90. doi:10.1007/s10592-019-01230-z

Cairns KM, Crowther MS, Nesbitt B, Letnic M (2022) The myth of wild dogs in Australia: are there any out there? *Australian Mammalogy* **44**, 67–75. doi:10.1071/AM20055

Cairns KM, Crowther MS, Parker HG, Ostrander EA, Letnic M (2023) Genome-wide variant analyses reveal new patterns of admixture and population structure in Australian dingoes. *Molecular Ecology* **32**, 4133–4150. doi:10.1111/mec.16998

Cairns KM, Letnic M, Ritchie EG, Adams JW, Crowther MS, Dickman CR, Fillios M, Pascoe J, Smith BP, Newsome TM (2025) Taxonomic tangles posed by human association: the urgent need for an evidence-based review of dingo and domestic dog taxonomy and nomenclature. *Australian Mammalogy* **47**, AM24052. doi:10.1071/AM24052

Claridge AW, Spencer RJ, Wilton AN, Jenkins DJ, Dall D, Lapidge SJ, Dickman CR (2014) When is a dingo not a dingo? Hybridisation with domestic dogs. In *Carnivores of Australia: Past, Present and Future*. (Eds AS Glen and CR Dickman) pp. 151–172. CSIRO Publishing, Collingwood.

Clutton-Brock J (1999) *A Natural History of Domesticated Mammals*. 2nd edn. Cambridge University Press.

Commonwealth of Australia (1999) *Environment Protection and Biodiversity Conservation Act 1999*. https://www.legislation.gov.au/Series/C2004A00485

Coppinger R, Coppinger L (2002) *Dogs: A startling new understanding of canine origin, behavior, and evolution*. 2nd edn. The University of Chicago Press.

Crowther MS, Fillios M, Colman N, Letnic M (2014) An updated description of the Australian dingo (*Canis dingo* Meyer, 1793). *Journal of Zoology* **293**, 192–203. doi:10.1111/jzo.12134

Crowther MS, Cairns KM, Van Eeden LM, Letnic M (2021) Introgression does not influence the positive ecological and functional role of dingo populations. *Australian Zoologist* **41**, 338–346. doi:10.7882/AZ.2020.016

Cursino MS, Harriott L, Allen BL, Gentle M, Leung LKP (2017) Do female dingo–dog hybrids breed like dingoes or dogs? *Australian Journal of Zoology* **65**, 112–119. doi:10.1071/ZO17005

Daniels MJ, Corbett L (2003) Redefining introgressed protected mammals: when is a wildcat a wild cat and a dingo a wild dog? *Wildlife Research* **30**, 213–218. doi:10.1071/WR02045

Dickman CR, Lunney D (2001) Last howl of the dingo: the legislative, ecological and practical issues arising from the kill-or-conserve dilemma. In *A Symposium on the Dingo* (Eds CR Dickman and D Lunney) pp. 95–107. Royal Zoological Society of New South Wales, Mosman, NSW.

Dobney K, Larson G (2006) Genetics and animal domestication: new windows on an elusive process. *Journal of Zoology* **269**, 261–272.

Donfrancesco V, Allen B, Appleby R, Behrendorff L, Conroy G, Crowther M, Dickman C, Doherty T, Fancourt B, Gordon C, Jackson S, Johnson C, Kennedy M, Koungoulos L, Letnic M, Leung L, Mitchell K,

Nesbitt B, Newsome T, Pacioni C, Phillip J, Purcell B, Ritchie E, Smith B, Stephens D, Tatler J, van Eeden L, Cairns K (2023) Understanding conflict among experts working on controversial species: a case study on the Australian dingo. *Conservation Science and Practice* **5**, e12900. doi:10.1111/csp2.12900

Elledge AE, Leung LKP, Allen LR, Firestone K, Wilton AN (2006) Assessing the taxonomic status of dingoes *Canis familiaris dingo* for conservation. *Mammal Review* **36**, 142–156. doi:10.1111/j.1365-2907.2006.00086.x

Field MA, Yadav S, Dudchenko O, Esvaran M, Rosen BD, Skvortsova K, Edwards RJ, Keilwagen J, Cochran BJ, Manandhar B, Bustamante S, Rasmussen JA, Melvin RG, Chernoff B, Omer A, Colaric Z, Chan EKF, Minoche AE, Smith TPL, Gilbert MTP, Bogdanovic O, Zammit RA, Thomas T, Aiden EL, Ballard JWO (2022) The Australian dingo is an early offshoot of modern breed dogs. *Science Advances* **8**, eabm5944. doi:10.1126/sciadv.abm5944

Hytten K (2009) Dingo dualisms: exploring the ambiguous identity of Australian dingoes. *Australian Zoologist* **35**, 18–27. doi:10.7882/AZ.2009.003

ICZN (1957) Opinion 451. Use of the plenary powers to secure that the specific name *dingo* Meyer, 1793, as published in the combination *Canis dingo*, shall be the oldest available name for the dingo of Australia (Class Mammalia). *Opinions and Declarations of the International Commission on Zoological Nomenclature* **15**, 329–338. doi:10.5962/p.149457

Jackson SM, Groves CP, Fleming PJ, Aplin KP, Eldridge MD, Gonzalez A, Helgen KM (2017) The wayward dog: is the Australian native dog or dingo a distinct species? *Zootaxa* **4317**, 201–224. doi:10.11646/zootaxa.4317.2.1

Jackson SM, Fleming PJ, Eldridge MD, Ingleby S, Flannery T, Johnson RN, Helgen KM (2019) The dogma of dingoes – taxonomic status of the dingo: a reply to Smith *et al. Zootaxa* **4564**, 198–212. doi:10.11646/zootaxa.4564.1.7

Jackson SM, Fleming PJ, Eldridge MD, Archer M, Ingleby S, Johnson RN, Helgen KM (2021) Taxonomy of the dingo: it's an ancient dog. *Australian Zoologist* **41**, 347–357. doi:10.7882/AZ.2020.049

Kerr R (1792) *The Animal Kingdom, or Zoological System, of the Celebrated Sir Charles Linnaeus: Class I, Mammalia*. J. Murray & R. Faulder, London. doi:10.5962/bhl.title.57940

Koungoulos L (2021) Domestication through dingo eyes: an Australian perspective on human–canid interactions leading to the earliest dogs. *Human Ecology* **49**, 691–705. doi:10.1007/s10745-021-00262-9

Kreplins TL, Stephens D, Laing N, Oskam C (2024) An assessment of dingo ancestry in camp dogs in Western Australia. *Australian Journal of Zoology* **72**, ZO24008. doi:10.1071/ZO24008

Macintosh NWG (1975) The origin of the dingo: an enigma. In *The Wild Canids: Their Systematics, Behavioural Ecology and Evolution*. (Ed. MW Fox) pp. 87–106. Van Nostrand Reinhold, New York.

Mazell P, Phillip A (1789) Dog of New South Wales. In *The Voyage of Governor Phillip to Botany Bay with an Account of the Establishment of the Colonies of Port Jackson and Norfolk Island*. pp. 274–275. Stockdale, London.

Meyer FAA (1793) *Systematisch-summarische Uebersicht der neuesten zoologischen Entdeckungen in Neuholland und Afrika: nebst zwey andern zoologischen Abhandlungen*. Dykische Buchhandlung, Leipzig.

Newsome AE, Corbett LK (1985) The identity of the dingo III. The incidence of dingoes, dogs and hybrids and their coat colours in remote and settled regions of Australia. *Australian Journal of Zoology* **33**, 363–375. doi:10.1071/ZO9850363

Newsome T (2021) Looking to the future: what next for the dingo? *Australian Zoologist* **41**, 643–653. doi:10.7882/AZ.2021.040

Newsome TM (2014) Makings of icons: Alan Newsome, the red kangaroo and the dingo. *Historical Records of Australian Science* **25**, 153–171. doi:10.1071/HR14013

Parr WC, Wilson LA, Wroe S, Colman NJ, Crowther MS, Letnic M (2016) Cranial shape and the modularity of hybridization in dingoes and dogs: hybridization does not spell the end for native morphology. *Evolutionary Biology* **43**, 171–187. doi:10.1007/s11692-016-9371-x

Parsons MH, Blumstein DT (2010) Familiarity breeds contempt: kangaroos persistently avoid areas with experimentally deployed dingo scents. *PLoS One* **5**, e10403. doi:10.1371/journal.pone.0010403

Probyn-Rapsey F (2015) Dingoes and dog-whistling: a cultural politics of race and species in Australia. *Animal Studies Journal* **4**, 55–77. doi:10.14361/transcript.9783839425183.fm

Ritchie E, Smith BP, Cairns KM, Takau S, Rassip W (2023) 'The boss of Country', not wild dogs to kill: living with dingoes can unite communities. *The Conversation* (2 October). https://theconversation.com/the-boss-of-country-not-wild-dogs-to-kill-living-with-dingoes-can-unite-communities-214212

Shipman P (2021) What the dingo says about dog domestication. *The Anatomical Record* **304**, 19–30. doi:10.1002/ar.24517

Smith B (2015) *The Dingo Debate: Origins, Behaviour and Conservation*. CSIRO Publishing, Clayton South.

Smith B, Cairns K, Adams J, Newsome T, Fillios M, Deaux E, Parr W, Letnic M, van Eeden L, Appleby R, *et al.* (2019) Taxonomic status of the Australian dingo: the case for *Canis dingo* Meyer, 1793. *Zootaxa* **4564**, 173–197. doi:10.11646/zootaxa.4564.1.6

Smith BP, Litchfield CA (2009) A review of the relationship between indigenous Australians, dingoes (*Canis dingo*) and domestic dogs (*Canis familiaris*). *Anthrozoos* **22**, 111–128. doi:10.2752/175303709X434149

Smith BP, Ritchie E, van Eeden L (2018) Why the WA government is wrong to play identity politics with dingoes. *The Conversation* (31 August). Available at https://theconversation.com/why-the-wa-government-is-wrong-to-play-identity-politics-with-dingoes-102344

Smith BP, Bradshaw C, Ritchie E, Adams JW, Cairns KM, Crowther M (2019) The dingo is a true-blue, native Australian species. *The Conversation* (7 March). https://theconversation.com/the-dingo-is-a-true-blue-native-australian-species-111538

Smith BP, Appleby RG, Cairns KM (2023) Reimagining the dingo: the 'Australian wolf' or just a feral dog? In *The Wolf: Culture, Nature, Heritage*. (Eds I Convery, O Nevin, E van Maanen, P Davis and K Lloyd) pp. 161–178. The Boydell Press, Suffolk, UK.

Stephens D, Wilton AN, Fleming PJ, Berry O (2015) Death by sex in an Australian icon: a continent-wide survey reveals extensive hybridization between dingoes and domestic dogs. *Molecular Ecology* **24**, 5643–5656. doi:10.1111/mec.13416

Souilmi Y, Wasef S, Williams MP, Conroy G, Bar I, Bover P, Mitchell KJ (2024) Ancient genomes reveal over two thousand years of dingo population structure. *Proceedings of the National Academy of Sciences of the United States of America* **121**, e2407584121. doi:10.1073/pnas.2407584121

van Eeden LM, Dickman CR, Newsome TM, Crowther MS (2019) What should we do with wild dogs? Taxonomic tangles and the management of dingo–dog hybridisation. *Australian Zoologist* **40**, 92–101. doi:10.7882/AZ.2018.031

vonHoldt BM, Brzeski KE, Wilcove DS, Rutledge LY (2017) Redefining the role of admixture and genomics in species conservation. *Conservation Letters* **11**, e12371. https://doi.org/10.1111/conl.12371

Weeks AR, Kriesner P, Bartonicek N, van Rooyen A, Cairns KM, Ahrens CW (2024) Genetic structure and common ancestry expose the dingo–dog hybrid myth. *Evolution Letters* **9**, 1–12.

Wood Jones F (1925) *The Mammals of South Australia: Part I. The Monotremes and the Carnivores*. Government Printer, Adelaide, SA.

Zeuner FE (1963) *A History of Domesticated Animals*. Harper & Row.

PART 5: DINGOES IN THE CROSSFIRE

Allen BL, West P (2013) Influence of dingoes on sheep distribution in Australia. *Australian Veterinary Journal* **91**, 261–267. doi:10.1111/avj.12075

Allen LR (2000) Measuring predator control effectiveness: reducing numbers may not reduce predator impact. *Proceedings of the Vertebrate Pest Conference* **19**, 284–289.

Allen LR, Fleming PJS (2004) Review of canid management in Australia for the protection of livestock and wildlife: potential application to coyote management. *Sheep & Goat Research Journal* **19**, 97–104.

Appleby R (2015a) Dingo-human conflict: attacks on humans. In: *The Dingo Debate*. (Ed. B Smith) pp. 131–159. CSIRO Publishing, Melbourne.

Appleby R (2015b) Dingo-human conflict: attacks on livestock. In *The Dingo Debate*. (Ed. B Smith) pp. 103–131. CSIRO Publishing, Melbourne.

Appleby R, Smith BP (2018) Do wild canids kill for fun? In: *Wild Animals and Leisure: Rights and Wellbeing*. (Eds N Carr and J Young) pp. 181–209. Routledge, New York.

Appleby R, Mackie J, Smith BP, Bernede L, Jones D (2017a) Human–dingo interactions on Fraser Island: an analysis of serious incident reports. *Australian Mammalogy* **40**, 146–156. doi:10.1071/AM16026

Appleby R, Smith B, Bernede L, Jones D (2017b) Utilising aversive conditioning to manage the behaviour of Fraser Island dingoes (*Canis dingo*). *Pacific Conservation Biology* **23**, 335–358. doi:10.1071/PC17017

Appleby R, Smith BP, Mackie J, Bernede L, Jones D (2017c) Preliminary observations of dingo responses to assumed aversive stimuli. *Pacific Conservation Biology* **23**, 295–301. doi:10.1071/PC17005

Appleby R, Smith BP, Jones D, Conroy G, Behrendorff L (2025) A population viability analysis of K'gari (Fraser Island) wongari (dingoes). *Australian Mammalogy* **47**, AM23009. doi:10.1071/AM23009

Bacon JS (1946) *The Australian Dingo, The King of the Bush*. J.S. Bacon, Sydney.

Ballard G, Fleming PJS, Meek PD, Doak S (2020) Aerial baiting and wild dog mortality in south-eastern Australia. *Wildlife Research* **47**, 99–105. doi:10.1071/WR18188

Behrendorff L (2017) A prickly subject: innovative handling of a difficult prey. *Australian Mammalogy* **40**, 294–296. doi:10.1071/AM17024

Behrendorff L (2018) Clever girl? An observation of innovative prey handling by a dingo (Canis dingo). *Pacific Conservation Biology* **24**, 194–197. doi:10.1071/PC17044

Behrendorff L (2021) Best-practice dingo management: six lessons from K'gari (Fraser Island). *Australian Zoologist* **41**, 521–533. doi:10.7882/AZ.2021.001

Behrendorff L, Leung LKP, McKinnon A, Hanger J, Belonje G, Tapply J, Jones D, Allen BL (2016) Insects for breakfast and whales for dinner: the diet and body condition of dingoes on Fraser Island (K'gari). *Scientific Reports* **6**, 23469. doi:10.1038/srep23469

Behrendorff L, King R, Allen BL (2023) Efficacy of management efforts to reduce food-related dingo–human interactions and conflict on K'gari (Fraser Island), Australia. *Animals (Basel)* **13**, 204. doi:10.3390/ani13020204

Brumm A (2022) Before Azaria: a historical perspective on dingo attacks. *Animals (Basel)* **12**, 1592. doi:10.3390/ani12121592

Carter J, Wardell-Johnson A, Archer-Lean C (2017) Butchulla perspectives on dingo displacement and agency at K'gari-Fraser Island, Australia. *Geoforum* **85**, 197–205. doi:10.1016/j.geoforum.2017.08.001

Conroy GC, Lamont RW, Bridges L, Stephens D, Wardell-Johnson A, Ogbourne SM (2021) Conservation concerns associated with low genetic diversity for K'gari–Fraser Island dingoes. *Scientific Reports* **11**, 9503. doi:10.1038/s41598-021-89056-z

Coppinger R, Feinstein M (2015) *How Dogs Work*. University of Chicago Press, Chicago.

Ecker S, Please PM, Maybery D (2016) Constantly chasing dogs: assessing landholder stress from wild dog attacks on livestock using quantitative and qualitative methods. *Australasian Journal of Environmental Management* **24**, 1–15.

Fleming PJ, Allen BL, Allen LR, Ballard G, Bengsen AJ, Gentle MN, Saunders GR (2014) Management of wild canids in Australia: free-ranging dogs and red foxes. In *Carnivores of Australia: Past, Present and Future*. (Eds AS Glen and CR Dickman) pp. 105–149. CSIRO Publishing, Collingwood.

Fleming PJS, Corbett LK, Harden R, Thomson PC (2001) *Managing the Impacts of Dingoes and Other Wild Dogs*. Bureau of Rural Sciences, Canberra.

Fleming PJS, Allen LR, Lapidge SJ, Robley A, Saunders GR, Thomson PC (2006) A strategic approach to mitigating the impacts of wild canids: proposed activities of the Invasive Animals Cooperative Research Centre. *Australian Journal of Experimental Agriculture* **46**, 753–762. doi:10.1071/EA06009

Forsyth DM, Woolnough AP, Nimmo DG, Ritchie EG, Kennedy MS, Pople AR, Watson I (2014) A comment on the influence of dingoes on the Australian sheep flock. *Australian Veterinary Journal* **92**, 461–462. doi:10.1111/avj.12277

Gong W, Sinden J, Braysher M, Jones R (2009) *The Economic Impacts of Vertebrate Pests in Australia*. Invasive Animals Cooperative Research Centre, Canberra.

Hewitt L (2009) 'Major economic costs associated with wild dogs in the Queensland grazing industry'. AgForce, Brisbane.

Howard TM, Thompson LJ, Frumento P, Alter T (2018) Wild dog management in Australia: an interactional approach to case studies of community-led action. *Human Dimensions of Wildlife* **23**, 242–256. doi:10.1080/10871209.2017.1414337

Kruuk H (1972) Surplus killing by carnivores. *Journal of Zoology* **166**, 233–244. doi:10.1111/j.1469-7998.1972.tb04087.x

MacNulty D, Mech LD, Smith D (2007) A proposed ethogram of large-carnivore predatory behavior, exemplified by the wolf. *Journal of Mammalogy* **88**, 595–605. doi:10.1644/06-MAMM-A-119R1.1

McLeod R (2004) *Counting the Cost: Impact of Invasive Animals in Australia*. Cooperative Research Centre for Pest Animal Control, Canberra.

McNeill AT, Luke K-PL, Goullet MS, Gentle MN, Allen BL (2016) Dingoes at the doorstep: home range sizes and activity patterns of dingoes and other wild dogs around urban areas of north-eastern Australia. *Animals (Basel)* **6**, 48. doi:10.3390/ani6080048

Newsome AE, Catling PC, Corbett LK (1983) The feeding ecology of the dingo II. Dietary and numerical relationships with fluctuating prey populations in south-eastern Australia. *Australian Journal of Ecology* **8**, 345–366. doi:10.1111/j.1442-9993.1983.tb01332.x

Parker MA (2006) Bringing the dingo home: discursive representations of the dingo by Aboriginal, colonial and contemporary Australians. PhD thesis, University of Tasmania.

Philip J (2019) The institutionalisation of poison: a historical review of vertebrate pest control in Australia, 1814 to 2018. *Australian Zoologist* **40**, 129–139. doi:10.7882/AZ.2018.025

Philip J (2021) A historical review of Australian aerial vertebrate pest control, targeting dingoes and wild dogs 1946–2019. *Australian Zoologist* **41**, 580–592. doi:10.7882/AZ.2020.011

Philip J (2022) The Dingo Barrier Fence: presenting the case to decommission the world's longest environmental barrier in the United Nations Decade on Ecosystem Restoration 2021–2030. *Biologia Futura* **73**, 9–27. doi:10.1007/s42977-021-00106-z

Pollock TI, Hunter DO, Hocking DP, Evans AR (2022) Eye in the sky: observing wild dingo hunting behaviour using drones. *Wildlife Research* **50**, 212–223. doi:10.1071/WR22033

Probyn-Rapsey F (2017) Eating dingoes. *Australian Zoologist* **39**, 39–42. doi:10.7882/AZ.2016.026

Purcell BV (2010) A novel observation of dingoes (*Canis lupus dingo*) attacking a swimming eastern grey kangaroo (Macropus giganteus). *Australian Mammalogy* **32**, 201–204. doi:10.1071/AM10001

Purcell BV, Glover A, Mulley RC, Close RL (2012) Euro-Australian culture and dilemmas within the science and management of the dingo, *Canis lupus dingo*. In *Science Under Siege: Zoology Under Threat*. (Eds P Banks, D Lunney and C Dickman) pp. 114–120. Royal Zoological Society of New South Wales, Mossman, NSW.

Shaw CH (1946) The Dingo. *The Bulletin* **67** (3463), 26 June, 28.

Shepherd NC (1981) Predation of red kangaroos, Macropus rufus, by the dingo, *Canis familiaris dingo* (Blumenbach) in north-western New South Wales. *Wildlife Research* **8**, 255–262. doi:10.1071/WR9810255

Smith B, Appleby R (2018) Promoting human-dingo co-existence in Australia: moving towards more innovative methods of protecting livestock rather than killing dingoes (*Canis dingo*). *Wildlife Research* **45**, 1–15. doi:10.1071/WR16161

Smith BP (2024) A dingo gold mine. In *Wild Science: Unexpected Encounters When Working in Nature*. (Ed. H Waudby) pp. 179–186. CSIRO Publishing, Melbourne.

Smith BP, Vague AL, Appleby RG (2018) Attitudes towards dingoes (*Canis dingo*) and their management: a case study from a mining operation in the Great Sandy Desert, Western Australia. *Pacific Conservation Biology* **25**, 308–321. doi:10.1071/PC18049

Smith CH, Jardine W (1839) *The Natural History of Dogs: Canidae or Genus Canis of Authors; Including Also the Genera Hyaena and Proteles*. W.H. Lizars, Edinburgh. doi:10.5962/bhl.title.39557.

Tapply J (2018) Contemporary dingo management on K'gari (Fraser Island, Great Sandy National Park) under the Queensland Parks and Wildlife Service. *Australasian Journal of Environmental Management* **25**, 119–131. doi:10.1080/14486563.2017.1422042

Thompson J, Shirreffs L, McPhail I (2003) Dingoes on Fraser Island: tourism dream or management nightmare. *Human Dimensions of Wildlife* **8**, 37–47. doi:10.1080/10871200390180136

Thompson L-J, Aslin H, Ecker S, Please P, Trestrail C (2013) *Social Impacts of Wild Dogs: A Review of Literature* (Prepared for AWI Ltd). ABARES, Canberra.

Thomson PC (1984) Dingoes and sheep in pastoral areas. *Journal of the Department of Agriculture, Western Australia, Series 4* **25**, 27–31.

Thomson PC (1992) The behavioural ecology of dingoes in north-western Australia. III. Hunting and feeding behaviour, and diet. *Wildlife Research* **19**, 531–541. doi:10.1071/WR9920531

van Eeden L, Smith B, Crowther M, Dickman C, Newsome T (2019) 'The dingo menace': an historic survey of graziers' management of an Australian carnivore. *Pacific Conservation Biology* **25**, 245–256. doi:10.1071/PC18031

Vucetich JA, Vucetich LM, Peterson RO (2012) The causes and consequences of partial prey consumption by wolves preying on moose. *Behavioral Ecology and Sociobiology* **66**, 295–303. doi:10.1007/s00265-011-1277-0

Waudby HP, Smith BP, Robinson GM, Petit S, Earl G (2021) Applying a social-ecological system framework to diagnose drivers of dingo management practices. *Australian Zoologist* **41**, 550–567. doi:10.7882/AZ.2020.036

Wallach AD, Ritchie EG, Read J, O'Neill AJ (2009) More than mere numbers: the impact of lethal control on the social stability of a top-order predator. *PLoS One* **4**, doi:10.1371/journal.pone.0006861.

Wallach AD, Johnson CN, Ritchie EG, O'Neill AJ (2010) Predator control promotes invasive dominated ecological states. *Ecology Letters* **13**, 1008–1018. doi:10.1111/j.1461-0248.2010.01492.x

Wallach AD, Ramp D, O'Neill AJ (2017) Causes of cattle mortality on a predator friendly station in central Australia. *Journal of Mammalogy* **98**, 45–52. doi:10.1093/jmammal/gyw156

Wicks S, Mazur K, Please P, Ecker S, Buetre B (2014) 'An integrated assessment of the impact of wild dogs in Australia'. ABARES Research Report No. 14.4, Canberra, Australia.

Zeven M, Cardilini AP, Hayley A, Borrie WT (2023) "Horrendous": people's experience of their dog's death from 1080 baiting. *Anthrozoos* **36**, 909–926. doi:10.1080/08927936.2023.2210438

PART 6: EMBRACING THE WILD

Allen LR, Sparkes EC (2001) The effect of dingo control on sheep and beef cattle in Queensland. *Journal of Applied Ecology* **38**, 76–87. doi:10.1046/j.1365-2664.2001.00569.x

Andelt WF, Phillips RL, Gruver KS, Guthrie JW (1999) Coyote predation on domestic sheep deterred with electronic dog-training collars. *Wildlife Society Bulletin* **27**, 12–18.

Ballard G, Fleming PJS, Meek PD, Doak S (2020) Aerial baiting and wild dog mortality in south-eastern Australia. *Wildlife Research* **47**, 99–105. doi:10.1071/WR18188

Boronyak L, Jacobs B (2023) Pathways to coexistence with dingoes across Australian farming landscapes. *Frontiers in Conservation Science* **4**, 1126140. doi:10.3389/fcosc.2023.1126140

Boronyak L, Quartermain E (2022) *Predator Smart Farming: Modernising Australia's Approach to Livestock Protection.* Humane Society International Australia.

Boronyak L, Jacobs B, Wallach A (2020) Transitioning towards human–large carnivore coexistence in extensive grazing systems. *Ambio* **49**, 1982–1991. doi:10.1007/s13280-020-01340-w

Boronyak L, Jacobs B, Wallach A, McManus J, Stone S, Stevenson S, Zaranek H (2022) Pathways towards coexistence with large carnivores in production systems. *Agriculture and Human Values* **39**, 47–64. doi:10.1007/s10460-021-10224-y

Boronyak L, Jacobs B, Smith B (2023) Unlocking lethal dingo management in Australia. *Diversity* **15**, 642. doi:10.3390/d15050642

Brink H, Purcell BV, Letnic M, Webster HS, Appleby RG, Jordan NR (2019) Pets and pests: a review of the contrasting economics and fortunes of dingoes and domestic dogs in Australia, and a proposed new funding scheme for non-lethal dingo management. *Wildlife Research* **46**, 365–377. doi:10.1071/WR19030

Campbell G, Emmott A, Pollock D, Traill BJ (2022) Can dingoes increase graziers' profits and help maintain Australia's rangelands? *The Rangeland Journal* **44**, 129–135. doi:10.1071/RJ22002

Dawson SJ, Kreplins TL, Kennedy MS, Wicks S, Fleming PA (2025) An economic analysis of cell-fencing in semi-arid rangelands. *The Rangeland Journal* **47**, RJ24023. doi:10.1071/RJ24023

Emmott A (2021) The Dingo as a management tool on a beef cattle enterprise in western Queensland. *Australian Zoologist* **41**, 459–466. doi:10.7882/AZ.2020.033

Gustavson CR, Garcia J, Hankins WG, Rusiniak KW (1974) Coyote predation control by aversive conditioning. *Science* **184**, 581–583.

Hoare R, King J (2023) The Law Fails to Reflect and Support the Dingo's Cultural Significance. *Environmental and Planning Law Journal* **39**(5), 450.

Hytten K (2009) Dingo dualisms: exploring the ambiguous identity of Australian dingoes. *Australian Zoologist* **35**, 18–27. doi:10.7882/AZ.2009.003

Jenkins DJ, Noad B (2003) *Guard Animals for Livestock Protection: Existing and Potential Use in Australia.* NSW Agriculture, Orange, NSW.

Newsome T (2021) Looking to the future: what next for the dingo? *Australian Zoologist* **41**, 643–653. doi:10.7882/AZ.2021.040

Pearson AI, Cobb ML (2025) Social license pressure and public support for the lethal control of dingoes in Australia. *bioRxiv*, 2025–01.

Philip J (2022) The Dingo Barrier Fence: presenting the case to decommission the world's longest environmental barrier in the United Nations Decade on Ecosystem Restoration 2021–2030. *Biologia Futura* **73**, 9–27. doi:10.1007/s42977-021-00106-z

Pollock D (2021a) Managing the unmanageable: reinstating the dingo for pastoral sustainability in Australian rangelands. *Proceedings of the Royal Society of Victoria* **133**, 27–31. doi:10.1071/RS21005

Pollock D (2021b) Rediscovering a role for dingoes on Wooleen Station. *Australian Zoologist* **41**, 480–486. doi:10.7882/AZ.2021.006

Pollock TI, Hunter DO, Hocking DP, Evans AR (2022) Eye in the sky: observing wild dingo hunting behaviour using drones. *Wildlife Research* **50**, 212–223. doi:10.1071/WR22033

Prowse TA, Johnson CN, Cassey P, Bradshaw CJ, Brook BW (2015) Ecological and economic benefits to cattle rangelands of restoring an apex predator. *Journal of Applied Ecology* **52**, 455–466. doi:10.1111/1365-2664.12378

Robley A, Lindeman M, Cook I, Moloney P, Woodford L (2015) *Dingo Semiochemicals: Towards a Non-lethal Control Tool for the Management of Dingoes and Wild Dogs in Australia.* Technical Report 263. Arthur Rylah Institute for Environmental Research, Heidelberg, Vic.

Ritchie E, Smith BP, Cairns K, Takau S, Rassip W (2023) 'The boss of Country', not wild dogs to kill: living with dingoes can unite communities. *The Conversation* (2 October). https://theconversation.com/the-boss-of-country-not-wild-dogs-to-kill-living-with-dingoes-can-unite-communities-214212.

Rossler ST, Gehring TM, Schultz RN, Rossler JL, Hawley JE, Wydeven AP (2012) Shock collars as a site-aversive conditioning tool for wolves (*Canis lupus*). *Wildlife Society Bulletin* **36**, 176–184.

Smith B, Appleby R (2015) Forging a new future for the Australian dingo. In *The Dingo Debate: Origins, Behaviour and Conservation.* (Ed. B Smith) pp. 301–315. CSIRO Publishing, Melbourne.

Smith B, Snijders L, Tobajas J, Whitehouse-Tedd K, van Bommel L, Pitcher B, St Clair C, Appleby R, Jordan N, Greggor AL (2022) Deterring and repelling wildlife. In: *Wildlife Research in Australia: Practical and Applied Methods.* (Eds B Smith, H Waudby, C Alberthsen and J Hampton) pp. 210–232. CSIRO Publishing, Melbourne.

Smith BP, Appleby RG (2018) Promoting human–dingo co-existence in Australia: moving towards more innovative methods of protecting livestock rather than killing dingoes (Canis dingo). *Wildlife Research* **45**, 1–15. doi:10.1071/WR16161

Smith BP, Appleby RG, Jordan NR (2021) Co-existing with dingoes: challenges and solutions to implementing non-lethal management. *Australian Zoologist* **41**, 491–510. doi:10.7882/AZ.2020.024

Smith BP, Jaques NB, Appleby RG, Morris S, Jordan NR (2020) Automated shepherds: responses of captive dingoes to sound and an inflatable, moving effigy. *Pacific Conservation Biology* **27**, 195–201. doi:10.1071/PC20022

Thomson P (1984) Dingoes and sheep in pastoral areas. *Journal of Agriculture Western Australia* **25**, 27–31.

Thomson P (1992) The behavioural ecology of dingoes in north-western Australia. III. Hunting and feeding behaviour, and diet. *Wildlife Research* **19**, 531–541. doi:10.1071/WR9920531

van Bommel L, Johnson CN (2012) Good dog! Using livestock guardian dogs to protect livestock from predators in Australia's extensive grazing systems. *Wildlife Research* **39**, 220–229. doi:10.1071/WR11135

van Bommel L, Johnson CN (2014) Protecting livestock while conserving ecosystem function: non-lethal management of wild predators. In *Carnivores of Australia: Past, Present and Future.* (Eds AS Glen and CR

Dickman) pp. 323–353. CSIRO Publishing, Collingwood.

van Bommel L, Johnson CN (2015) How guardian dogs protect livestock from predators: territorial enforcement by Maremma sheepdogs. *Wildlife Research* **41**, 662–672. doi:10.1071/WR14190

van Bommel L, Johnson CN (2017a) Olfactory communication to protect livestock: dingo response to urine marks of livestock guardian dogs. *Australian Mammalogy* **39**, 219–226. doi:10.1071/AM15049

van Bommel L, Johnson CN (2017b) Predation control. In *Advances in Sheep Welfare*. (Ed. J Blache) pp. 177–196. Woodhead Publishing, Cambridge, UK.

van Bommel L, Johnson CN (2024) Still a good dog! Long-term use and effectiveness of livestock guardian dogs to protect livestock from predators in Australia's extensive grazing systems. *Wildlife Research* **51**, WR23008. doi:10.1071/WR23008

van Bommel L, Magrath M, Coulson G, Johnson CN (2024) Livestock guardian dogs establish a landscape of fear for wild predators: implications for the role of guardian dogs in reducing human–wildlife conflict and supporting biodiversity conservation. *Ecological Solutions and Evidence* **5**, e12299. doi:10.1002/2688-8319.12299

van Eeden LM, Crowther MS, Dickman CR, Macdonald DW, Ripple WJ, Ritchie EG, Newsome TM (2018) Managing conflict between large carnivores and livestock. *Conservation Biology* **32**, 26–34. doi:10.1111/cobi.12959

van Eeden L, Dickman C, Crowther M, Newsome T (2021a) A Theory of Change for promoting coexistence between dingoes and livestock production. *Conservation Science and Practice* **3**, e304. doi:10.1111/csp2.304

van Eeden LM, Crowther MS, Dickman CR, Newsome TM (2021b) Wicked "wild dogs": Australian public awareness of and attitudes towards dingoes and dingo management. *Australian Zoologist* **41**, 467–479. doi:10.7882/AZ.2020.019

van Eeden LM, Dickman CR, Ritchie EG, Newsome TM (2017) Shifting public values and what they mean for increasing democracy in wildlife management decisions. *Biodiversity and Conservation* **26**, 2759–2763. doi:10.1007/s10531-017-1378-9

van Eeden LM, Eklund A, Miller JR, López-Bao JV, Chapron G, Cejtin MR, Treves A (2018) Carnivore conservation needs evidence-based livestock protection. *PLoS Biology* **16**, e2005577. doi:10.1371/journal.pbio.2005577

van Eeden LM, Newsome TM, Crowther MS, Dickman CR, Bruskotter J (2020) Diverse public perceptions of species' status and management align with conflicting conservation frameworks. *Biological Conservation* **242**, 108416. doi:10.1016/j.biocon.2020.108416

Walker BJ, Letnic M, Bucknall MP, Watson L, Jordan NR (2024) Male dingo urinary scents code for age class and wild dingoes respond to this information. *Chemical Senses* **49**, bjae004. doi:10.1093/chemse/bjae004

Wallach AD, Ritchie EG, Read J, O'Neill AJ (2009) More than mere numbers: the impact of lethal control on the social stability of a top-order predator. *PLoS One* **4**, e6861. doi:10.1371/journal.pone.0006861

Waudby HP, Smith BP, Robinson GM, Petit S, Earl G (2021) Applying a social-ecological system framework to diagnose drivers of dingo management practices. *Australian Zoologist* **41**, 550–567. doi:10.7882/AZ.2020.036

Index

www.ingramcontent.com/pod-product-compliance
Lightning Source LLC
LaVergne TN
LVHW070850210926
848482LV00016B/906
* 9 7 8 1 4 8 6 3 1 8 4 8 3 *